Lecture Notes in Networks and Systems 740

The series "Lecture Notes in Networks and Systems" publishes the latest developments in Networks and Systems—quickly, informally and with high quality. Original research reported in proceedings and post-proceedings represents the core of LNNS.

Volumes published in LNNS embrace all aspects and subfields of, as well as new challenges in, Networks and Systems.

The series contains proceedings and edited volumes in systems and networks, spanning the areas of Cyber-Physical Systems, Autonomous Systems, Sensor Networks, Control Systems, Energy Systems, Automotive Systems, Biological Systems, Vehicular Networking and Connected Vehicles, Aerospace Systems, Automation, Manufacturing, Smart Grids, Nonlinear Systems, Power Systems, Robotics, Social Systems, Economic Systems and other. Of particular value to both the contributors and the readership are the short publication timeframe and the world-wide distribution and exposure which enable both a wide and rapid dissemination of research output.

The series covers the theory, applications, and perspectives on the state of the art and future developments relevant to systems and networks, decision making, control, complex processes and related areas, as embedded in the fields of interdisciplinary and applied sciences, engineering, computer science, physics, economics, social, and life sciences, as well as the paradigms and methodologies behind them.

Indexed by SCOPUS, INSPEC, WTI Frankfurt eG, zbMATH, SCImago.

All books published in the series are submitted for consideration in Web of Science.

For proposals from Asia please contact Aninda Bose (aninda.bose@springer.com).

Sascha Ossowski · Pawel Sitek · Cesar Analide ·
Goreti Marreiros · Pablo Chamoso ·
Sara Rodríguez
Editors

Distributed Computing and Artificial Intelligence, 20th International Conference

Springer

Editors
Sascha Ossowski
ETSII/CETINIA
Universidad Rey Juan Carlos
Madrid, Madrid, Spain

Cesar Analide
Department of Informatics, Engineering School
University of Minho
Braga, Portugal

Pablo Chamoso
BISITE
University of Salamanca
Salamanca, Salamanca, Spain

Pawel Sitek
Politechnika Śląska Kielce
Kielce University of Technology
Santa Cruz, Poland

Goreti Marreiros
Departamento de Engenharia Informática
Polytechnic Institute of Porto
Porto, Portugal

Sara Rodríguez
BISITE
University of Salamanca
Salamanca, Salamanca, Spain

ISSN 2367-3370 ISSN 2367-3389 (electronic)
Lecture Notes in Networks and Systems
ISBN 978-3-031-38332-8 ISBN 978-3-031-38333-5 (eBook)
https://doi.org/10.1007/978-3-031-38333-5

This Springer imprint is published by the registered company Springer Nature Switzerland AG
The registered company address is: Gewerbestrasse 11, 6330 Cham, Switzerland

Preface

The technology transfer in fields such as distributed computing or artificial intelligence is still a challenge, and for that reason, this type of contributions will be specially considered in this symposium. Distributed computing performs an increasingly important role in modern signal/data processing, information fusion and electronics engineering (e.g., electronic commerce, mobile communications and wireless devices). Particularly, applying artificial intelligence in distributed environments is becoming an element of high added value and economic potential.

The 20th International Symposium on Distributed Computing and Artificial Intelligence 2023 (DCAI 2023) is a forum to exchange of ideas between scientists and technicians from both academic and business areas and is essential to facilitate the development of systems that meet the demands of today's society. Research on intelligent distributed systems has matured during the last decade, and many effective applications are now deployed. Nowadays, technologies such as Internet of things (IoT), industrial Internet of things (IIoT), big data, blockchain and distributed computing in general are changing constantly because of the large research and technical effort being undertaken in both universities and businesses. Most computing systems from personal laptops to edge/fog/cloud computing systems are available for parallel and distributed computing. This conference is the forum in which to present application of innovative techniques to complex problems in all these fields.

This year's technical program will present both high quality and diversity, with contributions in well-established and evolving areas of research. Specifically, 108 papers were submitted, by authors from 31 different countries (Algeria, Angola, Austria, Brazil, Burkina Faso, Canada, Croatia, Czechia, Denmark, Ecuador, Egypt, France, Germany, Greece, India, Israel, Italy, Japan, Moldova, Netherlands, Nigeria, Norway, Poland, Portugal, Serbia, South Africa, Spain, Tunisia, Turkey, USA and Zambia), representing a truly "wide area network" of research activity. The DCAI'23 technical program has selected 36 full papers in the main track and, as in past editions, it will be special issues in ranked journals such as Electronics, Sensors, Systems and Advances in Distributed Computing and Artificial Intelligence Journal. These special issues will cover extended versions of the most highly regarded works. Moreover, DCAI'23 Special Sessions have been a very useful tool to complement the regular program with new or emerging topics of particular interest to the participating community.

This symposium is organized by the LASI and Centro Algoritmi of the University of Minho (Portugal). We would like to thank all the contributing authors, National Associations (AEPIA, APPIA), the sponsors (AIR Institute), the funding supporting of the project "COordinated intelligent Services for Adaptive Smart areaS (COSASS), Reference: PID2021-123673OB-C33, financed by MCIN /AEI /10.13039/501100011033/FEDER,

UE., and finally, the local organization members and the program committee members for their hard work, which was essential for the success of DCAI'23.

June 2023

Sascha Ossowski
Pawel Sitek
Cesar Analide
Goreti Marreiros
Pablo Chamoso
Sara Rodríguez

Organization

Honorary Chairman

Sigeru Omatu	Hiroshima University, Japan

Program Committee Chairs

Sascha Ossowski	King Juan Carlos University, Spain
Pawel Sitek	Kielce University of Technology, Poland

Special Sessions Chairs

Rashid Mehmood	King Abdulaziz University, Saudi Arabia
Victor Alves	University of Minho, Portugal

Organizing Committee Chairs

Sara Rodríguez	University of Salamanca, Spain
Cesar Analide	University of Minho, Portugal
Goreti Marreiros	ISEP/GECAD, Portugal

Advisory Board

Yuncheng Dong	Sichuan University, China
Francisco Herrera	University of Granada, Spain
Kenji Matsui	Osaka Institute of Technology, Japan
Tan Yigitcanlar	Queensland University of Technology, Australia
Tiancheng Li	Northwestern Polytechnical University, China

Local Organizing Committee

Paulo Novais (Chair)	University of Minho, Portugal
José Manuel Machado (Co-chair)	University of Minho, Portugal

Hugo Peixoto	University of Minho, Portugal
Regina Sousa	University of Minho, Portugal
Pedro José Oliveira	University of Minho, Portugal
Francisco Marcondes	University of Minho, Portugal
Manuel Rodrigues	University of Minho, Portugal
Filipe Gonçalves	University of Minho, Portugal
Dalila Durães	University of Minho, Portugal
Sérgio Gonçalves	University of Minho, Portugal

Program Committee

Ana Almeida	ISEP-IPP, Portugal
Giner Alor Hernandez	Instituto Tecnologico de Orizaba, Mexico
Cesar Analide	University of Minho, Portugal
Luis Antunes	University of Lisbon, Portugal
Fidel Aznar Gregori	Universidad de Alicante, Spain
Olfa Belkahla Driss	University of Manouba, Tunisia
Orlando Belo	University of Minho, Portugal
Holger Billhardt	Universidad Rey Juan Carlos, Spain
Amel Borgi	ISI/LIPAH, Université de Tunis El Manar, Tunisia
Lourdes Borrajo	University of Vigo, Spain
Adel Boukhadra	National high School of Computer Science, Algeria
Edgardo Bucciarelli	University of Chieti-Pescara, Italy
Juan Carlos Burguillo	University of Vigo, Spain
Francisco Javier Calle	Departamento de Informática. Universidad Carlos III de Madrid, Spain
Rui Camacho	University of Porto, Portugal
Juana Canul Reich	Universidad Juarez Autonoma de Tabasco, Mexico
Carlos Carrascosa	GTI-IA DSIC Universidad Politecnica de Valencia, Spain
Sérgio Manuel Carvalho Gonçalves	University of Minho, Portugal
Luis Fernando Castillo Ossa	University of Caldas, Colombia
Rafael Corchuelo	University of Seville, Spain
Paulo Cortez	University of Minho, Portugal
Stefania Costantini	Dipartimento di Ingegneria e Scienze dell'Informazione e Matematica, Univ. dell'Aquila, Italy
Kai Da	National University of Defense Technology, China

Giovanni De Gasperis	DISIM, Università degli Studi dell'Aquila, Italy
Fernando De La Prieta	University of Salamanca, Spain
Carlos Alejandro De Luna-Ortega	Universidad Politecnica de Aguascalientes, Mexico
Raffaele Dell'Aversana	Università "D'Annunzio" di Chieti-Pescara, Italy
Worawan Diaz Carballo	Thammasat University, Thailand
Youcef Djenouri	LRIA_USTHB, Algeria
António Jorge Do Nascimento Morais	Universidade Aberta, Portugal
Dalila Durães	University of Minho, Portugal
Ramon Fabregat	Universitat de Girona, Spain
Ana Faria	ISEP, Portugal
Pedro Faria	Polytechnic of Porto, Portugal
Florentino Fdez-Riverola	University of Vigo, Spain
Alberto Fernandez	University Rey Juan Carlos, Spain
Peter Forbrig	University of Rostock, Germany
Felix Freitag	Universitat Politècnica de Catalunya, Spain
Alberto Freitas	University of Porto, Portugal
Toru Fujinaka	Hiroshima University, Japan
Francisco García-Sánchez	University of Murcia, Spain
Irina Georgescu	Academy of Economic Studies, Romania
Abdallah Ghourabi	Higher School of Telecommunications SupCom, Tunisia
Ana Belén Gil González	University of Salamanca, Spain
Arkadiusz Gola	Lublin University of Technology, Poland
Juan Gomez Romero	University of Granada, Spain
Evelio Gonzalez	Universidad de La Laguna, Spain
Angélica González Arrieta	Universidad de Salamanca, Spain
Alfonso Gonzalez-Briones	University of Salamanca, Spain
Carina Gonzalez-González	Universidad de La Laguna, Spain
Z. X. Guo	Sichuan University, China
Aurélie Hurault	IRITENSEEIHT, France
Elisa Huzita	State University of Maringa, Brazil
Gustavo Isaza	University of Caldas, Colombia
Patricia Jiménez	Universidad de Huelva, Spain
Bo Nørregaard Jørgensen	University of Southern Denmark, Denmark
Vicente Julian	Universitat Politècnica de València, Spain
Geylani Kardas	Ege University International Computer Institute, Turkey
Amin M. Khan	UiT The Arctic University of Norway, Norway
Naoufel Khayati	COSMOS Laboratory ENSI, Tunisia
Guenter Koch	Humboldt Cosmos Multiversity, Germany

Egons Lavendelis	Riga Technical University, Latvia
Tiancheng Li	Northwestern Polytechnical University, China
Weifeng Liu	Hangzhou Dianzi University, China
Ivan Lopez-Arevalo	Cinvestav Tamaulipas, Mexico
Ramdane Maamri	LIRE laboratory UC Constantine2- Abdelhamid Mehri Algeria, Algeria
Benedita Malheiro	Instituto Superior de Engenharia do Porto and INESC TEC, Portugal
Eleni Mangina	UCD, Ireland
Fábio Marques	University of Aveiro, Portugal
Goreti Marreiros	ISEP/IPP-GECAD, Portugal
Angel Martin Del Rey	Department of Applied Mathematics, Universidad de Salamanca, Spain
Fabio Martinelli	IIT-CNR, Italy
Ester Martinez	Universidad de Alicante, Spain
Philippe Mathieu	University of Lille, France
Kenji Matsui	Osaka Institute of Technology, Japan
Shimpei Matsumoto	Hiroshima Institute of Technology, Japan
Rene Meier	Lucerne University of Applied Sciences, Switzerland
José Ramón Méndez Reboredo	University of Vigo, Spain
Yeray Mezquita	University of Salamanca, Spain
Mohd Saberi Mohamad	Universiti Malaysia Kelantan, Malaysia
Jose M. Molina	Universidad Carlos III de Madrid, Spain
Paulo Moura Oliveira	UTAD University, Portugal
Paulo Mourao	University of Minho, Portugal
António J. R. Neves	University of Aveiro, Portugal
Jose Neves	University of Minho, Portugal
Julio Cesar Nievola	Pontifícia Universidade Católica do Paraná, PUCPR Programa de Pós Graduação em Informática Aplicada, Brazil
Nadia Nouali-Taboudjemat	CERIST, France
Paulo Novais	University of Minho, Portugal
José Luis Oliveira	University of Aveiro, Portugal
Sigeru Omatu	Hisroshima, Japan
Mauricio Orozco-Alzate	Universidad Nacional de Colombia, Colombia
Sascha Ossowski	University Rey Juan Carlos, Spain
Miguel Angel Patricio	Universidad Carlos III de Madrid, Spain
Juan Pavón	Universidad Complutense de Madrid, Spain
Reyes Pavón	University of Vigo, Spain
Pawel Pawlewski	Poznan University of Technology, Poland
Stefan-Gheorghe Pentiuc	University Stefan cel Mare Suceava, Romania

Tiago Pinto	IPP, Portugal
Julio César Ponce	Universidad Autónoma de Aguascalientes, Mexico
Juan-Luis Posadas-Yague	Universitat Politècnica de València, Spain
Jose-Luis Poza-Luján	Universitat Politècnica de València, Spain
Isabel Praça	GECAD/ISEP, Portugal
Radu-Emil Precup	Politehnica University of Timisoara, Romania
Mar Pujol	Universidad de Alicante, Spain
Francisco A. Pujol	Specialized Processor Architectures Lab, DTIC, EPS, University of Alicante, Spain
Araceli Queiruga-Dios	Department of Applied Mathematics, Universidad de Salamanca, Spain
Mariano Raboso Mateos	Facultad de InformáticaUniversidad Pontificia de Salamanca, Spain
Miguel Rebollo	Universitat Politècnica de València, Spain
Jaime A. Rincon	Universitat Politècnica de València, Spain
Ramon Rizo	Universidad de Alicante, Spain
Sergi Robles	Universitat Autònoma de Barcelona, Spain
Sara Rodríguez	University of Salamanca, Spain
Luiz Romao	Univille, Brazil
Gustavo Santos-Garcia	Universidad de Salamanca, Spain
Ichiro Satoh	National Institute of Informatics, Japan
Emilio Serrano	Universidad Politécnica de Madrid, Spain
Manuel Fernando Silva Rodrigues	University of Minho, Portugal
Nuno Silva	DEI & GECAD - ISEP - IPP, Portugal
Pedro Sousa	University of Minho, Portugal
Masaru Teranishi	Hiroshima Institute of Technology, Japan
Zita Vale	GECAD - ISEP/IPP, Portugal
Rafael Valencia-Garcia	Departamento de Informática y Sistemas. Universidad de Murcia, Spain
Miguel A. Vega-Rodríguez	University of Extremadura, Spain
Paulo Vieira	Insituto Politécnico da Guarda, Portugal
José Ramón Villar	University of Oviedo, Spain
Friederike Wall	Alpen-Adria-Universitaet Klagenfurt, Austria
Zhu Wang	XINGTANG Telecommunications Technology Co., Ltd., China
Li Weigang	University of Brasilia, Brazil
Bozena Wozna-Szczesniak	Institute Of Mathematics and Computer Science, Jan Dlugosz University in Czestochowa, Poland
Michal Wozniak	Wroclaw University of Technology, Poland
Michifumi Yoshioka	Osaka Pref. Univ., Japan

Agnieszka Zbrzezny	University of Warmia and Mazury, Faculty of Mathematics and Computer Science, Poland
Zhen Zhang	Dalian University of Technology, China
Yun Zhu	Shaanxi Normal University, China
André Zúquete	University of Aveiro, Portugal

Organizing Committee

Juan M. Corchado Rodríguez	University of Salamanca and AIR Institute, Spain
Fernando De la Prieta	University of Salamanca, Spain
Sara Rodríguez González	University of Salamanca, Spain
Javier Prieto Tejedor	University of Salamanca and AIR Institute, Spain
Ricardo S. Alonso Rincón	AIR Institute, Spain
Alfonso González Briones	University of Salamanca, Spain
Pablo Chamoso Santos	University of Salamanca, Spain
Javier Parra	University of Salamanca, Spain
Liliana Durón	University of Salamanca, Spain
Marta Plaza Hernández	University of Salamanca, Spain
Belén Pérez Lancho	University of Salamanca, Spain
Ana Belén Gil González	University of Salamanca, Spain
Ana De Luis Reboredo	University of Salamanca, Spain
Angélica González Arrieta	University of Salamanca, Spain
Angel Luis Sánchez Lázaro	University of Salamanca, Spain
Emilio S. Corchado Rodríguez	University of Salamanca, Spain
Raúl López	University of Salamanca, Spain
Beatriz Bellido	University of Salamanca, Spain
María Alonso	University of Salamanca, Spain
Yeray Mezquita Martín	AIR Institute, Spain
Sergio Márquez	AIR Institute, Spain
Andrea Gil	University of Salamanca, Spain
Albano Carrera González	AIR Institute, Spain

DCAI 2023 Sponsors

Contents

Time-Series Modeling for Intrusion Detection Systems

Konstantinos Psychogyios[1(✉)], Stavroula Bourou[1], Andreas Papadakis[1], Nikolaos Nikolaou[1], and Theodore Zahariadis[1,2]

[1] Synelixis Solutions S.A., 10 Farmakidou Av, 34100 Chalkida, Greece
psychogios@synelixis.com
[2] National and Kapodistrian University of Athens, 157 72 Athens, Greece
zahariad@uoa.gr

Abstract. The advent of computer networks and the Internet has drastically altered the means by which we share information & interact with each other. However, this technological advancement has also created room for malevolent behaviour where individuals exploit weak points with the intent of gaining access to confidential data, blocking activity etc. To this end, intrusion detection systems (IDS) are needed to filter malicious traffic and prevent common attacks. In the past, these systems relied on a fixed set of rules or comparison with previous attacks. However, with the increased availability of computational power & data, machine learning has emerged as a promising solution for this task. While many systems now use this methodology in real-time for a reactive approach to mitigation, we aim to explore the potential of configuring it as a proactive time series prediction. In this work, we delve into this possibility further. More specifically, we convert a classic IDS dataset to a time-series format and use predictive models to forecast forthcoming malign packets. The findings indicate that our model performs strongly, exhibiting accuracy that is within a 4% margin when compared to conventional real-time detection.

Keywords: Intrusion detection system · Cybersecurity · Machine learning · Time-series

1 Introduction

The universality of the Internet and computer networks has revolutionized the way we interact with each other enabling information sharing and collaboration on an unprecedented scale. However, this pervasive connectivity has also created new opportunities for malicious actors to exploit vulnerabilities and gain unauthorized access to sensitive information [1]. As a result, the importance of effective intrusion detection systems cannot be overstated, while the need for proactive notification is emerging [2]. IDS stands for a hardware device or software application used to monitor and detect suspicious network traffic and

S. Ossowski et al. (Eds.): DCAI 2023, LNNS 740, pp. 1–10, 2023.
https://doi.org/10.1007/978-3-031-38333-5_1

potential security breaches flagging malicious activity. This monitoring takes place at packet-level and thus such a system can distinguish malicious from benign packets. Traditionally, this component has been implemented as a firewall and later as a rule-based expert system. Due to the rise of ML in recent years [3–5], state-of-the-art approaches are based on ML technologies applied on data logs from IDS, to classify packets as suspicious or not [6,7].

These systems can be broadly categorized into two types: (i), classification-based [8] and (ii), anomaly-based [9]. Classification-based IDSs use machine learning algorithms to classify incoming data into different categories based on a set of features. Even though classification-based IDSs are effective in detecting known attacks, they can be less effective in identifying new and unknown attacks that have small correlation with the training dataset. On the other hand, anomaly detection-based approaches use statistical models and machine learning algorithms to establish a baseline of normal behavior and identify deviations from that baseline. Unlike classification-based IDSs, anomaly-based IDSs can detect unknown or novel attacks that have not been previously seen. However apart from this advantage, these models are known to produce significantly lower results when the forthcoming packets are of a previously seen class [10].

The methodologies discussed above suffer from a significant drawback, namely the process of identifying and categorizing anomalies takes place in real-time. To address this issue, the present study suggests a proactive intrusion detection system that can detect and isolate malevolent packets prior to their entry into the system, thus avoiding potential damage that could have occurred if action was taken too late. To achieve this, the proposed model utilizes a window of W preceding packets to predict the existence of an attack in the subsequent $W + T$ packets (prediction window), allowing for the classification of the upcoming behaviour in advance and enabling the application of security measures. To evaluate the effectiveness of this approach, we conducted experiments on the UNSW-NB15 dataset [11], a widely used benchmark for evaluating intrusion detection systems. Our model achieved an F1-score of 80% for the $T = 1$ case, which is only 4% lower than that of a classification IDS. Moreover, it attained an F1-score of 90% for predicting the existence of an attack in the next $T = 20$ packets.

Our main contributions can be summarized as follows:

- We extend and improve existing intrusion detection systems by implementing proactive prediction.
- We compare our approach with state-of-the-art current methodologies.
- We experiment with the configuration parameters (e.g. size of the input window) of our method thoroughly to provide insights.

2 Related Work

The field of IDS using machine learning has seen extensive research with new methods and datasets emerging frequently [12,13]. Maseer, Z. K. et. al. [14] evaluated many machine learning classification methods on the CIC-IDS2017

[15] dataset. Regarding the pre-processing steps, they conducted one-hot encoding and normalization. They also employed parameter tuning and k-fold cross validation for the training phase. The methods employed were the standard classification approaches, such as random forest, support vector machines, convolutional neural networks etc. for the task of binary classification. They measured accuracy (with binary accuracy, f1-score, precision etc.) and training/testing time. Results showed that KNN, Random forest and naive bayes achieve excellent results for the aforementioned metrics.

Imran, M. et. al. [16] evaluated custom autoencoder-based models against the KDD-99 [17] for the problem of multiclass classification. They developed a non-symmetric deep autoencoder, which was either used as a single model (NDAE) or in a stacked manner (S-NDAE). The term non symmetric refers to the architecture of the encoder and decoder models, which in this case are not similar (symmetric). They evaluated the performance of these models with common metrics, namely accuracy, precision etc. Results showed that this method achieves better metrics compared to different state-of-the-art approaches. Saba, T. et. al. [18] developed an intrusion detection model tested with the datasets of BoT-IoT [19] and NID[1]. The proposed model was a convolutional neural network. The BoT-IoT dataset was used for the task of multi-class classification, whereas NID was used for binary classification. Results showed that the proposed model was able to classify the packets with an accuracy of 95%. Tahri, R. et. al. [20] compared many articles that proposed IDS systems based on the UNSW-NB15 dataset and more specifically on the kaggle, 100.000 sample, version. In their survey, they found that random forest was the best performing model in most of the studies, reaching an accuracy of up to 98%, specificity of up to 98% and sensitivity of 94% for the task of binary classification.

Regarding approaches that address this problem as a time-series instance, Duque A. S. et al. [21] propose the use of machine learning-based intrusion detection techniques or analyzing industrial time series data. The paper evaluates three different algorithms, namely Matrix Profiles, Seasonal Autoregressive Integrated Moving Average (SARIMA), and LSTM-based neural networks, using an industrial dataset based on the Modbus/TCP protocol. The paper demonstrates that the Matrix Profiles algorithm outperforms the other models in detecting anomalies in industrial time series data, requiring minimal parameterization effort.

2.1 Findings

From the aforementioned state-of-the-art review, we see that most researches have implemented the intrusion detection system within the classification framework. Even though this approach is effective, it has limited application since the inference happens in real-time and not proactively. However, some researchers have tried to frame this problem as a time-series prediction with promising results. When it comes to the UNSW-NB15 dataset, most researchers choose

[1] https://www.kaggle.com/datasets/sampadab17/network-intrusion-detection.

an incomplete version of the dataset (approx. 100.000 samples instead of ~ 2.5 million samples). This dataset is tailored (unbalance is removed, only samples with strong correlation with the target label are kept) for machine learning applications and is not a representative version of real-world scenarios. For the reasons above, our work differs significantly since we: (i), use the whole version of the dataset to emulate real-world applications (ii) frame the intrusion detection system as a time-series problem (iii) conduct multiple experiments to thoroughly compare against the common approaches.

3 Dataset

The Dataset we used to validate our approach is UNSW-NB15 which was created by researchers at the University of South wales in Australia. This dataset was produced by the IXIA PerfectStorm where raw packets were gathered for common attacks such as Fuzzers, Analysis, Backdoors, and Denial Of Service (DoS). The main reason we chose this dataset is because it has been widely employed in research works and serves as a benchmark for comparison between different network intrusion detection systems.

The UNSW-NB15 dataset contains approximately 2.5 million samples of network packets and 49 corresponding features. From these features, we conclude with 44 after the pre-processing steps. The features include basic information such as source and destination IP addresses, as well as more advanced features such as the packet size, time to live, and protocol type. This dataset also contains time-based features, namely starting and ending time of a packet, that helps us sort and convert the samples to a time-series format. Concerning the labels (binary), the dataset consists of 2218760 benign packets and, 321283 malicious. Also, the mean time between two consecutive packets is 1 s. The mean number of packets involved in interactions between specific entities (IPs) in this dataset is 8167.

4 Methodology

In this section, we describe the proposed methodology for the proactive notification based on IDS logs. We describe the pre-processing steps as well as the model we used for the prediction. A detailed overview of the whole system can be viewed on Fig. 1.

The architecture consists of three main components: packet input, pre-processing and ML model.

4.1 Pre-processing

Before feeding the dataset to the model, we first convert it into the right format. Thus we initially **clean the data**, discarding NaN and corrupted values. The latter could be unrealistic attribute values, such as negative packet length or

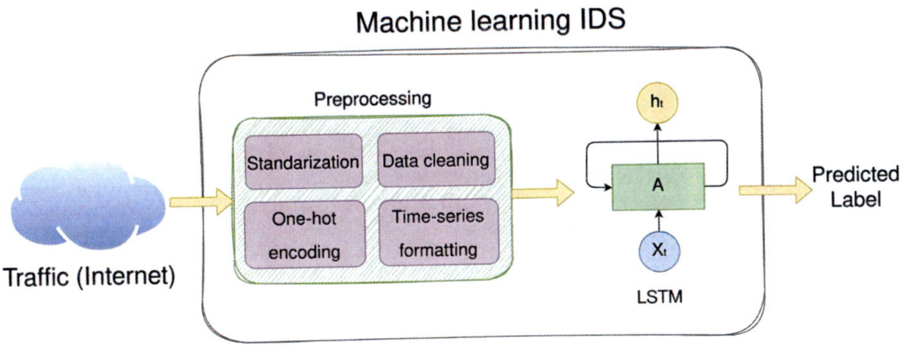

Fig. 1. Methodology for the proactive IDS.

unrealistic values too far from the feature mean (outliers). Such errors could be due to feature extraction tool error or transportation losses.

The second pre-processing step is **one-hot encoding**, since we have many categorical features that can take multiple values. This technique given a column with N different values, creates N different binary columns where 1 indicates the presence of this value in the current instance and 0 the absence. An example is the variable the categorical variable 'service', which indicates the type of packet and can be: 'dns', 'http', 'smtp', etc.

The third pre-processing step we implemented is **data scaling**, using Min-Max scaler. Scaling is the process of transforming the values of numeric variables in a dataset to a common scale. Since neural networks tend to calculate the distance between data points, we don't want features with larger scales dominate the ones with lower.

The final stage of pre-processing involves formatting the data into a **time-series format**. To accomplish this, we begin by sorting the dataset based on the time (Starting time) feature. We do not group the dataset based on IPs because it has been generated in the same network and thus consecutive packets are correlated even if they have not been recorded in the exact same location. Next using the sorted dataset, we generate windows comprising W time points and a *label*, where W is the size of the input window. Here, the initial W points serve as the input to the model, and the *label* is the target that must be predicted. We also emphasize again that for each of the W input points we keep all the features (not only the labels) formulating the problem of multivariate time-series forecasting. The *label* is calculated based on a value T and indicates the existence of an attack in the forthcoming $W + T$ packets. For example, if $W = 10$ and $T = 5$ and also the labels of the T packets were [0 1 0 0 1] the *label* of the time-series instance would be 1 because there is a one in the T vector. Otherwise, it would be 0. Also in such a case, we would use ten consecutive instances of the dataset as input for the model.

We also incorporate overlapping between the different windows to utilize the data to its fullest potential.

4.2 ML Model

The Machine learning model we employed for our task is Long Short Term Memory (LSTM) [22]. This is a type of recurrent neural network (RNN) that can preserve earlier information and carry it forward compared to standard previous approaches. To this end, a sliding window of size W is used where each item is a packet in a consecutive time step. Compared to vanilla RNNs, each LSTM cell is composed of gates, namely the (i), input gate (ii) forget gate and (iii) output gate. Concerning the mathematics of these gates we have:

$$i_t = sigmoid(W_i.[h_{t-1}, x_t] + b_i) \tag{1}$$

$$f_t = sigmoid(W_f.[h_{t-1}, x_t] + b_f) \tag{2}$$

$$o_t = sigmoid(W_o.[h_{t-1}, x_t] + b_o) \tag{3}$$

where i_t, f_t, and o_t are the activations of the input, forget, and output gates at time t, respectively. W_i, W_f, and W_o are weight matrices, b_i, b_f, and b_o are bias terms of the corresponding gates. In addition, h_{t-1} is the previous hidden state, and x_t is the current input.

The equation for updating the cell state is given by:

$$c_t = f_t * c_{t-1} + i_t * tanh(W_c.[h_{t-1}, x_t] + b_c) \tag{4}$$

where c_t is the current cell state, W_c is a weight matrix, b_c is a bias term, and $tanh$ is the hyperbolic tangent activation function (tanh).

Finally, the equation for computing the current hidden state is given by:

$$h_t = o_t * tanh(c_t) \tag{5}$$

where h_t is the current hidden state.

5 Experiments

This section outlines the experiments conducted to validate the approach presented in Sect. 4. We describe the experimental set-up, namely the methodology we used to validate our models. Then we show the used metrics and demonstrate results.

5.1 Experimental Set-Up

To utilize our dataset, we employ 5-Fold cross validation and average the results across all the 5 folds. We note that after the time-series formatting instead of individual samples we have windows of size W and thus the shuffling and splitting to folds is performed on such independent windows. To demonstrate the robustness of our proposed approach, we make a comparison with the standard binary classification IDS approach. We also want to show the effect of input window size to the results and thus exhibit the change in accuracy while the value of this variable is ascending.

Apart from experiments concerning the input window size, we also experimented for the prediction further into the future. Based on the aforementioned methodology, we still predict a single label, which is not always the next packet. For each window, we create a new label which indicates the existence of an attack in the next T time steps.

5.2 Metrics

To ensure the effectiveness of our predictive models, we evaluate their performance using F1-score, Precision, and Recall. These metrics are necessary due to the imbalanced nature of our dataset, where simple accuracy could produce inaccurate results. F1-score, Precision, and Recall provide a more accurate measure of the model's performance by accounting for the imbalanced dataset and ensure a fair evaluation.

5.3 Results

Impact of W on Model Performance
Firstly, we want to test the effect of window size regarding the model's accuracy. For this reason, we predict the label of the $W+1$ packet using a window size (W) of ascending size. We also want to test these cases against the classic, real-time binary classification to compare the efficiency of the two different approaches. For this experiment, the value of T if fixed having $T = 1$. With this value fixed, after the time-series formatting we end up with $2,540,033$ time-series objects having $2,540,043$ initial instances. This is expected since we have overlapping of time steps and we also have -10 instances for the first time-series object. The label distribution is teh same as the original dataset. Results can be viewed on Table 1.

Table 1. Results for ascending input window size.

Model	F1-score	Precision	Recall
Classification	0.84	0.90	0.78
Time-Series (W=1)	0.72	0.80	0.68
Time-Series (W=25)	0.74	0.84	0.68
Time-Series (W=50)	0.77	0.86	0.70
Time-Series (W=100)	0.80	0.88	0.73
Time-Series (W=200)	0.80	0.88	0.74

The present study reveals that the simple (real-time) classification scenario yields marginally better results compared to the time-series case, which is unsurprising given that making classifications using present variables is substantially easier than making predictions about future events. Specifically, the F1-score of the classification method is only 4% higher than that of the time-series model for $W = 100$, which is deemed acceptable. Additionally, the study indicates that an increase in the input window size enhances model accuracy. This is because utilizing more samples generally leads to improved predictions, by providing more data for analysis. Nevertheless, this also entails a greater computational burden, as a larger window entails more parameters for the model. More specifically, the study demonstrates that the metrics when W is increased from $W = 1$ to $W = 100$, with a 2–3% increase for each case. No further enhancement in the metrics is observed between $W = 100$ and $W = 200$, implying that all relevant information for the prediction has already been exploited considering that the lifetime of any interaction between an actor and the IDS is briefer). Consequently, the methodology described herein involves a natural compromise between computational complexity & accuracy, necessitating customization for each unique application.

Impact of T on Model Performance

To observe results related to predictions that cover a longer time frame (different values of T), one may consult Table 2 for a fixed window size $W = 50$.

Table 2. Results for ascending prediction horizon with label distributions.

Time steps	Labels	F1-score	Precision	Recall
T=1	$0:87\%, 1:13\%$	0.77	0.86	0.70
T=10	$0:75\%, 1:25\%$	0.85	0.89	0.82
T=20	$0:63\%, 1:37\%$	0.91	0.92	0.91
T=30	$0:54\%, 1:46\%$	0.93	0.92	0.94

The table reveals that the metrics exhibit improvement as the predictive horizon expands. This trend is expected since a larger value of T leads to more general predictions, with the focus being on detecting the presence of an attack in the upcoming T packets, rather than predicting the label of a particular packet. The most significant increase is observed between $T = 1$ and $T = 10$, where the F1-score improves by 8%, while precision and recall increase by 3% and 12%, respectively. As the value of T increases, the improvements in the metrics diminish, with only a 6% and 2% enhancement observed between the subsequent T values. We also display the label distributions for each of these cases. It is evident that with an increase in the value of T, the dataset achieves a greater balance, as one would anticipate due to the higher probability of an attack occurring within a larger window frame.

6 Conclusion and Future Work

Technological advances regarding the Internet and computer systems provided many advantages in numerous areas of human lives such as improved communication and information sharing. However, these advances have also paved the way for malicious actor exploitation through network attacks and capitalizing on vulnerabilities. This has led to the research and deployment of many intrusion detection systems that monitor network packets and detect malevolent traffic. In this study, we have re-examined the conventional approach of intrusion detection systems based on machine learning and have redefined the problem as a time-series prediction task. Our results demonstrate that our proposed methodology is capable of proactive operation rather than merely reacting to security breaches with an F1-score within a margin of 4% compared to the real-time approach. By embracing the time-series prediction approach, we acknowledge the dynamic nature of network threats and the need for a proactive defense strategy. This shift in perspective allows us to anticipate and counteract emerging attack techniques, thereby staying one step ahead of malicious actors. Through our research, we aim to contribute to the continuous improvement of intrusion detection systems, ensuring the safety and security of our interconnected digital world.

In the future, we aim to validate our approach further by employing more IDS datasets to further support our claims. We also plan to evaluate the performance of a model trained on one dataset but tested on another dataset with the same features to assess cross-dataset validation. Furthermore, we plan to test more complex and robust recurrent neural network architectures instead to assess performance increase compared to the vanilla LSTM. Lastly, we also plan to test our methodology in both centralized and federated learning paradigms since such a model would benefit from diverse data gathered from multiple devices.

Acknowledgements. This work was funded by the H2020 CyberSEAS project, contract no. 101020560, within the H2020 Framework Program of the European Commission.

References

1. Alshamrani, A., Myneni, S., Chowdhary, A., Huang, D.: A survey on advanced persistent threats: techniques, solutions, challenges, and research opportunities. IEEE Commun. Surv. Tutorials **21**(2), 1851–1877 (2019)
2. Khraisat, A., Gondal, I., Vamplew, P., Kamruzzaman, J.: Survey of intrusion detection systems: techniques, datasets and challenges. Cybersecurity **2**(1), 1–22 (2019). https://doi.org/10.1186/s42400-019-0038-7
3. Psychogyios, K., Velivassaki, T.H., Bourou, S., Voulkidis, A., Skias, D., Zahariadis, T.: GAN-driven data poisoning attacks and their mitigation in federated learning systems. Electronics **12**(8), 1805 (2023)
4. Psychogyios, K., Ilias, L., Ntanos, C., Askounis, D.: Missing value imputation methods for electronic health records. EEE Access **11**, 21562–21574 (2023)
5. Psychogyios, K., Ilias, L. and Askounis, D. : Comparison of missing data imputation methods using the Framingham heart study dataset. In: 2022 IEEE-EMBS International Conference on Biomedical and Health Informatics (BHI), pp. 1-5. IEEE (2022)

6. Halbouni, A., Gunawan, T.S., Habaebi, M.H., Halbouni, M., Kartiwi, M., Ahmad, R.: Machine learning and deep learning approaches for cybersecurity: a review. IEEE Access **10**, 19572–19585 (2022)
7. Anastasakis, Z., et al.: Enhancing cyber security in IoT systems using FL-based IDS with differential privacy. In: 2022 Global Information Infrastructure and Networking Symposium (GIIS), pp. 30–34. IEEE (2022)
8. Le, T.T.H., Oktian, Y.E., Kim, H.: XGBoost for imbalanced multiclass classification-based industrial internet of things intrusion detection systems. Sustainability **14**(14), 8707 (2022)
9. Hajisalem, V., Babaie, S.: A hybrid intrusion detection system based on ABC-AFS algorithm for misuse and anomaly detection. Comput. Netw. **136**, 37–50 (2018)
10. Pang, G., Shen, C., Cao, L., Hengel, A.V.D.: Deep learning for anomaly detection: a review. ACM Comput. Surv. (CSUR) **54**(2), 1–38 (2021)
11. Moustafa, N., Slay, J.: UNSW-NB15: a comprehensive data set for network intrusion detection systems (UNSW-NB15 network data set). In: 2015 Military Communications and Information Systems Conference (MilCIS), pp. 1–6. IEEE Australia (2015)
12. Thakkar, A., Lohiya, R.: A review on machine learning and deep learning perspectives of IDS for IoT: recent updates, security issues, and challenges. Arch. Comput. Methods Eng. **28**, 3211–3243 (2021)
13. Saranya, T., Sridevi, S., Deisy, C., Chung, T.D., Khan, M.A.: Performance analysis of machine learning algorithms in intrusion detection system: a review. Procedia Comput. Sci. **171**, 1251–1260 (2020)
14. Maseer, Z.K., Yusof, R., Mostafa, S.A., Bahaman, N., Musa, O., Al-rimy, B.A.S.: DeepIoT. IDS: hybrid deep learning for enhancing IoT network intrusion detection. Comput. Mater. Continua **69**(3), 3945–3966 (2021)
15. Sharafaldin, I., Lashkari, A.H., Ghorbani, A.A. : Toward generating a new intrusion detection dataset and intrusion traffic characterization. In: ICISSp, pp. 108–116 (2018)
16. Imran, M., Haider, N., Shoaib, M., Razzak, I.: An intelligent and efficient network intrusion detection system using deep learning. Comput. Electr. Eng. **69**, 107764 (2022)
17. Bay, S.D., Kibler, D., Pazzani, M.J., Smyth, P.: The UCI KDD archive of large data sets for data mining research and experimentation. ACM SIGKDD Explor. Newslett. **2**(2), 81–85 (2000)
18. Saba, T., Rehman, A., Sadad, T., Kolivand, H., Bahaj, S.A.: Anomaly-based intrusion detection system for IoT networks through deep learning model. Comput. Electr. Eng. **99**, 107810 (2022)
19. Koroniotis, N., Moustafa, N., Sitnikova, E., Turnbull, B.: Towards the development of realistic botnet dataset in the internet of things for network forensic analytics: Bot-IoT dataset. Future Gener. Comput. Syst. **100**, 779–796 (2019)
20. Tahri, R., Jarrar, A., Lasbahani, A., Balouki, Y.: A comparative study of machine learning algorithms on the UNSW-NB 15 dataset. In: TM Web of Conferences, vol. 48, p. 03002. EDP Sciences (2022)
21. Anton, S.D., Ahrens, L., Fraunholz, D., Schotten, H.D.: Time is of the essence: machine learning-based intrusion detection in industrial time series data. In: IEEE International Conference on Data Mining Workshops (ICDMW), pp. 1-6. IEEE (2018)
22. Hochreiter, S., Schmidhuber, J.: Long short-term memory. Neural Comput. **9**(8), 1735–80 (1997)

Estimation of Occlusion Region Using Image Completion by Network Model Consisting of Transformer and U-Net

Tomoya Matsuura(✉) and Takayuki Nakayama

Osaka Institute of Technology, 1-45, Chayamachi, Kita-ku, Osaka, Osaka, Japan
m1m23r36@st.oit.ac.jp, takayuki.nakayama@oit.ac.jp

Abstract. Object recognition from the images with occlusion was a difficult problem. In this paper, we attempted to solve this problem by image completion. In order to interpolate the hidden regions of objects from the surrounding image data, this paper proposed a new pix2pix type image generation network, in which the transformer was used instead of the convolution network in the generator. In this model, U-Net composed of the transformer blocks encodes the contextual relationships between the pixels, and the followed additional transformer block generate the interpolated image using them. Since the convolution was based on the global features, it could not interpolate the images if the missing regions were large. By replacing it with transformer, it was possible to analogize the missing regions from surrounding pixels. The effectiveness of the proposed method was confirmed by image interpolation experiments for several images with occlusions.

Keywords: GAN · Transformer · U-Net

1 Introduction

Recently, life support robots have been actively developed to solve the problem of a shortage of caregivers due to the declining birthrate and aging population. Autonomous mobile robots such as the life support robots must be able to sense their surroundings, make decisions, and move by themselves without human assistance. However, to recognize the objects from the image data is difficult since the brightness of lighting is not constant, and objects often overlap with other objects. Especially, recognition of such the hidden objects is difficult, and often causes the malfunction due to oversights or misrecognition of the objects. Thus, in this paper, we propose to improve propose a neural network to interpolate the hidden parts of objects from the partially appeared images.

2 Related Work

Image completion has been studied actively, and various type of neural networks have been developed [1]. Pix2Pix was proposed by Isora, et al. in 2017

S. Ossowski et al. (Eds.): DCAI 2023, LNNS 740, pp. 11–20, 2023.
https://doi.org/10.1007/978-3-031-38333-5_2

[2]. Pix2Pix is one of the Generative Adversarial Networks (GAN) in which the convolution-deconvolution network is used as the generator. Convolution is used in Pix2Pix to extract features by applying filters to images. Convolution-deconvolution network can regenerate the shape of the objects, but it is sometimes difficult to obtain the whole image due to the reduction of the image size. When convolution is used for image completion, the missing image is reconstructed based on the nearby pixels. Therefore, if the large part of the surrounding information is missing, the completion may not work. Also, it may not work well for images with fine textures and lines, because the convolution extracts the global features and cannot generate the images with much of local features. For these reasons, the quality of the complemented image is not sufficient when the missing region is large, to use it for the recognition of the objects with occlusion in the decision making of life support robots.

To improve the quality of complemented image, a mechanism to analogize missing information from the surrounding information is required. The method for analogizing missing information from surrounding information has been studied well natural language processing. In 2017, Vaswani et al. proposed Transformer [3]. The Transformer has a mechanism to extract the contextual relationship between the words and analogize the missing word from the surroundings. The Transformer includes a Multi-Head Attention mechanism and a Positional Encoding mechanism. Multi-Head Attention performs Self-Attention in parallel, and extract the important information as to the relationships between the multiple elements of time series data. Positional Encoding preserves the order information of the time-series data, which is lost in the neural network, by attaching positional information to the data.

In the field of the image processing, Vision Transformer (ViT) [4] applied the Transformer for image recognition. It extracts relationships between pixels in an image, just like words in natural language. Image data is divided into patches of a specific size determined in advance, vectorized, and input to the Transformer Encoder. The encoder performs Multi-Head Attention and finally classifies the image using Multilayer Perceptron (MLP). However, application to the image completion was not taken into account.

Swin Transformer [5], an improved version of ViT, reduces computational complexity and adjusts patch size. Patches are grouped into multiple windows, and the relevance of each window is calculated. The patch size is changed according to the layer, and different features of the region are taken into account.

There is Swin-Unet [6], which combines the structures of Swin Transformer and U-Net. The U-Net has a Skip Connection, which stores the down-sampled feature maps in the encoder. The stored data is used for up-sampling in the decoder. In the down-sampling encoder, features for the contextual relationship between the pixels are extracted. Decoder reconstructs the images from the encoded features. Using this scheme, Swin-Unet makes it possible to find image anomalies and segment them.

It is expected that Transformer can extract the features encoding the contextual relationship between the parts of images. So that the U-Net structure can provide clearer complemented images. Therefore, in this paper, we propose a GAN model using Transformer and U-Net in the generator.

3 Proposed Network

3.1 Summary

In this study, we propose a new network structure combining Transformer and U-Net. The encoder composed of the Transformers encodes the features concerned with the contextual relationship between the parts of images. The features extracted by the encoder are transferred to the Skip Connection and synthesized by the Decoder. Transferring information by skip connection compensates for the information loss occurring in the encoder and enables to generate the higher resolution output images.

3.2 ARISE2wise

The proposed network model (Fig. 1) aims to generate the whole image of an object from a part of the occlusion image. This model can be likened to "A region is enough to the wise." since it understands the whole image from a part of the image. Therefore, the proposed generative network model will be referred to as "ARISE2wise".

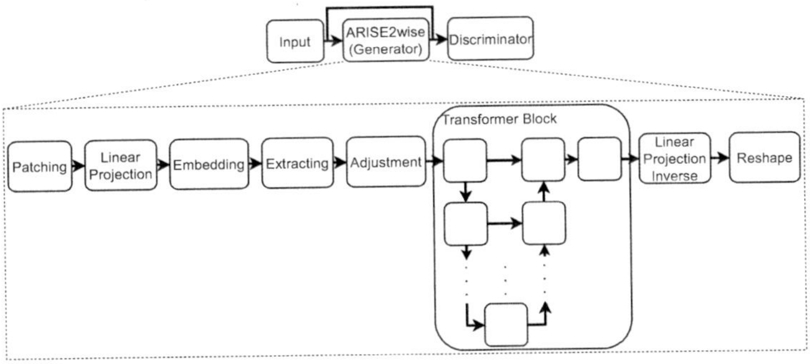

Fig. 1. The architecture of ARISE2wise.

The basic structure of ARISE2wise is GAN, where the generator, is composed of Transformer and U-Net. The images generated by generator from the input image partially hidden by the patch and the original input image are input to discriminator, which evaluates whether the generated image is identical to the input image or not. Patching Layer, Linear Projection Layer, and Embedding Layer patch and vectorize the data as in ViT, and then add position information. In Extracting Layer, patches are applied to tensor image data to hide the specific regions of images. In Adjustment Layer, the data extracted in the Extracting layer is expanded to the same size as the number of patches before extraction by a linear transformation. In Transformer Block, processing including Transformer and U-Net is performed. The details of Transformer Block are explained in the next section. Linear Projection Inverse Layer transforms the data to the same size as before vectorization. Reshape Layer transforms the data into a shape that can be treated as an image. The output is fed into Discriminator along with the input.

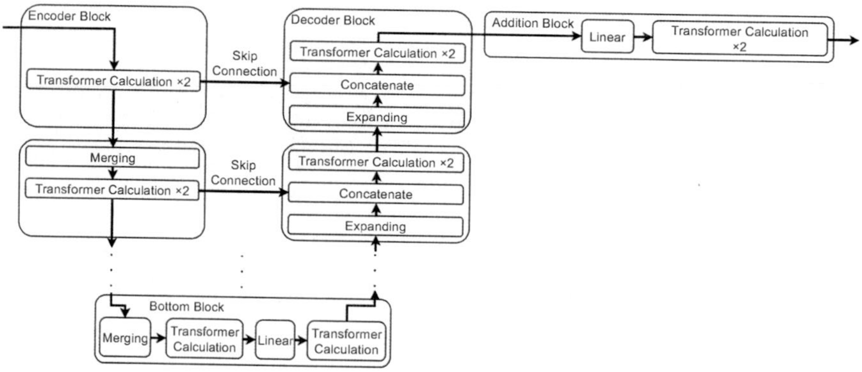

Fig. 2. Transformer Block

4 Transformer Block in ARISE2wise

4.1 Transformer Block

Transformer Block (Fig. 2) contains the same geometry as in U-Net, with the addition of Additive Block. When image completion was performed using U-Net that include Transformer, the output images were not consistent with the input images due to missing parts and other defects. Therefore, Transformer Block was added after the U-Net process, to reconstruct the features extracted in the encoder blocks. From the above, we added Addition Block after the U-Net structure.

4.2 Network Like U-Net

The network like U-Net consists of Encoder Block, Bottom Block, and Decoder Block. Encoder Block performs Merging and Transformer Calculation. Merging is Affine, which halves the amount of data by performing linear transformations. Transformer Calculation performs Multi-Head Attention and other processing similar to Transformer Encoder of ViT. In Bottom Block, Merging, Transformer Calculation, and Linear Transformations are performed. In Decoder

Block, Expanding and Transformer Calculation are performed. Expanding is Affine, which doubles the amount of data by performing linear transformation. The processed data is combined with Encoder Block data through the Skip Connection. The combined data is processed by Transformer Calculation for Multi-Head Attention and other processes.

4.3 Addition Block

When the output image was generated by simple U-Net from the patched image, it would contain the objects that did not exist in the original image or the objects whose shapes were collapsed. In other words, the output images were inconsistent with the input images. Therefore, in order to generate consistent images, Addition Block including the Transformer process should be added after the U-Net structure. In Addition Block, linear transformations and Transformer Calculation are computed. Linear Layer performs linear transformations. Transformer Calculation performs Multi-Head Attention and other processing similar to Transformer Encoder in ViT.

5 Image Completion Features

5.1 Datasets

CIFAR-10 [7] is used as the dataset. CIFAR-10 consists of 10 classes of image data, 50,000 as training data and 10,000 as test data. The image data are RGB images, and the image size is 32 pixels in height and 32 pixels in width.

5.2 Implementation

In this study, we consider a means to understand the whole image of an object from its parts without relying on language-based identification. In order to estimate the whole image from parts of an image (visual information), we deemed it necessary for the network model to possess an image completion function. Therefore, we use two types of network models, Pix2Pix and ARISE2wise, for the purpose of conducting image completion. For the two types of network models, the following two data processing methods are tried.

CENTER Inferring the central region from the edge region of the image.
EDGE Inferring the edge region from the central region of the image.

In each data processing method, the input is converted into the extracted image. In "Inferring the central region from the edge region of the image.", the data in the center of the input image is referred to and the number in the relevant area is converted to "1". In "Inferring the edge region from the central region of the image.", Pix2Pix refers to the data except the center of the input image and converts the value of the relevant area to "1". ARISE2wise extracts only the center of the input image. The size of the center region should be equivalent to 1/4 of the image size (1/2 in the vertical and horizontal direction).

The data input to Pix2Pix is converted from the number of dimensions (3,32,32) to the number of dimensions (512,16,16) by four times of Convolution. The output image is then converted to the number of dimensions (3,32,32) by four times of Deconvolution, and the output image and the correct image are input to the discriminator. Each network model is trained by feeding back the evaluation of the discriminator.

The data input to the Transformer Block undergoes transformer and merging processes in the Encoder Block. In this study, the Encoder Block is executed six times (N = 3), and the data is transformed until the number of patches is reduced to one digit. Next, in the Decoder Block, which is executed the same number of times as the Encoder Block, the data is transformed into the number of patches at the input by the Transformer and Expanding processes. After that, the data is processed twice (M = 2) in the Addition Block before being output as an image. Each network model is trained by inputting the output image and the input image (correct image) to the discriminator and receiving feedback from the evaluation.

In Pix2Pix and ARISE2wise, BCEWIthLogitsLoss and L1Loss are used as loss functions and Adam as optimization function. In Pix2Pix, Adam assumes a learning rate lr = 0.0002, betas = (0.5,0.999). In ARISE2wise, Adam assumes a learning rate lr = 0.001, betas = (0.9,0.999). Each network model is run for each data processing method with a batch size of 64 and 5000 epochs to obtain the loss function and output images.

Fig. 3. The results of image completion.

5.3 Result

The output images with image completion are evaluated and discussed (Fig. 3). Regarding the CENTER completion, the output images of Pix2Pix show colors and shapes that do not exist in Ideal. In the output images of ARISE2wise, the main object or background color is generated, although it is hazy. For the EDGE completion, the colors in the output images of Pix2Pix are so distorted that it is impossible to determine the shape of the object. The output images of ARISE2wise are hazy but do not produce any disconcerting colors or shapes.

Table 1. Similarity between ideals and outputs of each data processing method type.

Image	CENTER-Pix2Pix	CENTER-ARISE2wise	EDGE-Pix2Pix	EDGE-ARISE2wise
1	1254.06	124.07	1773.33	298.82
2	402.62	68.94	854.98	864.77
3	599.97	57.41	505.11	469.15
4	3618.93	172.29	2151.01	445.42
5	2913.09	51.97	2773.23	226.75

The similarity between the ideal and output images is evaluated and discussed (Table 1). Kullback-Leibler divergence (KL DIV) is used to calculate the similarity. The closer the value is to 0, the higher the similarity. In the completion of CENTER, the similarities of the output images of ARISE2wise are larger than those of Pix2Pix for all images. In the EDGE completion, the similarities of the output images of ARISE2wise are larger than those of Pix2Pix except for the second image from the top.

From the above, ARISE2wise has an advantage in the setting of this study.

6 Generation from Occlusion Images Using Completion

6.1 Datasets

Images of YCB objects [8–10] are used as the dataset. As for the objects to be used as the dataset, we arbitrarily select 10 classes from a total of 77 types of objects included in the YCB objects. Specifically, the 10 classes are: baseball, sugar box, dice, fork, hammer, mug, plate, scissors, turner, and sponge. We collect images taken by rotating the 3D data of YCB objects on Unity, and use 3888 images per class and 38880 images for all 10 classes as a dataset. The images used as the dataset are RGB images, and the image size is 128 pixels in height and 128 pixels in width. For the test data, multiple types of YCB objects are selected from the above YCB objects, and images are taken by arranging them in such a way that they overlap each other.

6.2 Implementation

ARISE2wise is used to generate a whole image of an object from an occlusion image. The generation is done by inferring the edge region from the central region of the image.

The data is converted into an extracted image in the preliminary stage of input to the Transformer Block. The data input to the Transformer Block undergoes transformer and merging processes in the Encoder Block. In this study, the Encoder Block is executed six times (N = 6), and the data is transformed until the number of patches is reduced to one digit. Next, in the Decoder Block, which is executed the same number of times as the Encoder Block, the data is transformed into the number of patches at the input by the Transformer and Expanding processes. After that, the data is processed twice (M = 2) in the Addition Block before being output as an image. Each network model is trained by inputting the output image and the input image (the correct image) to the discriminator and receiving feedback from the evaluation. In ARISE2wise, BCEWIthLogitsLoss and L1Loss are used for the loss functions, and Adam is used for the optimization function. Adam assumed learning rate lr = 0.001, betas = (0.9, 0.999). The network model is run for each data processing method with a batch size of 64 and 5000 epochs to obtain the loss function and output images (Fig. 4).

6.3 Result

The output images from the occlusion images are evaluated and discussed. In the output image of Turner, the generated image is smaller than the Ideal image, but the shape is similar and the roundness of the tip is reproduced. In the output image of Hammer, an overall image similar to the ideal shape is generated, although the size is different, and the curved surface is generated smoothly. In the output image of Mug, a semicircle, which does not exist in the ideal form, is generated on the side of the mug. In the output image of Mug&Plate, a semicircle that does not exist in the ideal shape is generated, and the dish included in the extraction region also exists in the output image. In the output image of Mug&Box, the lower part of the mug is generated larger than it should be, and the box included in the extracted region is also present in the output image.

The similarity between the ideal and output images is evaluated and discussed (Table 2). KL DIV is used to calculate the similarity. The closer the value is to 0, the higher the similarity. The output image of Mug has the highest similarity among the five types of images. The similarity is relatively low for Mug&Plate and Mug&Box, where there are multiple objects in the extraction region.

From the above, it is possible to estimate the hidden parts of objects in the image by image completion. However, if there are multiple objects in the extraction region, other objects than those to be estimated will remain in the output image. In addition, it is difficult to estimate the whole image if the

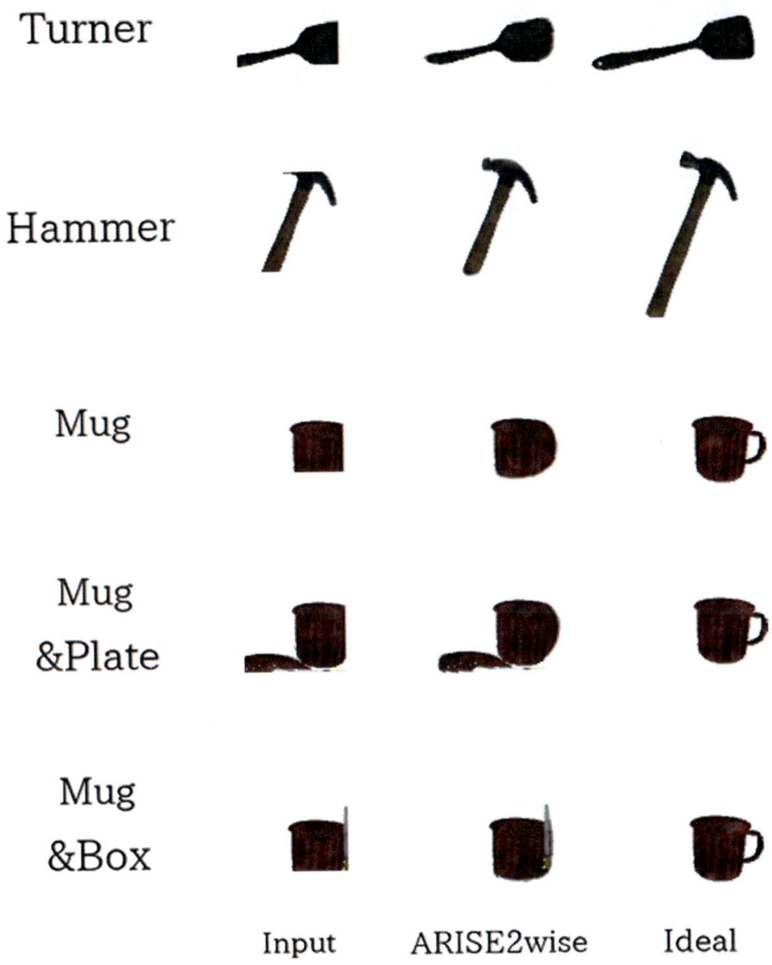

Fig. 4. The results generated from occlusion images using image completion.

Table 2. Similarity between ideals and outputs from the occlusion images.

Image	ARISE2wise
Turner	4862.46
Hammer	3593.59
Mug	3253.50
Mug&Plate	7100.61
Mug&Box	6697.65

extraction region does not contain any characteristic shape. In terms of similarity, the estimated shape does not always match the original shape and size, so that the similarity is high even if the shape is unnatural to the visual sense.

7 Conclusion

In this paper, we proposed a new GAN type model, which employs the Transformer U-Net as a generator, to realize the image complement from the image with occlusion. Since the model can extract the contextual relationship between the local image features by the Transformer network, the high-quality interpolation of hidden part of the image can be achieved using surrounding image data. To examine the effectiveness, we execute the image complement experiments for the image with missing central part and that with missing the peripheral part. Both experiments showed that the proposed network can generate the similar image to the original image with high quality. The effectiveness was also confirmed in the object restoration experiment from the image with occlusion. However, in some cases that the multiple objects are included in the region of interest or the characteristic shapes of the objects are hidden, image complement was failed. It is a future work to make it possible to extract the hidden objects accurately even from the images in which various objects overlap each other. Since the image complement from the occlusion image was successfully achieved in the case that the single object is included in the region of interest, it is expected that the proposed method can be applied to the object recognition and robot action planning.

References

1. Iizuka, S., Simo-Serra, E., Ishikawa, H.: Globally and locally consistent image completion. ACM Trans. Graph. (ToG) **36**(4), 1–14 (2017)
2. Isola, P., Zhu, J.Y., Zhou, T., Efros, A.A.: Image-to-image translation with generative adversarial networks. In: Proceedings of the IEEE Conference on Computer Vision and Pattern Recognition (2017)
3. Vaswani, A., et al.: Attention is all you need. arXiv:1706.03762. Submitted on 12 Jun 2017, Accessed 6 Dec 2017
4. Dosovitskiy, A., et al.: An image is worth 16x16 words: transformers for image recognition at scale. In: ICLR 2021 Conference Paper1909, Submitted on 22 Oct 2020, Accessed 3 Jun 2021
5. Liu, Z., et al.: Swin transformer: hierarchical vision transformer using shifted windows. arXiv:2103.14030, Submitted on 25 Mar 2021, Accessed 17 Aug 2021
6. Cao, H., et al.: Swin-Unet: Unet-like Pure Transformer for Medical Image Segmentation. arXiv:2105.05537, Submitted on 12 May 2021
7. Krizhevsky, A.: Learning multiple layers of features from tiny image, Technical report (2009)
8. Calli, B., Walsman, A., Singh, A., Srinivasa, S., Abbeel, P., Dollar, A.M.: Benchmarking in manipulation research: the YCB object and model set and benchmarking protocols. IEEE Rob. Autom. Mag. (2015)
9. Calli, B., et al.: Yale-CMU-Berkeley dataset for robotic manipulation research. Int. J. Rob. Res. **36**(3) (2017)
10. Calli, B., Singh, A., Walsman, A., Srinivasa, S., Abbeel, P., Dollar, A.M.: The YCB object and model set: towards common bench marks for manipulation research. In: Proceedings of the 2015 IEEE International Conference on Advanced Robotics (ICAR), Istanbul, Turkey (2015)

Operation of a Genetic Algorithm Using an Adjustment Function

Francisco João Pinto(✉)

Department of Study Center, Scientific Research and Advanced Training in Computer Systems and Comunication , Faculty of Engineering, Agostinho Neto University, University Campus of the Camama, S/N, Luanda, Angola
fjoaopinto@yahoo.es

Abstract. In this work we describe in some details the operation of a genetic algorithm (GA), using an adjustment function to compare solutions and determine which is the best. The three basic processes of GAs are: selection of solutions based on their adjustment or adquate to the environment, reproduction for genes crossover, and mutation, which allows random changes to occur in genes. Through these processes, GAs will find better and better solutions to a problem as species evolve to better adjustment their environments. A basic process of a GA begins by randomly generating solutions or "chromosomes" to the problem. Posteriorly, an iterative process is carried out in which, at each step, the good solutions are selected and the crossing between species is carried out. Occasionally we can have mutations on certain solutions. Through the selection of good solutions in the iterative process, the computer will develop better and better solutions. The results of our experiment show that there is a general improvement over the initial population, both in the total adjustment, as well as the medium and maximum.

Keywords: Genetic Algorithm · Adjustment Function · Artificial Intelligence

1 Introduction

The history of GAs begins in the 1940s, when scientists began trying to take inspiration from nature to create the branch of artificial intelligence (AI). Research developed further in the branches of cognitive research and in the understanding of reasoning and learning processes until the late 1950s, when it began to look for models of genetic systems that could generate candidate solutions to problems that were too difficult to solve computationally.

One of the first attempts to associate natural evolution with optimization problems was made in 1975, when Box presented his scheme of evolutionary operations. These were a method of systematically perturbing two or three control variables of an installation, analogous to what we understand today as mutation and selection operators [2]. Soon after, in the early 1960s, Bledsoe and Bremmerman began working with genes, using both binary and integer and real representations, and developing the precursors of combination operators.

S. Ossowski et al. (Eds.): DCAI 2023, LNNS 740, pp. 21–30, 2023.
https://doi.org/10.1007/978-3-031-38333-5_3

An attempt to use evolutionary processes to solve problems was made by I. Rechernberg in the first half of the 1960s, when he developed evolutionary strategies. This maintained a population of two individuals with chromosomes composed of real numbers at each instant, with one of the two chromosomes being a child of the other and generated through the exclusive application of the mutation operator. The process described by Rechenberg had broad theoretical foundations, and the mutation was applied from a Gaussian distribution of parameters and was used with practical success. Even though it does not include concepts that are currently widely accepted, such as larger populations and the crossover operator, Rechenberg's work can be considered pioneering, for having introduced evolutionary computation to practical applications [5].

In later works, evolutionary strategies addressed these flaws, being modified to include population and crossover operator concepts. The way they apply this operator is interesting because it includes the idea of using the average as an operator in addition to being able to involve many parents, ideas that can be applied to GAs when we use chromosomes with a continuous representation. However, despite not being the first researcher in the area, the one who would be called the father of GAs finally showed up at the end of the 60s, when John Holland "invents" GAs, although he focused eminently on discrete coding [8].

Holland formally studied the evolution of species and proposed a computational heuristic model that when implemented could offer good solutions to extremely difficult problems that were computationally insoluble up to that era. In 1975 Holland published his book, "Adaptation in Natural and Artificial Systems", in which he studied evolutionary processes, rather than designing new algorithms, as most people think. Holland's work presents GAs as a metaphor for evolutionary processes so that he could study adaptation and evolution in the real world by simulating it within computers. However, GAs have transcended the role originally envisioned by Holland and have become a tool of widespread use by computer scientists [5].

An interesting fact about Holland's work and his influence in the field of GA is that he originally used binary chromosomes, whose genes were just zeros and ones. This limitation was abolished by later researchers, but even today many scientists insist on using only the binary representation, even when there are others that may prove to be more suitable for solving the problem at hand.

Since then, genetic algorithms began to expand throughout the scientific community, generating a series of applications that could help solve extremely important problems that might not have been addressed otherwise. In addition to this scientific progress, there was also commercial development: in the 1980s commercial packages appeared using genetic algorithms [10].

In the 1980s, the progress of evolutionary algorithms and their popularization in the scientific community led to the emergence of the first conferences dedicated exclusively to these topics.

Nowadays, GAs have benefited greatly from interdisciplinarity. More and more computer scientists are looking for inspiration in other areas of research in order to absorb their ideas and make GAs more efficient and intelligent in problem solving [6].

2 Genetic Algorithms Overview

GAs are computer models based on biological genetics and evolution, and on the biological principle of "survival of the best". In the genetics of the life in which they are inspired, the two cells that unite to give rise to a descendant each contribute a genetic load in the 23 chromosomes they contain, numbered from 1 to 23. When reproduction occurs, chromosomes with the same number come together to form a new pair of chromosomes for the son. Each of the chromosomes in these pairs is made up of thousands or even millions of genes. A gene from the father and a gene from the mother form a pair of genes for the son that will help shape its characteristics. Sons normally inherit most of their genes from their parents. Occasionally, however, some of the gene values change due to unusual physical, chemical, or biological effects. Such changes are known as mutations. Over the generations, this mutation process is repeated and the result is that the individuals and genes that best adapt to the environment tend to remain, while the others tend to become extinct [1].

GAs are abstract models of natural genetics and the process of evolution. They include concepts such as chromosomes, genes, pairing or crossing of species, mutation, and evolution. The intention, however, is not to build computer models that reproduce this process from natural genetics, but rather to develop useful models that are easily implemented on computers by "borrowing" concepts from natural genetics.

The three basic processes of GAs are: selection of solutions based on their adjustment or adequate to the environment, reproduction for gene crossover, and mutation, that allows random changes to occur in genes. Through these processes, GAs will find better and better solutions to a problem as species evolve to better adjustment their environments. A basic process of a GA begins by randomly generating solutions or "chromosomes" to the problem. Posteriorly, an iterative process is carried out in which, at each step, the good solutions are selected and the crossing between species is carried out. Occasionally we can have mutations on certain solutions. Through the selection of good solutions in the iterative process, the computer will develop better and better solutions. This approach is applicable to many types of problems, such as optimization, or machine learning, as we will see [2].

The basic principle of the functioning of GAs is that a selection criterion will make, after many generations, the initial set of individuals generate more fit individuals. Most selection methods are designed to preferentially choose individuals with higher fitness scores, though not exclusively, in order to maintain population diversity.

3 Representation of Solutions in Genetic Algorithms

A GA begins with the design of a representation of a solution for a given problem. A solution in this context is any candidate value to be a final answer to the problem, regardless of whether this value provides a correct answer or not. For example, if we want to maximize the function $y = 5 - (x - 3)^3$, x = 1 would be one candidate solution, x = 2.5 would be another, while x = 3 is the correct solution.

Although the scheme used to represent solutions is up to the developer, one of the most common schemes is character strings. Consider a finite-length character string over a given alphabet, such as {0, 1}, {0, 1, 2}, or {A, B, Z}. The length of the chain with which the solutions will be represented will be chosen based on the alphabet used, and the amount of information that we want to represent in it. The broader the alphabet, the more information we can represent with each character. Therefore, fewer characters are needed to encode a specific amount of information. Suppose we represent each solution by a 12-bit string over the alphabet {0, 1}. This option could represent both solutions formed by a set of values of 12 variables or binary parameters, as well as solutions formed by 3 parameters using 4 bits for each one. In this case, each parameter could cover a range from 0000 to 1111, or, in decimal, from 0 to 15. In this context, a string is somewhat similar to a chromosome or set of chromosomes, and each parameter represented could be compared to a gene [1].

To see an application example, suppose that a company manufactures 4 products and the problem is to find the number of products that maximizes profit under certain circumstances. In this case we could represent each solution by 32-bit strings assigning the first 8 bits to represent the amount of product 1, the next 8 to represent product 2, and so on. Thus, (30, 10, 25, 30) would be one solution, (20, 20, 35, 40) another, etc.

Although the representation of solutions by means of character strings is the most frequent, it is not the only one, and the choice will depend on the problem to be solved. For example, for graph problems, the solution can be a graph which, in turn, can be represented by an adjacency matrix. For a genetic programming problem, each solution will be a computer program. In short, the most appropriate representation for each problem must be adopted [3].

In order to be able to compare solutions and determine which one is the best, an adjustment function is used that measures the proximity of a candidate solution to the objective of the problem. In the production chain example, profit itself could be used as an adjustment measure.

4 Basic Process of Operation of a Genetic Algorithm

Although there are numerous variations and extensions, the basic process of operation of a genetic algorithm is illustrated below, in which, at each step, a set of solutions called a population is generated.

Step 0. Initialization of the population.

0.1. Generate a set of solutions randomly.

Repeat the next 3 steps until the optimal solution is found or an end condition is satisfied

Step 1. Selection

1.1. Determine the values of the adjustment function for all solutions in the population.

1.2. Create a crossover group. Select solutions from the current population randomly and proportional to the value of its adjustment function. Solutions with better values will have a better chance of being chosen and will have to survive until the next generation. In this step, the concept of evolution based on the principle of natural selection is applied.

Step 2. Crossing of species.

2.1. Randomly choose two solutions. With a fixed crossover probability pc (e.g., pc = 0.7), randomly determine which crossovers will occur. If any crossing is made, go to step 2.2. Otherwise, make two descendent that will be exact copies of the two parents solutions, and go to step

2.2. Randomly select internal points (crossover points) of the solutions, and then swap the parts of the solution that found at those points. Carry out this second step for all the solutions obtained in step 1 that is, until the size of the new population reaches the size of the initial population, selecting a pair at random each time.

Step 3. Random mutation

3.1. With a fixed mutation probability and pm small (pm is typically chosen such that one mutation occurs for every thousand bit transfers across the crossing), randomly select a small portion of solutions to force some change (for example, change a bit from 0 to 1

The meaning of the crossover operations, which are performed in step 2, is as follows. Each solution can represent, depending on the problem we are dealing with, a set of values for the parameters, instructions to execute a particular task, etc. Parts of each solution may contain notions of importance or relevance. A GA exploits this information by reproducing the high-quality notions, and then crossing them between the best solutions. In short, it is about forcing a cross between the best representatives of the species to obtain a descendant that improves the quality of its parents [1].

Finally, the underlying idea of the third step of the algorithm is to model natural mutation to create a new species that could never otherwise be created by ordinary reproducing and crossing processes. In GAs, sometimes, after a certain number of iterations, most of the solutions can be so similar that no significant changes occur, even though they are far from the optimal solution. Through the process of mutation, the changed parts help to get the solution set out of the static configuration that may affect them.

Due to the GAs do not guarantee the optimal solution, the described process (steps 0–3) is repeated a certain number of times (for example 10), starting each time with different seeds for the generation of random numbers. Finally, the different solutions generated are compared, and the best one is chosen [9].

Let's see an example. A process workshop is a system made up of a series of workstations capable of manufacturing products. In this scenario, the objective of applying a genetic algorithm could be to determine an optimal schedule that maximizes profit, for example, by manufacturing the appropriate number of products, reducing production costs, downtime and inventory, or avoiding possible penalties due to delays in production or low quality of products. This is a difficult optimization problem since there are not only many complex regular factors, but also irregular factors such as a machine failure or a delay in supplies. The workshop schedule must be able to adjust quickly and flexibly to these changes. Due to of these complexities, the problem cannot be simply expressed by an elegant mathematical formula.

To exemplify the operation of the algorithm, let us consider a simple scenario in which the process workshop has 3 workstations. In each of them, only one product is manufactured. We will further simplify the problem and assume that a workstation either manufactures a certain amount of product (which we will represent as a 1), or it does not manufacture anything (which we will represent as a 0). In this context, therefore, each solution can be represented with a binary alphabet by a string of 3 characters. Suppose also that the value of the adjustment function is the binary number represented by the string. The problem is to find the string that gives the maximum adjustment value, i.e. string 111 for a maximum value of 7. In real situations, of course, the adjustment values are not so simply determined. The adjustment values may be determined by complex equations, simulation models, or by reference to observations made on experiments or real cases [1].

Next, we discuss how the algorithm works.

Step 0. Population initiation
We decided to use a population size of **n = 4**. Using random numbers, the following initial population is generated: {101}, {001}, {010}, {110}. Note that each solution is a string of length 3.

Iteration 1, Step 1. Reproduction. Generation of a crossing set.
For each solution $(string_i)$, determine the adjustment value, f_i, and the total adjustment. Equation (1) allows us to determine the total adjustment value (see Table 1).

$$F = \sum_i f_i. \tag{1}$$

Table 1. Adjustment values

i	$string_i$	f_i
1	101	5
2	001	1
3	010	2
4	110	6
		F = 14

2. Compute the probability of adjustment (or normalized adjustment),, and the expected count, with respect to the presence of a solution in the crossover set. Equation (2) allows us to calculate the adjustment probability, where f_i is the adjustment value and $\sum_i f_i$. is the total adjustment value, and Eq. (3) allows us to calculate the expected count, where n is the size of the population and p_i is the probability of adjustment.

$$p_i = \frac{f_i}{F} = \frac{f_i}{\sum_i f_i} \tag{2}$$

Expected Count = n·p_i,

3. Randomly select a new set of n strings (crossover set) whose average distribution is equal to the expected count distribution. In the example, from the set of 4 available

Table 2. Normalized adjustment

i	$string_i$	f_i	p_i	$n \cdot p_i$
1	101	5	0.357	1.428
2	001	1	0.071	0.284
3	010	2	0.143	0.572
4	110	6	0.429	1.716
	Total	**14**	1.000	4.000
	Average	**3.5**	0.250	1.000
	Maximum	**6**	0.429	1.716

strings, we will select an average of 1.716 strings as number 4, 1,428 strings as number 1, and so on for the rest. Since the number of generated strings has to be an integer, we can implement the selection process of n strings by assigning to each of them a certain range of random numbers that represent the probabilities of choosing that string. The distribution of these random numbers is chosen to be proportional to the probability distribution of the strings (a very common technique used in the Monte Carlo method—a simulation technique that uses random numbers). In our example, based on p_i we can assign the following ranges, (see Table 3):

Table 3. Random ranges

i	p_i	Random ranges
1	0.357	000–356
2	0.071	357–427
3	0.143	428–570
4	0.429	571–999

In this way, the probabilities pi are reproduced exactly, since each random number has a probability equal to 0.001 of being chosen. Thus, the probability of choosing one of the 357 numbers from 0 to 356 is 0.357, which is exactly the probability of choosing the string i. Once these ranges are established, we will only have to generate as many random numbers as the size n of the crossover set we want, and choose the strings associated with the values obtained for the random numbers.

Suppose that in our experiment we generated n = 4 random numbers which turned out to be 483, 091, 652, and 725. Based on these numbers, we select strings 3, 1, 4, and 4. The current count is (see Table 4):

And the generated crossing set is (see Table 5):

Table 4. Current count

$(Before)_i$	$(New)_i$	$string_i$
3	1	010
1	2	101
4	3	110
4	4	110

Table 5. Generated crossing set.

i	$string_i$	$n \cdot p_i$	Current count
1	101	1.428	1
2	001	0.284	0
3	010	0.572	1
4	110	1.716	2

Iteration 1. Step 2. Combination of species.

The following steps are executed for all solutions taking two solutions at a time:

1. A pair is randomly selected to cross it. We determine whether the crossing is effectively made based on a fixed crossing probability.In this case, we will assume that this probability is 1.
2. A crossover point is chosen randomly with uniform probability along the length of the string. Let L be the length of the strings, then the crossing point k is chosen within the interval [1, L−1].

Suppose that, in this case, strings 2 and 3 are selected to perform a crossover at crossover point k = 2, resulting in two new solutions (see Table 6):

Table 6. Next generation solutions (crossing point k = 2).

	Before crossing	Decendents
Solution1	10:1	100
Solution2	11:0	111

The other two remaining strings cross with k = 1, so the result end of the crossing process is (see Table 7):

Table 7. Crossing point k = 1

i	stringi	Cross with	k	New population
1	010	4	1	010
2	101	3	2	100
3	110	2	2	111
4	110	1	1	110

Iteration 1, Step 3. Random mutation.

As discussed, a common option is to use a crossover probability of one mutation for every 1000 bits transferred. In our case, only 12 bits have been transferred, so the probability of a mutation occurring in this iteration would be really low (0.012). Supose, however, that this is the case and the rightmost bit of string2 is chosen for mutation so that its value becomes 101.

Iteration 2 Step 1. Reproduction.

The process of reproduction, crossing and mutation would be repeated a number of times. We will show only part of the second iteration of step 1 or reproduction (see Table 8).

Table 8. Reproduction

i	$string_i$	f_i
1	010	2
2	101	5
3	111	7
4	110	6
	Total	**20**
	Average	**5**
	Maximum	**7**

We can observe how a general improvement is produced in the initial population both in the total, medium and maximum adjustment. In this experiment, the optimal solution (111 = 7) is already reached in this second iteration, so it is not necessary to carry out more iterations. In general, the process would be repeated more times. Normally, the solutions would get better and better until eventually, hopefully, the optimal solution would be found.

5 Conclusions

In this paper we investigate in detail the operation of GA, using an adjustment function in order to compare the solutions and determine which is the best. The results show that there is a general improvement over the initial population in the total, medium and maximum adjustment. We also can conclude that GAs are a part of an area of intelligent computing, called bio-inspired computing, that draws inspiration from nature to find solutions for problems that are too complex to be solved by traditional techniques. Moreover, GAs solve computational problems by modeling the process of natural evolution, carrying out reproduction and mutation processes, until finding suitable solutions to the problem. However, we can say that, as a search tool, GAs are extremely efficient, finding good solutions to problems that might not be solvable otherwise, in addition to being extremely simple to implement and modify.

References

1. Bonillo, V.M., Betanzos, A.A., Canosa, M.C., Berdiñas, B.G., Rey, E.M.: Fundamentos de Inteligencia Artificial, Capítulo 4, páginas 87–93, 97 e 98, 111–115, Universidad de la Coruña-España (2000)
2. Goldberg, D.E.: Genetic Algorithm in Search, Optimization and Machine Learning. Adilson-Wesley (1989)
3. Goldber, D.E.: Real-coded genetic algorithms, virtual alphabets and blocking, Department of general engineering, Illinois, EUA (1990)
4. Pinto, F.J.: Structure and operation of a basic genetic algorithm. In: 13th International Conference on Distributed Computing and Artificial Intelligent, vol. 474, pp. 53–59, Springer, Spain (2016). https://doi.org/10.1007/978-3-319-40162-1_6
5. Rechenberg, I.: Evolutionsstrategici optimierung technischer system nach prinzipien der der biologischen evolution. Verlag, Alemanha (1973)
6. Linden, R.: Algoritmos genéticos-3ª edição. Ciência Moderna, Lida, Rio de Janeiro (2012)
7. Alender, J.T.: On optimal size of genetic algorithms. In: Proceding CompEuro1996, Computer System and Software Engeneering 6th Annual European Computer Conference, pp. 65–70 (1996)
8. Coffin, D., Smith, R.E.: Linkage learning in estimation of distribution algorithms. In: Linkage in Evolutionary Computation. Studies in Computational Intelligence, vol. 157, pp. 141–156 (2008).https://doi.org/10.1007/978-3-540-85068-7_7, ISBN 978–3–540–85067–0
9. Pavai, G., Geetha, T.V.: New crossover operators using dominance and co-dominance principles for faster convergence of genetic algorithms. Soft. Comput. **23**(11), 3661–3686 (2018). https://doi.org/10.1007/s00500-018-3016-1
10. Shir, O.M.: Niching in evolutionary algorithms. In: Rozenberg, G., Bäck, T., Kok, J.N. (eds.) Handbook of Natural Computing. Springer, Heidelberg, pp. 1035–1069 (2012). https://doi.org/10.1007/978-3-540-92910-9_32

CUBA: An Evolutionary Consortium Oriented Distributed Ledger Byzantine Consensus Algorithm

Cyril Naves Samuel[1,2](✉), François Verdier[1], Severine Glock[2], and Patricia Guitton-Ouhamou[2]

[1] Université Côte d'Azur, LEAT/CNRS UMR 7248, Sophia Antipolis, France
cyril-naves.samuel@etu.univ-cotedazur.fr,
francois.verdier@univ-cotedazur.fr

[2] Renault Software Factory, Sophia Antipolis, France
{cyril-naves.samuel,severine.glock,patricia.guitton-ouhamou}@renault.com

Abstract. We propose a consortium-based distributed ledger (Blockchain) consensus algorithm overcoming the classical problems of Byzantine Fault Tolerant (BFT) consensus algorithms. The identified issues concerning Byzantine algorithms are scalability, performance, and attack resilience. These factors inspire us to conceive our novel consensus algorithm CUBA. CUBA expands to Contesting Utilitarian Byzantine Agreement which evaluates and valorizes each consensus action as a Utilitarian metric of the gamified participants in the network. The obtained utilitarian metrics are used as feedback to reorganize the network for faster performance of the network consensus or for being resilient to the malicious activity noticed. This consensus protocol is designed to sustain or increase the Utilitarian happiness in a Byzantine environment of identified participants for the network's liveness, safety, performance, and scalability. Evaluation results show an improved throughput, scalability, and malicious resilience compared to Proof of Authority protocols like PBFT, IBFT, and QBFT, as well as comparable to Clique for consortium Distributed Ledger networks.

Keywords: Distributed Consensus · Distributed Ledger BFT Algorithm · Blockchain · Byzantine Fault Tolerance · Performance · Scalability · Consortium

1 Introduction

Byzantine Fault Tolerant consensus algorithms [1] for Distributed Ledger Technology (Blockchain) in general shows a transaction throughput performance drop in scalability and consistency due to the following factors i) Communication Complexity and ii) Fork Issues which cause inconsistency of chains or state between the participating nodes. The classical blockchain BFT algorithms of Practical Byzantine Fault Tolerance (PBFT), Istanbul Byzantine Fault Tolerance (IBFT), Quorum Byzantine Fault Tolerance (QBFT), and Clique are

S. Ossowski et al. (Eds.): DCAI 2023, LNNS 740, pp. 31–43, 2023.
https://doi.org/10.1007/978-3-031-38333-5_4

affected by the above-mentioned drawbacks. In our consensus protocol CUBA, we overcome the earlier mentioned constraints to render the blockchain system more scalable and failure tolerant and maintain the consistency of the chain. Through quorum-based consensus message exchange, the protocol lightens the communication complexity and awards. It calculates each participant node's score according to protocol actions, making it resilient to malicious actions. We explain our proposal with a state of art optimization algorithms around BFT and later explain CUBA followed by its evaluation in a blockchain simulation against PBFT, IBFT, QBFT, and clique.

2 State of Art

The various methodologies or propositions to improve the Byzantine Fault Tolerant consensus protocols' scalability, performance, and resilience are discussed hereafter, in line with the Consistency, Availability, and Partition Tolerance Theorem. The **Reduced Communication Complexity** methodology aims to reduce the communication bottleneck through various techniques but introduces additional problems. Single Phase Protocol and Linear Communication Protocols like Zyzzyva [2] have a single phase of PRE-PREPARE message from a primary. Then it expects the replicas to acknowledge the message. But this assumption is strong as the network is always unpredictable, leading to liveness and consistency issues. These protocols elect a single proposer, which is optimistic without considering failure scenarios. They suffer communication overhead in fallback cases of view change and necessitating leadership selection. Also, the protocol of Clique [3], and Aura [4] follow the same pattern of a single chosen leader, which assumes the replicas work without any issue and are prone to chain fork problems or network deadlock if two leaders propose a block at the same height.

On the other hand, the **Randomised BFT approach** Protocols can be either for randomized selection of transactions as in HoneyBadger BFT protocol [5] or randomized selection of validator committee from a total set of validators in the case of Proteus BFT. These randomization protocols offer better performance and scalability for 100–200 nodes. A random generation can be faulty, and it should respect the properties of unbiasedly, Reliable, Verifiable, decentralization, and unpredictability. A public centralized random generation system called Randomness Beacon by the National Institute of Standards and Technology exists. Still, a decentralized ledger can be compromised if it relies on a centralized actor. Also, the factor of randomness in the sense of BFT Protocols cannot guarantee deterministic termination. It can only achieve a termination with a high probability, like in the case of Ben-Or's protocol [6].

Similarly, we also explore several other works around algorithm optimization for fuzzy controllers and last-mile delivery [14–16]. The proposed approaches are oriented towards collective decision-making and optimizing the steps needed for faster consensus. They employ the techniques of adaptive feedback for improvising the algorithm at each execution. Our CUBA algorithm for blockchain consensus inspires us to follow a feedback approach at the end of each block height interval for reorganizing and optimizing the network for faster finalization of blocks.

3 System Model

We design our protocol from a consortium perspective as private enterprises like Renault, Mercedes-Benz, Porsche, IBM, or other Original Equipment Manufacturers prefer this type of blockchain network to have a balance between transparency and privacy while creating inter-organizational synergies [7]. But we can very well extend this protocol to a much higher participant network of private or public nature. Our network definition comprises limited and well-identified participants with an established reputation. We assume that if there are f dishonest or faulty nodes in a network of N nodes, the consensus protocol needs to be tolerant for the condition $f \leq (N-1)/3$. Based on public key infrastructure, we consider a set of K-designated Node participants in the consortium network identified by $N_0 \ldots N_K$. The participants send and receive transactions that need a common agreement on the transaction order, validity, and immutability. The transactions agreed upon are, in turn, to be stored as a Block structure. We consider Transaction represented by T and Block B of size M containing a set of transactions $\{T_0 \ldots T_M\}$. The Blocks are enchained as a blockchain structure of height S represented as $\{B_0, B_1\{H(B_0)\} \ldots B_S H(B_{S-1})\}$ where H is the Hash Function digest of the Block data which is usually the hash of the previous block to be included in the current block.

Network Considerations: We assume an asynchronous network where the messages may be delayed but eventually turn synchronous, delivering the message in an unknown but finite time interval bound of Global Stabilisation Time δ (GST). This model respects the safety and liveness properties as the network can eventually stabilize within a limitation of some byzantine nodes. The nodes broadcast transactions, including signature and other application data, through a reliable channel that tries to rebroadcast the transactions unless delivered successfully. The nodes need to follow the extended Atomic broadcast protocol by the following properties of [8], all eventually delivering within a GST δ:

1. **Validity:** If an honest node broadcasts a Transaction T_i then it is eventually delivered.
2. **Agreement:** If an honest node broadcasts a Transaction T_i then all the nodes deliver the transaction.
3. **Total Order:** If honest nodes N_i and N_j deliver Transactions T_a and T_b, then N_i deliver the transaction T_a before T_b if and only if N_j deliver the transaction T_a before T_b.

Consensus Intuition: In the consortium network C, where each Node N_K possesses an intellectual knowledge π_K about others, each node strives, along with others, in maintaining the utilitarianism or in other words, consistency, availability, liveness and resilience of the network. This can be ensured by attributing an effective score to each node based on his knowledge and actions of the node in the past. Consider a Block containing a set of transaction emitters Te who has signed and issued the transactions represented as a set $\{Te_0, Te_1 \ldots Te_n\}$. Then this block has to be validated for consensus, taking into account the following:

1. If there is only a unique block at the destined index i
2. if a valid node proposes the block and provides a signed hash digest.
3. If the block upon consensus attracts the necessary votes or approvals from other benign nodes.
4. If the block is propagated and added to the index by all the nodes without any conflict or forks ensuring a single persistent blockchain, and
5. If all the above actions are performed, i.e., communication or response in an expected time interval δ individually to ensure the stability and liveness of the network.

All these individual actions can be attributed to a utilitarian score U_a, U_b, U_c where a, b, c are considered actions enumerated above like block proposal, consensus message votes, or block as well as message propagation. These actions are then cumulatively managed in the network for each node as each node's utilitarian UT_i score. This score is also maintained in the negative sense if the above-expected actions are not performed or ignored by a node like M_a, M_b, M_c where M represents misbehavior and a, b, c reprises the former notation significance. The negative scoring mechanism ensures a sense of discipline and competition as it can decrease the score of any supposedly utilitarian node. The Effective Utilitarian Score EU_n, n indicating the Block height or index level in the chain, is measured as:

$$\sum_{n=0}^{k} EU_n = U_a^n + U_b^n + U_c^n - M_a^n - M_b^n - M_c^n$$

Here the effective utilitarian score EU_n consolidates the positive actions score subtracted from the negative actions score. An effective utilitarian score signifies utilitarian health in the network, as higher values signify positive utilitarianism, and a lower score signifies the presence of malicious participants hampering the network.

Consensus Finalisation: If an honest node or participant N_A adds a block B_m containing a set of valid transactions $\{T_1 \ldots T_K\}$ after approval from other participants at a particular index M in the blockchain then no other node N_B can add another block B_n to the same index at any point of time [9] even in the presence of forks subject to eventual resolution. In addition to the basic requirement of Consistency, Availability, and Partition Tolerance in any distributed consensus protocol, we add other desirable properties like Transaction Finalisation Time, Scalability, Consensus Decentralization measures, and resilience capacity of the protocol against impending malicious attacks.

4 CUBA Consensus

In this section, we explain the CUBA consensus protocol represented in Fig. 1 starting from the transaction processing until the network evolution for self-optimization based on the utilitarian behavior of each node in the network.

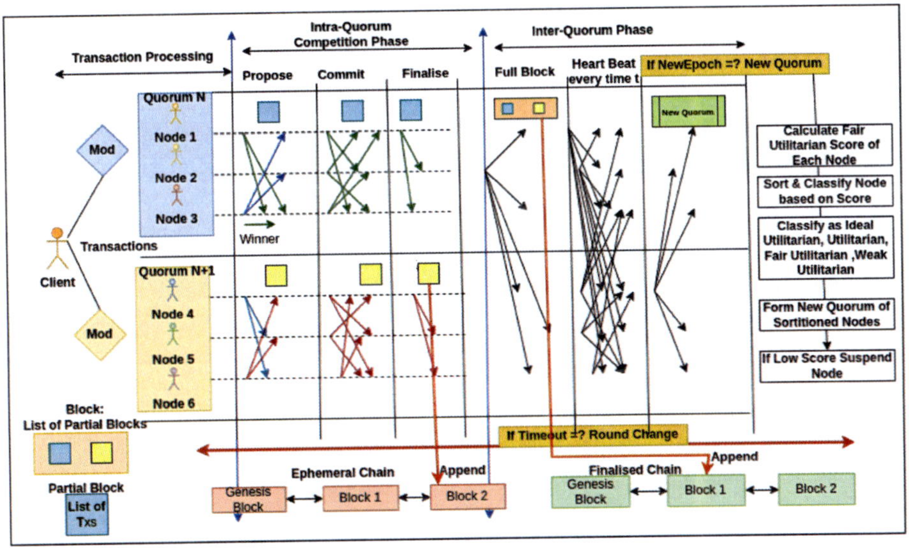

Fig. 1. CUBA Consensus Protocol

4.1 Consensus Network Organisation

In the Consortium Network C, the K participants are organized as a set of R Quorums $\{Q_1, Q_2, \ldots .Q_{R,}\}$. Each Quorum can accommodate members N_i belonging to more or less similar Net Effective Utilitarian Score threshold τ at the end of each Epoch. The Quorum organization facilitates two primary purposes 1) Lesser message communication complexity within a quorum rather than the entire network 2) Quorum can be used for grouping and allocating nodes belonging to similar utilitarian score levels. Quorum Membership changes for every epoch to optimize the network in removing malicious or faulty nodes. Each Quorum can accommodate until a Quorum Size σ up to the total no of Quorums in the network R to award a node N_i where $i \leq K$, membership for a new Epoch ϵ, then it needs to be classified into Ideal Utilitarian Υ_I, Utilitarian Υ, Fair Utilitarian Υ_F and Weak Utilitarian Υ_W depending on its utilitarian score in the descending order accumulated over epochs. Then for the Genesis Epoch ϵ_O, the classification would be a random distribution of nodes N_i as we don't have any prior Evaluation or Utilitarian State to consider. For Subsequent

Epochs, ϵ would be evaluated based on the τ for the preceding epoch. Quorum can communicate or pass messages in two ways, in Inter-Quorum Message M_I and Intra-Quorum Message M_ι.

Data Structure Definition: The fundamental unit in our protocol is Partial Block $P_{k,l}$ where k is the Block Height and l is the index position within a block or Quorum Index. Each $P_{k,l}$ is proposed by a Quorum l after the Intra-Quorum consensus for a given round k. It consists of a set of transactions self-assigned as part of its quorum membership. It also has the other identifiers of Timestamp, Previous Block State hash, current partial block hash, and proposer's signature. After the Intra-Quorum consensus, it will be placed inside a Block container B_k of Ephemeral Blockchain α_ψ for the k,l index and broadcasted to the consortium network. B_k needs the partial blocks to be proposed by each Quorum. As the partial blocks are received for the Block structure, the Temporal Hashing of the partial blocks received is updated. It is more to ascertain the state of an unfinalized container block containing a set of partial blocks before being updated with another partial block. This is done to maintain a history of state progression, as we always need non-repudiation in the network. After all the partial blocks are received for a Block k, it undergoes the secondary consensus of Inter-Quorum finalizing the block on the Finalized Blockchain β_ω. The Ephemeral Blockchain α_ψ is an intermediate chain with multiple blocks, and partial blocks can be proposed in parallel without waiting for a preceding block.

4.2 Transaction Processing

In this section, we look at the starting point of the protocol, which is the transaction processing by the blockchain network. As represented in Fig. 1, each of the client C_g where $g \in \mathbb{N}$ submits a transaction T_χ where $\chi \in \mathbb{N}$ to the network which is forwarded to the nodes. Each node rebroadcasts the transaction to the whole network in the desired network topology of Fully Connected, Ring Lattice, or Watts Strogatz. Then the process of Intra-Quorum and Inter-Quorum happens simultaneously for multiple index partial blocks and blocks in a pipelined nature.

Intra-Quorum Phase: The Intra-Quorum phase consensus operates in sub-phases of PREPARE, COMMIT, and FINALISE as illustrated in Fig. 1. Its working is as follows:

1. Each Quorum Q_r upon receiving the transactions in its pool up to a partial block size κ proposes a partial block $P_{k,l}$ if the node is a partial block proposer based on the codepoint on the hash of Block height K. To create a gamified protocol the competitor proposer is also chosen among the quorum members to propose a similar Partial Block $P'_{k,l}$.
2. During the PROPOSE sub-phase, the block proposed by the proposer or competitor is propagated in the Quorum network. $P_{k,l}$, which achieves the threshold of 2/3 votes in the COMMIT phase, is confirmed. During the FINALISE

sub-phase, the finalizer proposer is chosen to add the partial block to a block container and then broadcast it to the quorum network. If the finalized block is not achieved within a timeout Δ, then a Fulfiller is selected to finalize the block. In all the cases of Intra-Quorum and Inter-Quorum phases, when there is a blockchain state deadlock either in the ephemeral or finalized chain, a ROUNDCHANGE phase is initiated. This will entail a Quorum Reorganisation, and the nodes which have failed or behave maliciously are removed for optimization in the subsequent round.

Inter-Quorum Phase: In the previous phase of Intra-Quorum consensus (detailed in Sect. 4.2), individual blocks were placed in the block container $\mathrm{B}_{\mathrm{c,k}}$. Then the temporal hash is updated in the container as and when all the indexes of a partial block in the block B_{K}. When a ρ partial blocks are received in the ephemeral chain α_{ψ} equivalent to the number of quorums in the network, a full block proposer is chosen to propose the block B_{K}. Block B_{K} is formed considering the last temporal hashing state by the block proposer. In case the proposer fails to complete the block, then after a timeout of η then, a fulfiller is chosen. Fulfiller proposes the block B'_{K}, which is updated. In case of conflict between the blocks B_{K} and B', then the time is compared to resolve the conflict. The block confirmed is stored in the finalized blockchain β_{ω}. In case of non-fulfillment of blocks, then a RoundChange phase is initiated as a default view change for Quorum Reorganisation.

4.3 Network Evolution

Continuing the explanation from Inter-Quorum where a Block B_{K} is either formed by the block proposer or fulfilled, it is propagated throughout the network. Each Node N_{i} maintains an internal register of the utilitarian score derived from each Block finalized. The post-mortem on the Block can be used to attribute the utilitarian score as explained as follows:

1. Each Block is composed of partial blocks wherein each partial block has several attributes for differentiating its utilitarian score among all nodes. First, the Inter-Block time coefficient is calculated concerning the previous block as it measures the real effort maintained in the network. Then this coefficient is utilized as a multiplication factor for each positive or negative score.
2. The positive score in Block B_{K} can be attributed to an action if it improves the utilitarian score as the following:
   (a) **Partial Block Proposal Winner** if a proposer succeeds in the PROPOSE and COMMIT sub-phase for the concerned Quorum.
   (b) **Commit Win** for those who participated by votes in reaching the desired threshold faster than the competitor partial block.
   (c) **Heart Beat** score for those who have sent their ping message to all the nodes to establish their liveness metric.

   The negative score for B_{K} can be derived as the actions in the opposite sense as listed before:

(a) **Partial Block Proposal Loser:** if a proposer loses in the competition race against a peer in the Quorum as a penalization.
(b) **Commit Loser:** for those expected to endorse a winning partial block but failed, as noted by their vote signatures.
(c) **Heart Beat Missed:** score for those who failed benignly or malicious behavior to cascade their liveness message for fixed frequency interval.
(d) **Malicious:** score, if the block or partial block propagated, invalidates with forged signatures, hash, or votes.

Each score is consolidated at the block level multiplied by the inter-block time coefficient as outlined in Algorithm: 1. Each score is based on the node's previous disposition of Ideal Utilitarian, Utilitarian, Fair Utilitarian, or Weak Utilitarian. The getFairnessScore method calculates the fairness score to give more weightage scores to weak participants and lesser weightage to more benign participants. This is to encourage the node to recover if it fails intermittently. The net score is calculated as a difference between the positive and negative scores.

For a current Epoch E, before forming the next Epoch E+1, we need an effective understanding of the Utilitarian Score across nodes with varying epochs. The essence of the forgetting coefficient component is to give more weightage to recent epochs and lesser weightage to past epochs. This will gradually descend if we move from the recent epoch to the genesis epoch. The values of the forgetting coefficient are calculated concerning the current epoch. It is calculated by iterating for each previous epoch since the current epoch. The calculated forgetting coefficient for each previous epoch is applied by multiplying each with EffectiveScore obtained in the previous section. This reduces or augments the relevance based on the score obtained in earlier or recent epochs.

Node Utilitarian Classification: Then we classify the nodes based on the utilitarian score until the recent epoch and reorganize the Quorum formation, which works as follows:

1. The Nodes are sorted in descending order based on the Utilitarian Score until the recent epoch, then split into four equal intervals of nodes based on their rank. The First interval of nodes is marked and placed into Ideal Utilitarian List, Second Interval into Fair Utilitarian, the Third into Utilitarian, and the fourth into Weak Utilitarian. As obvious by their nomenclature, the Weak Utilitarians are the nodes susceptible to failures and have continuously degraded the network's performance.
2. In case a node has been in the Weak Utilitarian list for Z previous epochs, then it is suspended for an upcoming epoch as a kind of penalization and remediation at the same time. Heart_Message is expected of a node every θ frequency for inclusion in the classification and the proposition of the new quorum. If the heartbeat message is not received, it is also suspended for the Epoch in progress. The suspension factor works on the transparent Utilitarian Score and Heart Beat Message to enforce the vitality of the blockchain network.

```
// Input: New Block B_K
Function UtilitarianScoreUpdate(B_K):
    // Inter-Block Time Coefficient
    Inter_Block_Time_Coefficient ← B_K.blocktime - B_K-1.blocktime;
    // Normalising to milliseconds
    Time_Coefficient ← Time_Coefficient/100000;
    for EachPartialBlock In B_K.PartialBlocks do
        Utilitarian_Proposal_Win ← Utilitarian_Proposal_Win +
          Mapping<B_K,<EachPartialBlock.PartialBlockProposer , PartialBlock_Proposal_win*
          Time_Coefficient +
          getFairnessScore(CurrentEpoch,Node Identifier) >>;
        Utilitarian_Proposal_Loss ← Utilitarian_Proposal_Loss +
          Mapping<B_K,<EachPartialBlock.PartialBlockCompetitor , PartialBlock_Proposal_loss*
          Time_Coefficient >>;
        Utilitarian_Commit_Win ← Utilitarian_Commit_Win +
          Mapping<B_K,<EachPartialBlock.PartialBlockCommiter , PartialBlock_Commit_win*
          Time_Coefficient +
          getFairnessScore(CurrentEpoch,Node Identifier) >>;
        Utilitarian_Commit_Loss ← Utilitarian_Commit_loss +
          Mapping<B_K,<EachPartialBlock.PartialBlockCommitLoser , PartialBlock_Commit_loss*
          Time_Coefficient >>;
    end
```

Algorithm 1: Algorithm for Calculating the final fair Utilitarian Score of Nodes

Quorum Reorganisation: Then the final action to be performed as perform of the Quorum Reorganisation works as follows:

1. For each new Epoch E, a selected Quorum Proposer based on past Epochs of performing Intra-Quorum, Inter-Quorum consensus, Utilitarian Calculation, Fairness, and Inter-Block Coefficient, Forgetting Quotient and finally, we sort each classified node into a Quorum. Blockchain network requires ρ quorums wherein each quorum needs to be filled with classified nodes.
2. The Quorums are filled with each up to σ number of nodes in their order of ascending ranks. This will organize the Ideal Utilitarian, Utilitarian, Fair Utilitarian, and Weak Utilitarian in homogeneous groups of each quorum. A minute heterogeneity can be observed since there cannot be even allocation of quorums to nodes of the same utilitarian classification due to the suspension or failure of nodes. The newly formed QUORUM_MESSAGE<E_ε> with quorum reorganizations for the new Epoch is proposed by the Quorum Proposer to the other peers in the network for evolution improving optimization and resilience.

5 Consensus Evaluation

We analyze the CUBA protocol to understand its performance and limitations in line with the above perspective. CUBA works in an eventual synchrony model by progressing the epochs where a new set of ρ quorums, each having σ members inside a quorum. As the progression of the finalized chain is based on the block height and epoch, it is resistant to clock skewness to an extent. But in case of unexpected anomalies of partial block fulfillment timeout, full block fulfillment timeout, and round change timeout, the protocol depends on clock synchronizations. This can lead to an intermittent liveness issue until the node recovers within a global timeout δ. So essentially, our protocol cannot offer strong consistency but an eventual consistency model. Also, in case of an extended issue of liveness, a round change is invoked where the quorums are reorganized, and failure nodes can be suspended for certain epochs rendering the network resilient. It is evident by the heartbeat message and the inter-block time coefficient a utilitarian score that can diminish the overall score of a failure node. Our protocol exhibits average availability, eventual consistency, and partition tolerance. Let us understand the protocol by following questions:

1. Why not high availability? In the protocol, we can sustain until 2f+1 node for f failure nodes in the overall network as we need benign nodes to progress, or we would repeatedly fall to the round-change phase. This makes the system reliant on the minimum 2f+1 nodes for safety and liveness to render the blockchain available normally. A highly available system would be immune against any threshold of bad actors we cannot offer in our byzantine setting.
2. How CUBA makes a tradeoff between partition tolerance and eventual consistency? A blockchain protocol should mandatorily be partition tolerant either a priori like Practical Byzantine Fault Tolerance (PBFT) [10] protocol where there is no possibility of forks or a posteriori like Clique [3], Authority Round (Aura) protocol [4] which uses a fork resolution algorithm like GHOST (Greediest Heaviest Observed Subtree) in addition to resolving the blockchain partitions. In CUBA, we avoid the presence of forks as there is a checkpoint mechanism of Quorum reorganization every epoch interval, which detects the drift and auto-corrects to prevent the adverse scenario. Organization in terms of multiple quorums, each limited to a particular index of the partial block, cannot create a long chain to produce a fork without consensus from the benign participants. In addition, The intra-quorum phase has three sub-phases, making it eventually consistent with an acceptable timeout. The corner case of more than 2f+1 malicious nodes can create a partition that renders the system prone to byzantine failure, a proper limitation within the threshold.

5.1 Experimental Evaluation

We implement the CUBA protocol using Java Object Oriented language, where the consensus algorithms are tested in a blockchain simulator. The Simulator implemented behaves as a blockchain protocol including the components of 1) Network, 2) Cryptography, 3) Blockchain and Transaction Payload Object, 4) Peer-to-Peer Socket Communication 5) Storage Pools for Transaction, Message, and Consensus handling. Along with CUBA, the other Proof of Authority consensus algorithms of PBFT, Clique, Istanbul Byzantine Fault Tolerance (IBFT), and Quorum Byzantine Fault Tolerance (QBFT) [12] are implemented for comparison bench-marking. The infrastructure for testing the above implementation

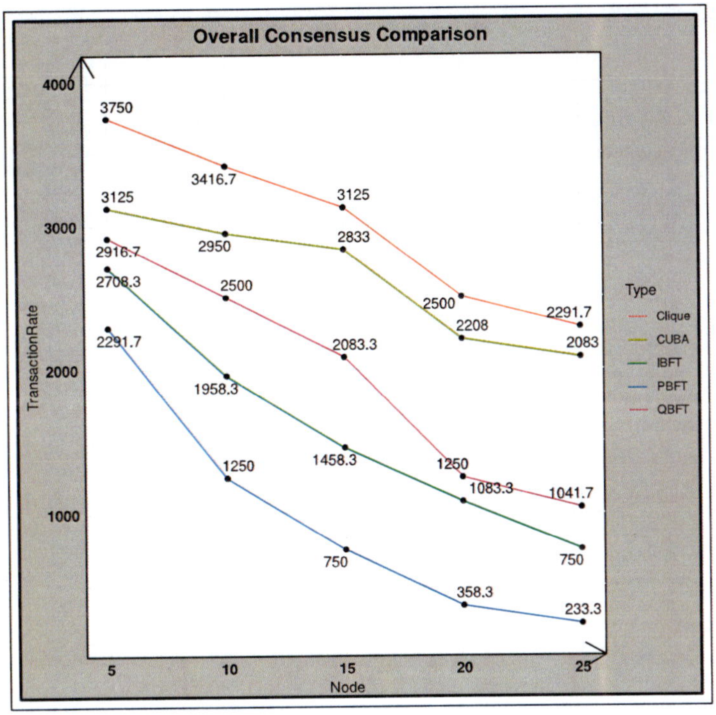

Fig. 2. CUBA Consensus Protocol Comparative Evaluation

is on TAS Cloud, whose data centers are based in Sophia Antipolis, Nice, France. The Simulator is built into a JAVA archive ported to an Apache Tomcat Docker Container Image, which is pushed to DockerHub. Then the Kubernetes YAML script is generated to deploy Pods containing the earlier-built docker containers. Then the launch script for the test starting the transmission of the transactions is launched from the test orchestrator machine. We represent the results in a consolidated Fig. 2, which compares CUBA against classical BFT algorithms of Clique, IBFT, PBFT, and QBFT consensus implementations. The implementation of the CUBA protocol and cloud test suite is public. The simulator is available at https://github.com/scyrilnaves/article-cuba. The result analysis can be discussed as follows:

1. Clique shows a relatively smooth drop in scalability but not a steep drop as it's more of a proposed block propagated to the entire network. But this single phase can introduce the presence of forks if there is a noticeable latency in the network [11,13].
2. PBFT with 4 phases of PRE-PREPARE, PREPARE, COMMIT, and ROUND CHANGE augments the message complexity in proportion to the number of nodes. IBFT is a down-phased version of PBFT with PREPARE, COMMIT, and ROUND CHANGE only in case of liveness issue, which is

better than the former but can be prone to duplicate block proposition at the same height, which is already acknowledged [12] which needs to solved by block locking.
3. QBFT is similar to IBFT but scales better as the block-locking mechanism is removed and replaced by round change.

Exceptionally for 15 Nodes, it sustains the throughput performance for 6 Quorums but slips around 2000 at the same setting. This is relatively better than PBFT, IBFT, and QBFT due to having message communication only within the quorum members. Also, the simultaneous proposition of multiple partial blocks and blocks for different block heights parallelizes the consensus with higher confirmation rates. It also can track the consensus actions and reorganize them for better performance, as faults can be detected based on the accumulated scores. CUBA performs slightly lesser than Clique, but the former has assured finalized blockchain and the presence of no forks. The Clique with a single block proposal phase has higher throughput due to linear message complexity, but it leads to chain consistency issues. So the transaction can never be considered finalized in Clique, which affects the blockchain security, but CUBA has a single finalized consistent chain. In addition, the protocol's ability to reorganize and identify the byzantine nodes optimizes the throughput further and resilience. CUBA has a higher performance than its BFT peers due to the reason of:

1. Reduced message complexity as communication is within the quorums of selected members.
2. Pipe-lining through multiple partial blocks and blocks proposition by the different quorums simultaneously in stages of ephemeral chain and finalized chain improves the protocol.
3. Resilience of the protocol by evolution based on the utilitarian score optimizes the network.

6 Conclusion

This paper proposes a BFT consensus algorithm based on the action outcome of each node participation in a consortium network. We detail the two phases of Intra-Quorum and Inter-Quorum for partial blocks and block finalization in the ephemeral and finalized blockchain. We further investigated the experimental implementation in a public cloud environment and measured its scalability. We compared the results bench-marked against classical protocols and the current state of art BFT protocols with promising results. The CUBA protocol shows more acceptable throughput conditions with the recovery due to quorum optimization at the end of each epoch. The protocol offers better scalability in a consortium network than PBFT, IBFT, and QBFT, except for Clique, which is susceptible to high forks. In the case of advertent behavior, the nodes can ensure resilience by ensuing suspension and reorganizing the network and ensuring the liveness and performance of the network.

References

1. Lamport, L., Shostak, R., Pease, M.: The Byzantine generals problem. ACM Trans. Program. Lang. Syst. https://doi.org/10.1145/357172.357176
2. Kotla, R., Alvisi, L., Dahlin, M., Clement, A., Wong, E.: Zyzzyva: speculative Byzantine fault tolerance. ACM Trans. Comput. Syst. **27** (2010). https://doi.org/10.1145/1658357.1658358
3. Clique PoA protocol and Rinkeby PoA testnet. https://github.com/ethereum/EIPs/issues/225. Accessed 4 Apr 2023
4. OpenEthereum Aura - Authority Round. https://openethereum.github.io/Aura
5. Jalalzai, M., Busch, C., Richard, G.: Proteus: a scalable BFT consensus protocol for blockchains. In: 2019 IEEE International Conference on Blockchain (Blockchain), pp. 308–313 (2019)
6. Bano, S., et al.: SoK: consensus in the age of blockchains. In: Proceedings of the 1st ACM Conference on Advances in Financial Technologies, pp. 183–198 (2019). https://doi.org/10.1145/3318041.3355458
7. Kaula, Inc. and AZAPA Co., Ltd.: Kaula together founded Automotive BlockChain Consortium (ABCC) toward creation of new mobility services. https://kaula.jp/wp/wp-content/uploads/2018/08/180719-ABCC-PR-en-2.pdf
8. Sedlmeier, P., Schleger, J., Helm, M.: Atomic Broadcasts and Consensus: A Survey (2020). https://www.net.in.tum.de/fileadmin/TUM/NET/NET-2020-11-1/NET-2020-11-1_19.pdf
9. Vukolić, M.: The quest for scalable blockchain fabric: proof-of-work vs. BFT replication. In: Camenisch, J., Kesdoğan, D. (eds.) iNetSec 2015. LNCS, vol. 9591, pp. 112–125. Springer, Cham (2016). https://doi.org/10.1007/978-3-319-39028-4_9
10. Castro, M., Liskov, B.: Practical Byzantine fault tolerance. In: Proceedings of the Third Symposium on Operating Systems Design and Implementation, pp. 173–186 (1999)
11. Ekparinya, P., Gramoli, V., Jourjon, G.: The attack of the clones against proof-of-authority. arXiv. abs/1902.10244 (2020)
12. Saltini, R., Hyland-Wood, D.: IBFT 2.0: a safe and live variation of the IBFT blockchain consensus protocol for eventually synchronous networks (2019)
13. Samuel, C., Glock, S., Verdier, F., Guitton-Ouhamou, P.: Choice of Ethereum clients for private blockchain: assessment from proof of authority perspective. In: 2021 IEEE International Conference On Blockchain And Cryptocurrency (ICBC), pp. 1–5 (2021)
14. Preitl, Z., Precup, R.-E., Tar, J., Takács, M.: Use of multi-parametric quadratic programming in fuzzy control systems. Acta Polytechnica Hungarica **3**(1), 1–17 (2006)
15. Maestro, J.A., Rodriguez, S., Casado, R., Prieto, J., Corchado, J.M.: Comparison of efficient planning and optimization methods of last mile delivery resources. In: Gao, H., J. Durán Barroso, R., Shanchen, P., Li, R. (eds.) BROADNETS 2020. LNICST, vol. 355, pp. 163–173. Springer, Cham (2021). https://doi.org/10.1007/978-3-030-68737-3_11 ISBN 978-3-030-68737-3
16. Precup, R.E., David, R.C., Roman, R.C., Petriu, E.M., Szedlak-Stinean, A.I.: Slime Mould algorithm-based tuning of cost-effective fuzzy controllers for servo systems. Int. J. Comput. Intell. Syst. **14**(1), 1042–1052. https://doi.org/10.2991/ijcis.d.210309.001. ISSN 1875-6883

From Data to Action: Exploring AI and IoT-Driven Solutions for Smarter Cities

Tiago Dias[1,2](✉), Tiago Fonseca[1], João Vitorino[1,2], Andreia Martins[1,2], Sofia Malpique[1,2], and Isabel Praça[1,2]

[1] School of Engineering, Polytechnic of Porto (ISEP/IPP), 4249-015 Porto, Portugal
{tiada,calof,jpmvo,teles,pique,icp}@isep.ipp.pt

[2] Research Group on Intelligent Engineering and Computing for Advanced Innovation and Development (GECAD), 4249-015 Porto, Portugal

Abstract. The emergence of smart cities demands harnessing advanced technologies like the Internet of Things (IoT) and Artificial Intelligence (AI) and promises to unlock cities' potential to become more sustainable, efficient, and ultimately livable for their inhabitants. This work introduces an intelligent city management system that provides a data-driven approach to three use cases: (i) analyze traffic information to reduce the risk of traffic collisions and improve driver and pedestrian safety, (ii) identify when and where energy consumption can be reduced to improve cost savings, and (iii) detect maintenance issues like potholes in the city's roads and sidewalks, as well as the beginning of hazards like floods and fires. A case study in Aveiro City demonstrates the system's effectiveness in generating actionable insights that enhance security, energy efficiency, and sustainability, while highlighting the potential of AI and IoT-driven solutions for smart city development.

Keywords: security · internet of things · smart city · machine learning

1 Introduction

The growth and urbanization of cities worldwide, associated with increasing population expectations, has led to a multitude of complex challenges across various domains, such as transportation, public safety, energy consumption, and infrastructure maintenance and management. These challenges can result in significant negative impacts on both the environment and citizens' standard of living, making them into a compelling impetus for the efforts of city planners, policymakers, engineers, and the public at large. Against this backdrop, the emergence of smart cities, powered by the use of advanced Information and Communication Technologies (ICT), such as the Internet of Things (IoT), Artificial Intelligence (AI), Machine Learning (ML), and data analytics, promise to unlock cities' potential to become more sustainable, intelligent, efficient, and ultimately livable for their inhabitants.

Despite its novelty and innovative nature, several cities around the world have already begun to implement smart city solutions [1]. For instance, Barcelona, Spain, has established a comprehensive smart city platform [2] that includes various solutions, such as

S. Ossowski et al. (Eds.): DCAI 2023, LNNS 740, pp. 44–53, 2023.
https://doi.org/10.1007/978-3-031-38333-5_5

smart parking, waste management, and air quality monitoring, all aimed at improving urban life for citizens. In Singapore [3], the Smart Nation initiative aims to harness technology to improve the quality of life of residents by enhancing transportation, healthcare, and public safety. In Portugal, the city of Aveiro is one of the city's leading these innovations and has made significant investment towards becoming a smarter city. The city has established the Aveiro Tech City Living Lab (ATCLL) [4], which is a research platform that serves as a testing ground for smart city solutions.

This work introduces the Intelligent City Management System (ICMS), the result of our participation in the first edition of the Aveiro Tech City hackathon, promoted by the municipality of Aveiro, that aimed to further enhance the capabilities and services of the city's management platform. ICMS is designed to enhance city management through a scalable and intuitive AI-powered system that integrates multiple data analysis and prediction dashboards. The system impact is evaluated by a live case study using real-world data derived from the ATCLL environment and additional IoT sensors available during the hackathon.

2 State-of-the-Art

The concept of a "smart city" has emerged as a response to the economic, social, and political challenges faced by post-industrial societies at the outset of the new millennium. This idea involves the strategic use of digital technologies to develop innovative solutions for urban communities. Therefore, the primary objective is to address the challenges encountered by urban society, including environmental pollution, demographic change, healthcare, the financial crisis, and resource scarcity [5].

Novel advancements in IoT are a key enabler for smart city applications [6], being responsible for generating an enormous quantity of data [7]. Indeed, in [8] Allam et al., have put forth a novel Smart City framework that integrates AI technology and urban systems, with a primary focus on enhancing urban sustainability and liveability. The authors contend that technology ought to serve as a fundamental cornerstone of Smart Cities, where Big Data can be derived from various domains through IoT. AI is proposed as an underlying feature capable of processing, analyzing, and interpreting the generated data. Moreover, to evaluate the electricity consumption patterns in Iran, Ghadamiet al. employed machine learning techniques and implemented dynamic strategies to foster citizen participation in renewable energy generation, informed by expert knowledge. The authors utilized a combination of an Artificial Neural Network and statistical analysis to develop a Decision Support System [9].

Another IoT applications for smart cities are related to vehicular traffic data, which represents one of the most vital data sources in a typical smart city. Effective analysis of this data can yield significant benefits for both citizens and governments. Neyestani et al. proposed a Mixed-Integer Linear Programming model for the traffic behavior of Plug-in Electric Vehicles, which can be integrated as a sub-module in various other studies such as operation and planning, thereby providing decision makers with valuable insights in urban environments [10, 11].

In this context, the ATCLL an open platform for developing, testing, and demonstrating innovative concepts, products, and services related to the urban environment.

It includes an advanced communication infrastructure and an urban data management and analytics platform that can collect, process, and analyze data from various sources. The platform offers opportunities for anyone, or any organizations interested in devising novel solutions for the predicaments encountered in contemporary urban settings [12].

ATCLL integrates a communication infrastructure and sensing platform, which comprises an array of smart lamp posts that are equipped with both traffic radars and video cameras. Additionally, the platform integrates buses and other vehicles that are fitted with devices that collect and transmit data. Furthermore, sensors are deployed throughout the city to monitor the number of people present in different zones, as well as to measure environmental quality and other relevant factors. The seamless integration of these components creates a comprehensive and sophisticated technological ecosystem that enables the collection and analysis of vast amounts of data, providing new and innovative ways to address the challenges faced by modern cities. Overall, the ATCLL is a cutting-edge initiative that combines technology, research, and innovation to create a living laboratory for urban development [4, 12].

3 Proposed Solution

The literature shows that smarter cities generate an enormous flow of information which can be useful to keep track, improve and solve issues inherent to the city. Ultimately, its progression and development provide its inhabitants with better life quality. However, infrastructure costs, privacy and security issues and interoperability of multiple systems can be an embargo to achieve this goal. As such, this work attempts to facilitate and leverage the implementation of smart devices installed across Aveiro city to create an Intelligent City Management System (ICMS).

The proposed system is an AI-powered comprehensive system that integrates multiple data analysis and prediction dashboards to provide a single point of management for the city. The system provides a holistic view of various sectors of the city, with analytics and forecasting capabilities that allow city managers to make decisions quickly and effectively, improving efficiency and resource allocation. To enable its use across different cities, the system is highly scalable and configurable.

The proposed system is divided into four different components: (i) City Security and Safety (CSS), (ii) City Energy Management (CEM), (iii) City Infrastructure Maintenance (CIM), and (iv) City Management Dashboard (CMD). In this Representational State Transfer (REST) architecture, each component is considered a REST API and their communication is based on HTTP request. The authors decided to follow this architecture, as it allows for very low coupling of the components that represent different city management fields, making the system highly scalable, reliable, portable, and maintainable (Fig. 1).

Each component is implemented in Python and provides a real-time analysis of several aspects of the city, using data gathered from the ATCLL. The following subchapters define the problem, the goal, and the implementation strategy of each city management component.

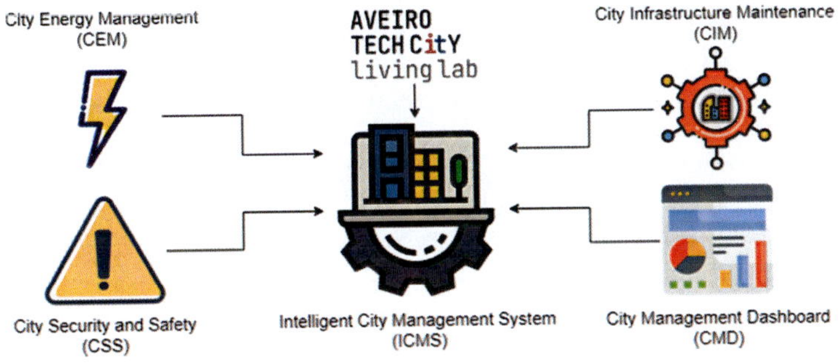

Fig. 1. ICMS architecture.

3.1 Use Case 1: Security and Safety

Ensuring the security and safety of both vehicles and pedestrians is of paramount importance in modern cities, as it not only protects the well-being of individuals but also contributes to the overall livability and sustainability of the city.

The CSS component focuses on improving road and pedestrian safety, by analyzing data provided by smart posts spread across Aveiro city. As described in Sect. 2, these smart posts are equipped with multiple sensors, of which cameras and speed radars, that are strategically placed to capture pedestrian and vehicle circulation in the same area. The authors considered the use of this information relevant to monitor in real-time driver and pedestrian safety. The premise of this correlation is that a driver should adapt his/her driving behavior depending on the number of pedestrians in a certain zone, since the probability of an accident that compromises safety of those around is much higher. The goal of this integrated component is to provide intelligently organized data and assist on the decision-making process of security implementations regarding the city's public highways. CSS works similarly to an expert system, as it is capable of correlating information using user defined rules, however, it further expands on it by being capable of performing feature computation. These features and rules should be managed by the city's security decision-makers.

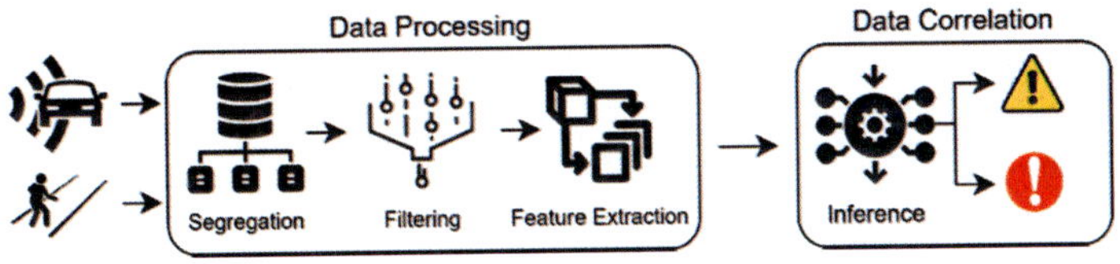

Fig. 2. CSS pipeline overview.

As described in Fig. 2, this use case is divided into the data processing and data correlation phases. The firstly the data consumed is segregated by smart post, to ensure correctness of the correlations. Then, the radar data that is unrelated to heavy or light vehicles are discarded. Lastly, for each smart post, the speed average and the number of pedestrians features are computed for the same time frame, which corresponds to the cadency configured, to aggregate the occurrences in each zone.

The second phase takes the processed intelligible data and correlates it according to rules defined by the decision-makers. The resulting correlation is then classified by the rules as warning or danger depending on the severity of the violation. Since the data that is correlated always belong to the same place, the city's security decision-makers can visualize where the violations are occurring and take into consideration a frequency level that is presented to decide whether security measures should be taken.

Even though this implementation only considers information regarding radar sensors and pedestrian count, other information can be included in the rules, as long as it is captured by the smart posts. The addition of other information, such as the existence of walkways or not can be included to attain a more fine-grained security analysis between pedestrians and vehicles.

3.2 Use Case 2: Energy Management

The global environmental changes and ongoing energy crisis have amplified the need for efficient energy management in urban environments. As cities around the world strive to become smarter cities, they are actively exploring ways to optimize their energy consumption and reduce their carbon footprints. In this context, City Energy Management (CEM) has emerged as a crucial component in the design and operation of our smart city platform (Fig. 3).

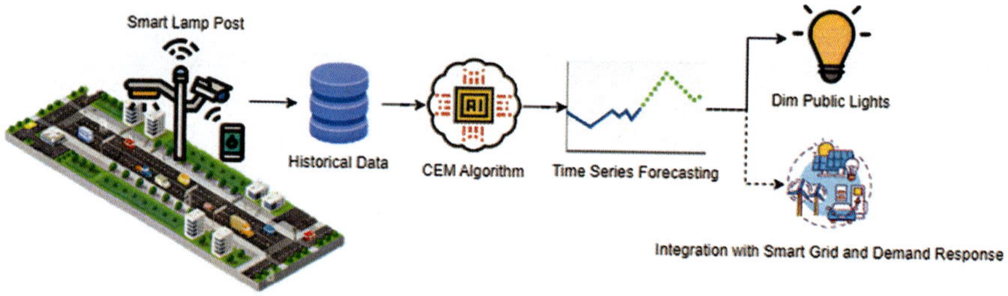

Fig. 3. CEM pipeline overview.

Our solution utilizes the ATCLL smart lamp posts, which are equipped with a variety of sensors and cameras. Given their historical data on the number of identified pedestrians, vehicles, and other moving objects in each street, our algorithm is designed to predict the number of movements likely to occur on the street in the next 24 h, accounting for the differences between workdays and weekends. Based on these predictions, CEM can provide recommendations on when to dim public lighting in specific streets. Alternatively, if the lights do not support dimming, CEM could advise shutting off half of the lamp posts in the street. This approach to public lighting enables a city to reduce its energy consumption while maintaining public safety.

Moreover, we highlight the possibility of integrating CEM with smart energy communities and intelligent demand response strategies, such as [13]. This can bring synergistic advantages because utility providers can effectively optimize and schedule flexible energy resources and energy storage across the city, leading to a reduction in costs and peak demand. By forecasting the absence of people on several streets at night, the system can dim the public lights in an event of peak grid consumption, acting as a smart regulating reserve mechanism. Participation in such mechanisms can even generate revenue for the city. Consequently, CEM not only acts as an isolated smart-city solution, but it can also be part of the creation of a more efficient and robust energy grid for urban areas, while incentivizing the use of renewable energy.

Regarding the specifically designed AI time-series forecasting algorithm, its implementation process consisted of four steps: collection and preprocessing of historical data, feature engineering, model selection, and model training. These are the steps that permit our algorithm to learn the patterns and relationships between various factors that influence the number of pedestrians, vehicles, and objects on the streets of a city. First in the data collection and preprocessing step, we collect historical data from the smart posts in Aveiro, which includes pedestrian and vehicle counts, as well as weather conditions, day of the week, holidays, time of day, and local events. The data is then preprocessed to remove any outliers, missing values, and inconsistencies.

Next, during the feature engineering step, we extract relevant features from the preprocessed data, which include temporal features (e.g., hour of the day, day of the week), weather features (e.g., temperature, humidity), and event-based features (e.g., holidays, local events). These features are crucial for improving the accuracy of our model. Finally, we selected and trained our machine learning model for time series forecasting using the preprocessed data and features, adjusting hyperparameters as necessary to minimize the error in the forecasts.

3.3 Use Case 3: Infrastructure Maintenance

A city's infrastructure is essential for its efficient functioning, and regular maintenance is crucial to ensure its longevity. However, regular wear and tear and other factors can cause these infrastructures to fail, often leading to costly maintenance issues. Therefore, monitoring the infrastructure of a city is crucial to its development but the identification of maintenance issues can be challenging and time-consuming.

As part of the ICMS platform, CIM (Fig. 4) attempts to automate and improve the monitorization of Aveiro by leveraging its smart public transportation to efficiently monitor in a distributed way the city's infrastructure, resorting to live-image capturing and computer vision to detect infrastructure defects, which in turn are reported in real-time to the city's infrastructure engineers, allowing them to make data-driven decisions to ensure maintenance of the infrastructure.

The You Only Look Once (YOLOv5) [14] algorithm is utilized within the component to perform object detection of maintenance issues and the beginning of hazards, using a live feed by the smart public transportation. The algorithm was trained using three annotated datasets, the Pothole Object Detection Dataset [15], the Roadway Flooding Image Dataset [16] and the FIRE Dataset [17], to detect the occurrence of potholes, floods, and fires, which are three concerning aspects for the city of Aveiro.

The CIM execution pipeline consists of analysing the captured images of the city relying on the employed YOLOv5 model. The algorithm detects the city's infrastructure defects, highlighting them with bounding boxes and assigning them a confidence score between 0 and 1. A higher confidence score reflects worse conditioning of the

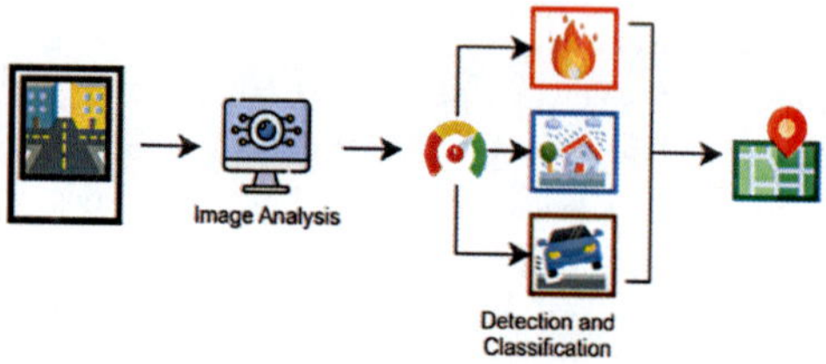

Fig. 4. CIM pipeline overview.

infrastructure and therefore requires more immediate attention. Lastly, the coordinates of detected issues are presented along with the highlighted images to the city's infrastructure engineers in an interactive map, so that they can be remotely analysed to decide which actions should be taken.

4 Case Study

An empirical case study was carried out to assess the feasibility and reliability of the proposed solution for the city of Aveiro. The organizing committee of the Aveiro Tech City hackathon provided two months of recorded data from the ATCLL, so the first month could be used for training of ML models, and the second for a holdout evaluation. ICMS was calibrated to the characteristics of the existing smart infrastructure and fine-tuned to the data readings of the city's IoT sensors. Then, the capabilities of the system were demonstrated live in the final stage of the hackathon.

Regarding the first use case, Security and Safety, the ratio between the number of speeding vehicles and the number of pedestrians, per hour of the day, can be visualized for each street of Aveiro equipped with the smart infrastructure. For the analyzed month, the ratio exceeded the allowed threshold in several streets and therefore further speed reduction mechanisms like speed humps and rumble strips are required to compel drivers to reduce speed. For instance, this can be noticed in a street where the threshold was slightly exceeded on a weekly basis (Fig. 5). Additionally, in this street and some nearby streets, there was a significant spike by the end of the month, possibly due to an event occurring in that area of the city with many pedestrians and vehicles in circulation. It could be valuable to correlate the sensor data with information about ongoing events in the city, to better distinguish between sporadic spikes and areas where drivers exhibit dangerous behavior on a regular basis.

Regarding the second use case, Energy Management, the number of pedestrians, vehicles, and other moving objects can be visualized for each street. To distinguish between day and night times, the latter have a darker grey background. Furthermore, the hours when no activity was registered on a street are highlighted in green blocks to indicate that power consumption could be reduced in those hours by shutting off public equipment or dimming public lighting. For instance, these blocks where efficiency could be improved can be noticed almost every night in Aveiro (Fig. 6).

Additionally, based on the historical data provided in the first month, the proposed solution can predict the number of movements likely to occur on the street in the next 24 h at a time, enabling a forecasting of the best hours to apply energy saving measures

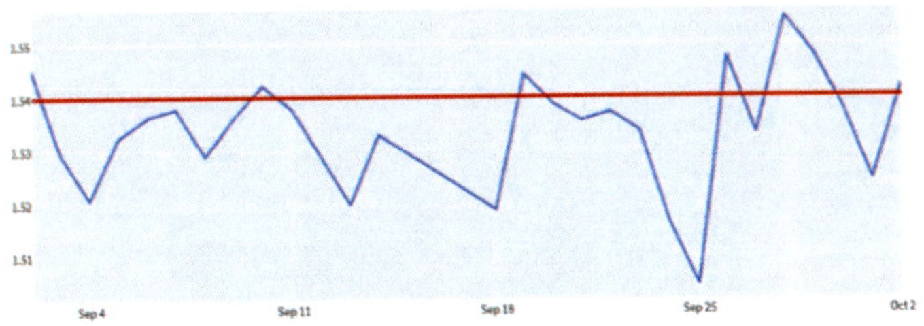

Fig. 5. CSS threshold analysis.

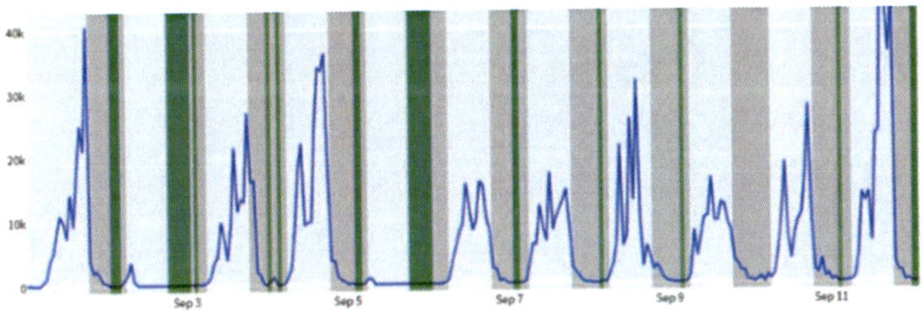

Fig. 6. CEM block identification.

(Fig. 7). These predictions achieved a good generalization throughout the second month of data, which demonstrates that cost savings could be achieved with more intelligent management of public lighting. If such algorithms were trained with an entire year of data, they could be improved to account for special holidays and events that may affect the hours of activity in different areas of the city.

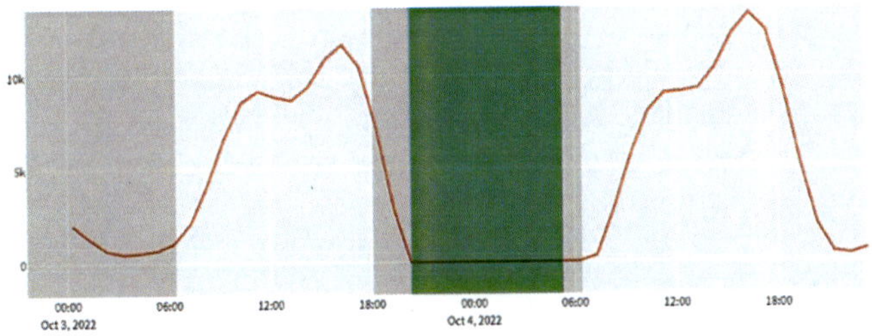

Fig. 7. CEM activity forecasting.

Regarding the third use case, Infrastructure Maintenance, every maintenance issue was created as an occurrence with location coordinates to be displayed in the interactive map of the ATCLL. For instance, in one of the main roundabouts of the city, a pothole was detected in a vehicle that was simulating the route of a bus, checking if it could be used as a mobile camera platform. The pothole was automatically assigned a confidence score of 0.41, which indicates that it is not an urgent issue, but it is still relevant for the municipality to fix it (Fig. 8). It is pertinent to note that the live feed of the camera was analyzed in real-time and immediately discarded afterwards, so only the frames where a public maintenance issue was detected were stored. Further cybersecurity measures like the anonymization of license plates are essential to comply with privacy regulations in smart city solutions that rely on camera feeds.

Fig. 8. CIM issue detection.

5 Conclusion

This work addressed several possible applications of AI to smart cities, in the context of the first edition of the Aveiro Tech City hackathon. The proposed system, ICMS, provides a data-driven approach to three use cases: (i) analyze traffic information to reduce the risk of traffic collisions and improve driver and pedestrian safety, (ii) identify when and where energy consumption can be reduced to improve cost savings, and (iii) detect maintenance issues like potholes in the city's roads and sidewalks, as well as the beginning of hazards like floods and fires.

By harnessing the power of AI and IoT, the proposed system can be significantly beneficial to the security, energy efficiency, and sustainability of a smart city. Further research efforts must be made to develop smart city solutions capable of tackling the environmental challenges of urban environments, so cities like Aveiro can provide more security and a better quality of life to their citizens.

Acknowledgements. The authors would like to thank the University of Aveiro, Instituto de Telecomunicações and Câmara Municipal de Aveiro for organizing the event and proving the city data utilized in this work.

This work has received funding from UIDB/00760/2020 and from UIDP/00760/2020.

References

1. Shamsuzzoha, A., Niemi, J., Piya, S., Rutledge, K.: Smart city for sustainable environment: a comparison of participatory strategies from Helsinki, Singapore and London. Cities **114**, 103194 (2021). https://doi.org/10.1016/j.cities.2021.103194
2. Rous, B.: The ACM digital library. Commun. ACM **44**(5), 90–91 (2001). https://doi.org/10.1145/374308.374363

3. Cavada, M., Tight, M.R., Rogers, C.D.F.: 14 - A smart city case study of Singapore—is Singapore truly smart? In: Anthopoulos, L. (ed.) Smart City Emergence, pp. 295–314. Elsevier (2019). https://doi.org/10.1016/B978-0-12-816169-2.00014-6
4. "Aveiro Tech City Living Lab." https://aveiro-living-lab.it.pt/citymanager. Accessed 23 Apr 2023
5. Ghazal, T.M., et al.: IoT for Smart Cities: machine learning approaches in smart healthcare—a review. Future Internet **13**(8) (2021). https://doi.org/10.3390/fi13080218
6. Bellini, P., Nesi, P., Pantaleo, G.: IoT-enabled smart cities: a review of concepts, frameworks and key technologies. Appl. Sci. **12**(3) (2022). https://doi.org/10.3390/app12031607
7. Ullah, Z., Al-Turjman, F., Mostarda, L., Gagliardi, R.: Applications of artificial intelligence and machine learning in smart cities. Comput. Commun. **154** (2020). https://doi.org/10.1016/j.comcom.2020.02.069
8. Allam, Z., Dhunny, Z.A.: On big data, artificial intelligence and smart cities. Cities **89**, 80–91 (2019). https://doi.org/10.1016/j.cities.2019.01.032
9. Ghadami, N., et al.: Implementation of solar energy in smart cities using an integration of artificial neural network, photovoltaic system and classical Delphi methods. Sustain. Cities Soc. **74**, 103149 (2021). https://doi.org/10.1016/j.scs.2021.103149
10. Neyestani, N., Damavandi, M.Y., Shafie-khah, M., Catalão, J.P.S.: "Modeling the PEV traffic pattern in an urban environment with parking lots and charging stations. IEEE Eindhoven PowerTech **2015**, 1–6 (2015). https://doi.org/10.1109/PTC.2015.7232637
11. Arasteh, H., et al.: Iot-based smart cities: a survey. In: 2016 IEEE 16th International Conference on Environment and Electrical Engineering (EEEIC), pp. 1–6 (2016). https://doi.org/10.1109/EEEIC.2016.7555867
12. Rito, P., et al.: Aveiro tech city living lab: a communication, sensing and computing platform for city environments. IEEE Internet Things J. 1 (2023). https://doi.org/10.1109/JIOT.2023.3262627
13. Fonseca, T., Ferreira, L.L., Landeck, J., Klein, L., Sousa, P., Ahmed, F.: Flexible loads scheduling algorithms for renewable energy communities. Energies (Basel) **15**(23) (2022). https://doi.org/10.3390/en15238875
14. YOLOv5 Docs - Ultralytics YOLOv8 Docs. https://docs.ultralytics.com/yolov5/. Accessed 23 Apr 2023
15. "Pothole Object Detection Dataset." https://public.roboflow.com/object-detection/pothole. Accessed 23 Apr 2023
16. Roadway Flooding Image Dataset—Kaggle. https://www.kaggle.com/datasets/saurabhshahane/roadway-flooding-image-dataset. Accessed 23 Apr 2023
17. FIRE Dataset—Kaggle. https://www.kaggle.com/datasets/phylake1337/fire-dataset. Accessed 23 Apr 2023

Adaptive Learning from Peers for Distributed Actor-Critic Algorithms

Chandreyee Bhowmick(✉), Jiani Li, and Xenofon Koutsoukos

Institute of Software Integrated Systems, Vanderbilt University, Nashville, TN 37209, USA
{chandreyee.bhowmick,xenofon.koutsoukos}@vanderbilt.edu

Abstract. Training distributed reinforcement learning models over a network of users (or agents) has great potential for many applications in distributed devices such as face recognition, health tracking, recommender systems, and smart homes. Cooperation among networked agents by sharing and aggregating their model parameters can benefit considerably the learning performance. However, agents may have different objectives and unplanned cooperation may lead to undesired outcomes. Therefore, it is important to ensure that cooperation in distributed learning is beneficial especially when agents receive information from unidentifiable peers. In this paper, we consider the problem of training distributed reinforcement learning models and we focus on distributed actor-critic algorithms because they are used successfully in many application domains. We propose an efficient adaptive cooperation strategy with linear time complexity to capture the similarities among agents and assign adaptive weights for aggregating the parameters from neighboring agents. Essentially, a larger weight is assigned to a neighboring agent that performs a similar task or shares a similar objective. The approach has significant advantages in situations when different agents are assigned different tasks and in the presence of adversarial agents. Empirical results are provided to validate the proposed approach and demonstrate its effectiveness in improving the learning performance in single-task, multi-task, and adversarial scenarios.

Keywords: Resilient learning · Multi-agent reinforcement learning · Adaptive aggregation

1 Introduction

Distributed learning has received increasing attention due to the growth of machine learning applications in multi-agent systems such as networks of mobile phone users, wearable devices, and smart homes [1,2]. Multiple agents operate in a distributed and cooperative manner to perform a learning task. For example, to learn the user behavior in a network of smart wearable devices, the collected data at each device differ among the users and it is natural to learn separate models. However, people exhibit similar behaviors due to which relatedness among models commonly exists and cooperation among agents could be leveraged to improve the learning performance [3]. The potential improvement of the learning performance based on cooperative reinforcement

S. Ossowski et al. (Eds.): DCAI 2023, LNNS 740, pp. 54–64, 2023.
https://doi.org/10.1007/978-3-031-38333-5_6

learning (RL) agents has been studied in [4]. Cooperation can be achieved using a central server (i.e., federated learning [5]) or by adopting a fully-distributed scheme where agents communicate with their neighbors, which is the focus of this paper, because of scalability and robustness against single-point-of-failure. Collaborative multi-agent reinforcement learning (MARL) has been used in modeling recommender systems [6], traffic networks [7] and distributed economic dispatch in smart grids [8].

MARL studies may consider the case where agents learn in independent Markov Decision Processes (MDPs) or in a shared MDP [9,10]. In this paper, we are interested in the scenario where agents learn in similar but *independent* MDPs. In this case, the agents make individual decisions and their actions do not influence each other. Agents interact with their immediate neighbors and exchange their parameters without sharing their data [8]. Related approaches include a distributed implementation of Q-learning called QD-learning where every agent collaboratively updates tabular Q-values [11], a distributed method for policy evaluation with linear value function approximation [4], and a distributed actor-critic algorithm to learn a policy that performs well on average for a set of tasks [12].

Multi-Task Learning (MTL) is used extensively in distributed multi-agent systems. Aggregation functions significantly impact its performance [13,14]. For example, typical approaches based on averaging [4] or consensus [9] improve the performance when the agents share a common objective. However, such methods lead to undesired outcomes when agents are involved in a multi-task network [15], similar observation is obtained in the presence of adversarial agents [16]. Other aggregation functions such as coordinate-wise median, trimmed mean [17], or geometric median that are considered more resilient, may also not perform well in such cases. Thus, it is crucial to ensure that cooperation in distributed learning is beneficial, especially when agents learn from unidentifiable peers. In general, adaptive aggregation methods are preferred over standard techniques in such scenarios since they can capture the similarity among agents. Kullback-Leibler (KL) divergence and Wasserstein distance [18,19] can be used for this purpose, however, they measure how one probability distribution is different from another, and are not applicable to deterministic policies.

The main contribution of this paper is a novel approach for adaptive learning from peers for distributed actor-critic algorithms. Agents make individual decisions, share actor and critic parameters with neighbors, and update them by aggregating the parameters received from their neighbors. Actor-critic algorithms are considered here given its recent advances in solving many challenging problems [9,20,21]. We propose an adaptive approach to aggregate the model parameters from neighbors to leverage the similarity among agents and improve learning performance, which is applicable to both stochastic and deterministic policies. We consider how the performances of two agents are related by fitting the model of one agent to another agent's data. If the two agents have similar goals, then their models should be able to fit well to each other's data distributions. We use sampled batch data for this purpose, under the assumption that the batch data is randomly sampled and unbiased [22]. The losses associated with critic and actor networks are used to measure the similarity between the agents. The aggregation weights can be derived by solving an optimization problem for minimizing these losses while combining parameters from neighbors by fitting the agent's own data to the neighbor's model. We evaluate the proposed method for three state-of-the-art actor-critic algorithms: Deep Deterministic Policy Gradient (DDPG), Soft Actor-Critic (SAC), and Twin Delayed DDPG (TD3). The evaluation results show that the adaptive aggregation significantly improves the learning performance as measured by the average (long-term) return compared to the non-cooperative case as well as baselines with other aggregation functions.

2 Distributed Actor-Critic in Multi-agent Networks

Markov Decision Processes (MDPs) are typically used to characterize the RL process. An MDP can be represented as $\langle \mathcal{S}, \mathcal{A}, P, R \rangle$, where $\mathcal{S}$ and $\mathcal{A}$ denote the state and action spaces respectively, $P(s'|s,a) : \mathcal{S} \times \mathcal{A} \times \mathcal{S} \to [0,1]$ is the state transition probability from state s to the next state s' determined by action a, and $R(s,a) : \mathcal{S} \times \mathcal{A} \to \mathbb{R}$ is the reward function defined as $R(s,a) = \mathbb{E}\left[r_{t+1}|s_t = s, a_t = a\right]$, with r_{t+1} being the immediate reward received at time t. An agent's action is characterized by a stochastic policy $\pi : \mathcal{S} \times \mathcal{A} \to [0,1]$ representing the probability of choosing action a at state s. Deterministic policies are represented by $\pi : \mathcal{S} \to \mathcal{A}$. Multiple agents cooperating with each other are represented by a graph. Consider a network of N connected agents modeled by an undirected graph $\mathcal{G} = (\mathcal{V}, \mathcal{E})$ where the set of nodes $\mathcal{V}$ represents the agents and the set of edges $\mathcal{E}$ represents the interactions between them. An edge $(l,k) \in \mathcal{E}$, where $l, k \in \mathcal{V}$, signifies that agents k and l exchange information with each other. The neighborhood of agent k is the set of agents that it directly interacts with including itself, and is denoted as $\mathcal{N}_k = \{l \in \mathcal{V} | (l,k) \in \mathcal{E}\} \cup \{k\}$.

In this paper, we consider a group of N reinforcement learning agents operating in similar but *independent* MDPs. The environment for agent k is modeled as $\mathcal{M}_k = \langle \mathcal{S}, \mathcal{A}, P^k, R^k \rangle$, where $k \in \{1, 2, \ldots, N\}$. The state and action spaces are same for every agent, but the transition probabilities and reward functions may be different. The expected time-average return of policy π for agent k is defined as

$$J_k(\pi) = \lim_{T \to \infty} \frac{1}{T} \sum_{t=0}^{T-1} \mathbb{E}(r_{t+1}^k) = \sum_{s \in \mathcal{S}} d_\pi^k(s) \sum_{a \in \mathcal{A}} \pi(s,a) R^k(s,a) \tag{1}$$

where $d_\pi^k = \lim_{t \to \infty} P^k(s_t = s|\pi)$ is the stationary distribution of the Markov chain under policy π for agent k. The action-value associated with a state-action pair (s,a) under policy π for agent k is defined as $Q_\pi^k(s,a) = \sum_t \mathbb{E}\left[r_{t+1}^k - J_k(\pi)|s_0 = s, a_0 = a, \pi\right]$. It is assumed that $Q_\pi^k(s,a)$ can be parameterized as $Q^k(s,a;w_k)$ with parameters w_k, and the policy π can be parameterized as π_{θ_k} with parameter θ_k. For the group of agents, the objective is to cooperatively learn the optimal policy that maximizes the objective:

$$\max_{\theta_1, \ldots, \theta_N} \left\{ \frac{1}{N} \sum_{k=1}^{N} J_k(\theta_k) \right\}. \tag{2}$$

Each agent executes an independent actor-critic algorithm with an aggregation step. Due to data privacy, the agents do not share their data, instead they update their model parameters as a weighted average of the neighbors' model parameters to solve the optimization problem (2). A typical actor-critic algorithm based on action-value function approximation with model aggregation [9] can be represented as:

$$\underbrace{\begin{aligned} \mu_{t+1}^k &= (1-\beta_{w,t}^k)\cdot \mu_t^k + \beta_{w,t}^k \cdot r_{t+1}^k, \\ \tilde{w}_t^k &= w_t^k + \beta_{w,t}^k \cdot \delta_t^k \cdot \nabla_w Q_t^k(w_t^k), \\ w_{t+1}^k &= \sum_{l\in\mathcal{N}_k} c_t(k,l)\cdot \tilde{w}_t^l. \end{aligned}}_{\text{critic step}} \quad \text{and} \quad \underbrace{\begin{aligned} \tilde{\theta}_t^k &= \theta_t^k + \beta_{\theta,t}^k \cdot Q_t^k(w_t^k)\cdot \psi_t^k, \\ \theta_{t+1}^k &= \sum_{l\in\mathcal{N}_k} b_t(k,l)\cdot \tilde{\theta}_t^l \end{aligned}}_{\text{actor step}}$$

Here the action-value function is denoted as $Q_t^k(w) \triangleq Q^k(s_t^k, a_t^k; w)$, the action-value temporal difference (TD) error is defined as $\delta_t^k \triangleq r_{t+1}^k - \mu_t^k + Q_t^k(w_t^k) - Q_t^k(w_t^k)$, the gradient of the log of the policy is given by $\psi_t^k \triangleq \nabla_\theta \log \pi_{\theta_t^k}(s_t^k, a_t^k)$, and $\beta_{w,t}^k, \beta_{\theta_t}^k > 0$ are the step-sizes. The critic step operates at a faster time scale to estimate the action-value function $Q_t^k(w_t^k)$ under policy $\pi_{\theta_t^k}$. The actor step improves the policy by gradient ascent at a slower rate. The row-stochastic aggregation matrices are defined as $C_t = \left[c_t(k,l)\right]_{N\times N}$ and $B_t = \left[b_t(k,l)\right]_{N\times N}$, where $c_t(k,l)$ and $b_t(k,l)$ are the weights assigned by agent k to agent l at time t for the aggregation of the critic and actor parameters, respectively. Our objective here is to find the optimal adaptive aggregation weights that solve (2) using a distributed algorithm. The proposed solution addresses not only single-task networks but also multi-task scenarios and networks with adversarial agents that share malicious information with their neighbors.

3 Adaptive Learning in Distributed Actor-Critic Algorithms

Here we present the design of adaptive aggregation weights by fitting an agent's data into the neighbor's model. The better an agent's model fits to the data distribution of another agent, the stronger the model is related to the underlying task of the other. Actor-critic algorithms typically use neural networks for actor and critic networks. These networks are associated with loss functions, which are measures of how close the neural network parameters are to the optimal values. For example, in DDPG [23], the actor function $\mu(s|\theta)$ specifies the policy by a deterministic mapping of a state to a specific action. Let B be the size of the sampled minibatch of data at a time step, where (s_i, a_i, r_i, s_{i+1}) is the ith sample of the batch. The critic parameters are updated by minimizing the mean-squared temporal difference (TD) error, so the critic loss is defined as

$$L_{k,t}^w(w) = \frac{1}{B}\sum_{i=1,\ldots,B} (y_i^k - Q_i^k(w))^2, \tag{3}$$

where the critic is parameterized by w. In this case,

$$y_i^k = r_i^k + \gamma Q'^k(s_{i+1}, \mu'^k(s_{i+1}|\theta_t'^k); w_t'^k),$$

with Q'^k and μ'^k being the target value function and target policy function of agent k, parameterized by $w_t'^k$ and $\theta_t'^k$, respectively. In DDPG, the actor parameters are updated by maximizing the expected action-value $\mathbb{E}[Q(s, \mu(s))]$, or minimizing $-\mathbb{E}[Q(s, \mu(s))]$. The actor loss $L^{\boldsymbol{\theta}}_{k,t}(\boldsymbol{\theta})$ is defined as

$$L^{\boldsymbol{\theta}}_{k,t}(\boldsymbol{\theta}) = \frac{1}{B} \sum_{i=1,\ldots,B} (-Q^k_i(s_i, \mu^k(s_i|\boldsymbol{\theta})), \tag{4}$$

when the actor is parametrized by $\boldsymbol{\theta}$ and μ^k is the policy function of agent k. These two loss functions, $L^w_{k,t}(w)$ and $L^{\boldsymbol{\theta}}_{k,t}(\boldsymbol{\theta})$, are used in the design of the adaptive aggregation weights. We derive the aggregation weights by solving an optimization problem with objective to minimize the losses while aggregating the parameters from the neighbors. The objective function is the sum of the losses from the neighboring agents computed using the parameters of the neighbor and data points of the agent itself. Thus, the optimization problem of agent k for critic aggregation is formulated as

$$\begin{aligned} &\min_{C_k} \sum_{l\in\mathcal{N}_k} \frac{1}{2} c_t^2(k,l) L^w_{k,t}(\tilde{w}^l_t) \\ &\text{subject to} \sum_{l\in\mathcal{N}_k} c_t(k,l) = 1, \quad c_t(k,l) \geq 0, \quad c_t(k,l) = 0 \text{ for } l \notin \mathcal{N}_k, \end{aligned} \tag{5}$$

where $C^k_t = \left[c_t(k,1)\ c_t(k,2) \cdots c_t(k,N)\right]^T$ is the vector of aggregation weights for agent k (the kth row of the critic aggregation matrix C_t)[1]. The critic loss is scalar and non-negative as the critic uses mean-squared TD-error[2]. After incorporating the constraint $\sum_{l\in\mathcal{N}_k} c_t(k,l) = 1$, the Lagrangian of (5) is given by

$$\mathcal{L}(c_t(k,l), \lambda) = \sum_{l\in\mathcal{N}_k} \frac{1}{2} c_t^2(k,l) L^w_{k,t}(\tilde{w}^l_t) + \lambda\Big(1 - \sum_{l\in\mathcal{N}_k} c_t(k,l)\Big),$$

where λ is the Lagrange multiplier. Now, taking the gradient of the Lagrangian w.r.t. $c_t(k,l)$ and equating it to zero, we get

$$c_t(k,l) L^w_{k,t}(\tilde{w}^l_t) - \lambda = 0, \quad \forall l \in \mathcal{N}_k. \tag{6}$$

This yields $c_t(k,l) = \frac{\lambda}{L^w_{k,t}(\tilde{w}^l_t)}$, $\forall l \in \mathcal{N}_k$. Using this in the constraint, we get $\lambda \sum_{l\in\mathcal{N}_k} \frac{1}{L^w_{k,t}(\tilde{w}^l_t)} = 1$, Thus, the Lagrange multiplier is given as $\lambda = \frac{1}{\sum_{l\in\mathcal{N}_k} L^w_{k,t}(\tilde{w}^l_t)^{-1}}$. Substituting this in (6), we get the optimal aggregation weights as

$$c_t(k,l) = \begin{cases} \frac{L^w_{k,t}(\tilde{w}^l_t)^{-1}}{\sum_{p\in\mathcal{N}_k} L^w_{k,t}(\tilde{w}^p_t)^{-1}} & \text{for } l \in \mathcal{N}_k, \\ 0 & \text{for } l \notin \mathcal{N}_k. \end{cases} \tag{7}$$

---

[1] A fraction of $\frac{1}{2}$ is introduced to the objective function for the simplification of the solution.

[2] If in a particular algorithm, the loss of the critic can be negative, we can include a softmax layer to $L^w_{k,t}(\cdot)$ in the objective function.

The optimal weights for actor aggregation are obtained in a similar manner. However, since the actor loss $L^{\theta}_{k,t}(\cdot)$ can have negative values, we apply a softmax layer to the optimization function, which is given as

$$\begin{aligned} &\min_{B_k} \sum_{l\in\mathcal{N}_k} \frac{1}{2} b_t^2(k,l) \exp(-L^{\theta}_{k,t}(\tilde{\theta}^l_t)) \\ &\text{subject to} \sum_{l\in\mathcal{N}_k} b_t(k,l) = 1, \quad b_t(k,l) \geq 0, \quad b_t(k,l) = 0 \text{ for } l \notin \mathcal{N}_k. \end{aligned} \tag{8}$$

The optimal weights for actor aggregation are then derived as

$$b_t(k,l) = \begin{cases} \dfrac{\exp(-L^{\theta}_{k,t}(\tilde{\theta}^l_t))}{\sum_{p\in\mathcal{N}_k} \exp(-L^{\theta}_{k,t}(\tilde{\theta}^p_t))} & \text{for } l \in \mathcal{N}_k, \\ 0 & \text{for } l \notin \mathcal{N}_k. \end{cases} \tag{9}$$

Algorithm 1. MARL with adaptive model aggregation.

Input: initial values of the parameters $\mu^i_0, \tilde{w}^i_0, w^i_0, \theta^i_0$, the initial states s^i_0, and step-sizes $\{\beta^i_{w,t}\}_{t\geq 0}$ and $\{\beta^i_{\theta,t}\}_{t\geq 0}$ for $i \in [1, 2, \cdots, N]$.
Each agent $i \in [1, 2, \cdots, N]$ executes action $a^i_0 \sim \pi_{\theta^i_0}(s_0, \cdot)$.
Initialize the step counter $t \leftarrow 0$.
repeat
  for all $i \in [1, 2, \cdots, N]$ **do**
    Local update: update $\tilde{w}^i_t$ (critic update) and $\tilde{\theta}^i_t$ (actor update).
    Send $\tilde{w}^i_t$ and $\tilde{\theta}^i_t$ to the neighbors.
  for all $k \in [1, 2, \cdots, N]$ **do**
    Critic aggregation step: Compute critic losses $L^{w}_{k,t}(\tilde{w}^l_t)$ and aggregation weights $c_t(k,l)$ using (7), $\forall l \in \mathcal{N}_k$. Aggregate critic parameters as $w^k_{t+1} = \sum_{l\in\mathcal{N}_k} c_t(k,l) \cdot \tilde{w}^l_t$.
    Actor aggregation step: Compute actor losses $L^{\theta}_{k,t}(\tilde{\theta}^l_t)$ and aggregation weights $b_t(k,l)$ using (9), $\forall l \in \mathcal{N}_k$. Aggregate actor parameters as $\theta^k_{t+1} = \sum_{l\in\mathcal{N}_k} b_t(k,l) \cdot \tilde{\theta}^l_t$.
  Increment the iteration counter $t \leftarrow t + 1$.
until convergence

MARL with the proposed adaptive aggregation is outlined in Algorithm 1. We now characterize the time complexity of the adaptive aggregation step. To find the weights in (7) and (9), we need to compute the batch loss for each neighbor with the sampled data. As evident from Eqs. (3) and (4), when the batch-size B is constant, the computational time is linear in the dimension of the model parameters and neighborhood size. Thus, the time complexity of the aggregation step is $O(d \cdot |\mathcal{N}_k|)$ for agent k, where $d = d_w$ for the critic and $d = d^{\theta}$ for the actor, d_w and d_{θ} being the dimensions of critic and actor network parameters. This can be combined with the time complexity of standard actor-critic algorithm [24] to find the overall complexity of Algorithm 1.

4 Evaluation

We evaluate[3] the proposed approach using three state-of-the-art off-policy actor-critic algorithms – DDPG [23], TD3 [21], and SAC [20] – for the MuJoCo continuous control tasks HalfCheetah, Walker2d, Ant, and Reacher [25] through the OpenAI Gym interface [26]. The goal of the evaluation is to quantify how the proposed approach improves the overall learning performance under different scenarios. For all examples, we use the ADAM optimizer [27], $\gamma = 0.99$, the learning rate is 0.001, and the batch-size is 256. The neural network architectures and remaining hyper-parameters are the same as in the original algorithms [20,21,23]. Agents learn in independent environments and they start with random environment seeds in OpenAI Gym. They exchange the model (actor/critic) parameters with their neighbors and aggregate them after each learning episode. The target neural networks (time-delayed copies of their original neural networks that slowly track the learned models) are updated using a soft update with $\tau = 0.001$ [23]. We compare the proposed approach with the following baselines: (i) no cooperation, (ii) aggregation using averaging, (iii) aggregation using median. In addition to the simple case when all the agents perform the same task (*single-task scenario*), we consider a *multi-task scenario* where agents are divided in two groups and each group is assigned a different learning task. To demonstrate resilience in the presence of adversaries, an *adversarial scenario* is considered where all agents perform the same task but some of them are adversarial and they send malicious information to their neighbors.

We consider a network of $N = 8$ agents with average connectivity $\frac{1}{N}\sum_{k=1}^{N}|\mathcal{N}_k| = 5.75$. A summary of the results for TD3, DDPG, and SAC is given in Table 1, Table 2, and Table 3, respectively[4]. The results are obtained when the agents are trained for 10^5 steps for the task Reacher, and 2×10^5 steps for the other tasks. Note that DDPG does not converge for the task Ant and is not listed. Also, to emphasize the discrepancies between normal and adversarial agents, we consider an enlarged reward for the task Reacher during training (not in the evaluation), that is, the new reward is 10 times the original reward given in OpenAI Gym. Without cooperation, the simulation takes 30–50 sec (different tasks have different training time) for each training episode (one episode has 1000 training steps) for 8 networked agents running in 8 threads using our Python code on an NVIDIA GeForce GTX 1080Ti GPU. Cooperation using averaging, median, and adaptive aggregation does not increase the training time notably.

We find that the adaptive aggregation improves the overall learning performance in all considered scenarios and tasks. In contrast, cooperation using the average and median-based aggregation may lead to worse learning performance when agents are performing multiple tasks or some of the agents are under attack. Specifically, in single-task scenario of using TD3 for Ant, we observe a 3 times improvement in the average return. For multi-task scenario, the agents perform two different tasks: HalfCheetah (agents 0–3) and Walker2d (agents 4–7). In this case, we observe about 44% improvement in the average return over the non-cooperative case for the three algorithms,

[3] The simulation code is available in https://github.com/cbhowmic/resilient-adaptive-RL.

[4] Maximum value for each task is in bold fonts; $\pm$ corresponds to a single standard deviation over the network.

Table 1. Max average return for TD3.

Scenario	Task	Noncoop	Average	Median	AdaLearn(Ours)
Single-task	HalfCheetah	6706.49 ± 774.00	$\mathbf{10350.08 \pm 82.14}$	9626.16 ± 80.44	$\mathbf{10249.95 \pm 84.18}$
	Walker2d	2701.53 ± 1290.04	4273.22 ± 73.09	3745.96 ± 143.28	$\mathbf{4731.30 \pm 468.54}$
	Ant	1044.14 ± 177.80	$\mathbf{3347.62 \pm 87.41}$	2802.76 ± 67.93	$\mathbf{3153.94 \pm 51.30}$
	Reacher	-6.31 ± 0.24	$\mathbf{-5.85 \pm 0.03}$	$\mathbf{-5.87 \pm 0.06}$	$\mathbf{-5.95 \pm 0.06}$
Multi-task	HalfCheetah+Walker2d	4507.75 ± 1289.29	3314.79 ± 2336.07	2901.95 ± 1923.04	$\mathbf{6708.19 \pm 2184.79}$
Adversarial	HalfCheetah (25% attacked)	6596.01 ± 622.46	348.31 ± 191.54	9283.08 ± 69.68	$\mathbf{9771.80 \pm 63.28}$
	HalfCheetah (50% attacked)	6467.72 ± 704.65	510.94 ± 47.32	3873.77 ± 117.46	$\mathbf{9488.09 \pm 192.31}$
	HalfCheetah (only 1 normal)	$\mathbf{5963.01 \pm 0.00}$	608.72 ± 0.00	-2.50 ± 0.00	$\mathbf{5891.17 \pm 0.00}$
	Walker2d (25% attacked)	2648.39 ± 803.81	415.75 ± 300.81	4348.52 ± 54.55	$\mathbf{4398.11 \pm 545.71}$
	Walker2d (50% attacked)	2906.32 ± 1596.26	157.65 ± 19.40	1115.38 ± 47.98	$\mathbf{5345.96 \pm 163.44}$
	Walker2d (only 1 normal)	1467.60 ± 0.00	-16.81 ± 0.00	-9.35 ± 0.00	$\mathbf{4458.48 \pm 0.00}$
	Ant (25% attacked)	1171.83 ± 390.21	990.60 ± 5.12	2696.77 ± 73.02	$\mathbf{3054.00 \pm 49.51}$
	Ant (50% attacked)	892.05 ± 11.00	982.06 ± 3.37	1697.30 ± 32.95	$\mathbf{2744.55 \pm 236.27}$
	Ant (only 1 normal)	1055.09 ± 0.00	878.62 ± 0.00	952.24 ± 0.00	**2232.09pm0.00**
	Reacher (25% attacked)	-6.42 ± 0.38	-12.92 ± 1.51	-6.36 ± 0.08	$\mathbf{-6.10 \pm 0.08}$
	Reacher (50% attacked)	-6.52 ± 0.12	-14.71 ± 2.82	-11.89 ± 0.52	$\mathbf{-6.14 \pm 0.16}$
	Reacher (only 1 normal)	$\mathbf{-6.85 \pm 0.00}$	-62.43 ± 0.00	-25.73 ± 0.00	$\mathbf{-7.17 \pm 0.00}$

Table 2. Max average return for DDPG.

Scenario	Task	Noncoop	Average	Median	AdaLearn(Ours)
Single-task	HalfCheetah	4728.48 ± 432.39	6968.39 ± 65.85	6637.61 ± 69.86	$\mathbf{7734.40 \pm 97.89}$
	Walker2d	1598.72 ± 437.28	$\mathbf{3278.27 \pm 184.70}$	1041.99 ± 122.27	$\mathbf{3219.94 \pm 189.01}$
	Reacher	-6.09 ± 0.18	$\mathbf{-5.86 \pm 0.02}$	-6.19 ± 0.11	$\mathbf{-5.87 \pm 0.02}$
Multi-task	HalfCheetah+Walker2d	3302.70 ± 1779.20	2933.86 ± 1597.24	1058.48 ± 346.85	$\mathbf{5365.06 \pm 2538.90}$
Adversarial	HalfCheetah (25% attacked)	4899.95 ± 258.43	-52.64 ± 29.20	6045.27 ± 49.68	$\mathbf{7668.68 \pm 836.77}$
	HalfCheetah (50% attacked)	4876.08 ± 184.65	-71.81 ± 43.45	1939.30 ± 90.08	$\mathbf{6361.74 \pm 727.24}$
	HalfCheetah (only 1 normal)	$\mathbf{4742.29 \pm 0.00}$	375.79 ± 0.00	-47.44 ± 0.00	$\mathbf{4353.34 \pm 0.00}$
	Walker2d (25% attacked)	1233.25 ± 174.05	319.87 ± 78.40	1542.28 ± 278.48	$\mathbf{2647.08 \pm 236.87}$
	Walker2d (50% attacked)	1774.96 ± 506.91	-7.49 ± 4.30	1123.09 ± 33.55	$\mathbf{2422.36 \pm 757.20}$
	Walker2d (only 1 normal)	1237.56 ± 0.00	-10.31 ± 0.00	-15.30 ± 0.00	$\mathbf{1644.62 \pm 0.00}$
	Reacher (25% attacked)	-6.18 ± 0.22	-16.99 ± 2.31	-6.07 ± 0.06	$\mathbf{-5.92 \pm 0.16}$
	Reacher (50% attacked)	$\mathbf{-6.24 \pm 0.24}$	-102.32 ± 6.59	-9.12 ± 0.20	$\mathbf{-6.62 \pm 0.17}$
	Reacher (only 1 normal)	$\mathbf{-6.03 \pm 0.0}$	-109.00 ± 0.00	-13.45 ± 0.00	$\mathbf{-6.10 \pm 0.00}$

whereas average and median-based aggregation result in worse performance than the non-cooperative case. To simulate the adversarial scenario, we consider that the adversarial agents send perturbed parameters using the Fast Gradient Sign Method (FGSM) [28], where the multiplier of the perturbation is 0.005. The average returns are calculated only for the normal agents, and we find that aggregation using averaging results in inferior performance than the non-cooperative case in all examples. Median based aggregation fails when more than half of the agents are adversarial. In contrast, the proposed method achieves better performance even in the case of an increased number of adversarial agents (about 50%–160% improvement in the average return when $\leq$50% agents are adversarial). When 7 out of 8 agents are under attack, there is only one normal agent in the network and cooperation using the proposed method results in a similar performance as the non-cooperative case. Detailed plots for all scenarios and

Table 3. Max average return for SAC.

Scenario	Task	Noncoop	Average	Median	AdaLearn(Ours)
Single-task	HalfCheetah	6399.85 ± 673.81	9614.02 ± 107.70	9588.57 ± 79.58	$\mathbf{9931.12 \pm 125.80}$
	Walker2d	3019.59 ± 718.71	$\mathbf{3662.94 \pm 60.70}$	3289.44 ± 93.99	$\mathbf{3656.56 \pm 33.73}$
	Ant	1979.03 ± 348.19	3720.35 ± 62.03	3146.27 ± 67.65	$\mathbf{3776.56 \pm 64.48}$
	Reacher	-6.33 ± 0.29	-6.50 ± 0.07	-6.50 ± 0.12	$\mathbf{-6.20 \pm 0.13}$
Multi-task	HalfCheetah+Walker2d	3938.91 ± 1632.00	3973.77 ± 1337.77	2958.70 ± 1853.09	$\mathbf{4857.24 \pm 3433.90}$
Adversarial	HalfCheetah (25% attacked)	6163.24 ± 823.36	261.09 ± 157.26	8616.53 ± 89.98	$\mathbf{8742.69 \pm 63.35}$
	HalfCheetah (50% attacked)	5755.53 ± 659.42	1003.12 ± 349.58	3725.91 ± 82.78	$\mathbf{7885.69 \pm 39.13}$
	HalfCheetah (only 1 normal)	$\mathbf{6710.47 \pm 0.00}$	-5.97 ± 0.00	668.58 ± 0.00	$\mathbf{6666.83 \pm 0.00}$
	Walker2d (25% attacked)	3396.36 ± 440.52	315.70 ± 62.68	2781.80 ± 63.98	$\mathbf{3531.32 \pm 36.73}$
	Walker2d (50% attacked)	$\mathbf{3132.66 \pm 812.49}$	3.16 ± 4.63	959.56 ± 10.18	$\mathbf{3115.22 \pm 47.73}$
	Walker2d (only 1 normal)	2284.98 ± 0.00	1.25 ± 0.00	-6.22 ± 0.00	$\mathbf{2695.72 \pm 0.00}$
	Ant (25% attacked)	2198.53 ± 254.48	994.86 ± 0.91	3022.71 ± 71.98	$\mathbf{3578.94 \pm 54.02}$
	Ant (50% attacked)	1803.09 ± 258.45	999.50 ± 1.15	1114.05 ± 56.17	$\mathbf{3483.97 \pm 70.63}$
	Ant (only 1 normal)	1383.55 ± 0.00	692.08 ± 0.00	576.88 ± 0.00	$\mathbf{1750.23 \pm 0.00}$
	Reacher (25% attacked)	-6.60 ± 0.72	-14.09 ± 1.10	-6.22 ± 0.05	$\mathbf{-6.17 \pm 0.02}$
	Reacher (50% attacked)	$\mathbf{-6.35 \pm 0.21}$	-31.41 ± 13.13	$\mathbf{-6.47 \pm 0.09}$	-6.50 ± 0.26
	Reacher (only 1 normal)	$\mathbf{-6.24 \pm 0.00}$	-70.99 ± 0.00	-101.27 ± 0.00	-6.59 ± 0.00

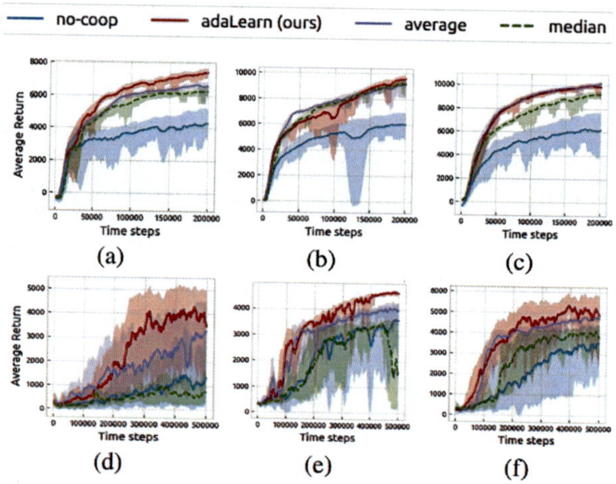

Fig. 1. Training curves for single-task scenario: (a)–(c) HalfCheetah, (d)–(f) Walker2d.

tasks are not included due to space constraints. Learning curves for HalfCheetah and Walker2d tasks in single-task scenario are given in Fig. 1 as a reference[5]. We performed the same experiments on a network of 20 agents, with small connectivity of 4.80. Similar observations were obtained on this network, as the adaptive method shows improved performance over baselines, demonstrating scalability of the proposed algorithm. Due to space constraints, the results for the larger network are not included here.

[5] The solid lines in the plots show the average return of the agents and the shaded area represents its range.

5 Conclusion

We consider distributed MARL using actor-critic algorithms where agents aggregate the model parameters from their neighbors to improve the learning performance. We propose an efficient cooperation strategy which assigns adaptive weights to neighbors by measuring the similarities among agents according to losses associated with the critic and actor networks. Evaluation results show that cooperation by promoting similarity among agents improves the learning performance. The performance improvement is observed in various scenarios and it is more prominent when the agents are performing different tasks and when some of the agents are adversarial. Our results indicate that adaptive learning from peers can improve performance in distributed RL algorithms. The proposed approach can be used in other distributed machine learning algorithms where determining similarity between the tasks of various agents is important.

References

1. Anguita, D., Ghio, A., Oneto, L., Parra, X., Reyes-Ortiz, J.L.: A public domain dataset for human activity recognition using smartphones. In: 21st European Symposium on Artificial Neural Networks, ESANN, Bruges, Belgium, 24–26 April 2013 (2013)
2. Chen, Y., Qin, X., Wang, J., Yu, C., Gao, W.: FedHealth: a federated transfer learning framework for wearable healthcare. IEEE Intell. Syst. **35**(4), 83–93 (2020)
3. Sayed, A.H., Tu, S.-Y., Chen, J., Zhao, X., Towfic, Z.J.: Diffusion strategies for adaptation and learning over networks: an examination of distributed strategies and network behavior. IEEE Signal Process. Mag. **30**(3), 155–171 (2013)
4. Macua, S.V., Chen, J., Zazo, S., Sayed, A.H.: Distributed policy evaluation under multiple behavior strategies. IEEE Trans. Autom. Control **60**(5), 1260–1274 (2014)
5. McMahan, H.B., Moore, E., Ramage, D., Agüera y Arcas, B.: Federated learning of deep networks using model averaging. CoRR, abs/1602.05629 (2016)
6. Afsar, M.M., Crump, T., Far, B.: Reinforcement learning based recommender systems: a survey. ACM Comput. Surv. **55**(7), 1–38 (2022)
7. Prabuchandran, K.J., Hemanth Kumar, A.N., Bhatnagar, S.: Multi-agent reinforcement learning for traffic signal control. In: 17th International IEEE Conference on Intelligent Transportation Systems (ITSC), pp. 2529–2534. IEEE (2014)
8. Liu, W., Zhuang, P., Liang, H., Peng, J., Huang, Z.: Distributed economic dispatch in microgrids based on cooperative reinforcement learning. IEEE Trans. Neural Netw. Learn. Syst. **29**(6), 2192–2203 (2018)
9. Zhang, K., Yang, Z., Liu, H., Zhang, T., Basar, T.: Fully decentralized multi-agent reinforcement learning with networked agents. In: International Conference on Machine Learning, pp. 5872–5881. PMLR (2018)
10. Lowe, R., Wu, Y., Tamar, A., Harb, J., Abbeel, P., Mordatch, I.: Multi-agent actor-critic for mixed cooperative-competitive environments. In: Advances in Neural Information Processing Systems, vol. 30 (2017)
11. Kar, S., Moura, J.M.F., Poor, H.V.: QD-learning: a collaborative distributed strategy for multi-agent reinforcement learning through Consensus + Innovations. IEEE Trans. Signal Process. **61**(7), 1848–1862 (2013)
12. Macua, S.V., Tukiainen, A., Hernández, D.G.-O., Baldazo, D., de Cote, E.M., Zazo, S.: Diff-DAC: distributed actor-critic for multitask deep reinforcement learning. arXiv preprint arXiv:1710.10363 (2017)

13. Yan, D., et al.: Multi-task deep reinforcement learning for intelligent multi-zone residential HVAC control. Electr. Power Syst. Res. **192**, 106959 (2021)
14. Zhang, Q., et al.: Multi-task fusion via reinforcement learning for long-term user satisfaction in recommender systems. In: Proceedings of the 28th ACM SIGKDD Conference on Knowledge Discovery and Data Mining, pp. 4510–4520 (2022)
15. Nassif, R., Vlaski, S., Richard, C., Chen, J., Sayed, A.H.: Multitask learning over graphs: an approach for distributed, streaming machine learning. IEEE Signal Process. Mag. **37**(3), 14–25 (2020)
16. Konstantinov, N., Lampert, C.: Robust learning from untrusted sources. In: International Conference on Machine Learning, pp. 3488–3498. PMLR (2019)
17. Lin, Y., Gade, S., Sandhu, R., Liu, J.: Toward resilient multi-agent actor-critic algorithms for distributed reinforcement learning. In: 2020 American Control Conference (ACC), pp. 3953–3958. IEEE (2020)
18. Blundell, C., Cornebise, J., Kavukcuoglu, K., Wierstra, D.: Weight uncertainty in neural network. In: International Conference on Machine Learning, pp. 1613–1622. PMLR (2015)
19. Shui, C., Abbasi, M., Robitaille, L.-É., Wang, B., Gagné, C.: A principled approach for learning task similarity in multitask learning. In: IJCAI (2019)
20. Haarnoja, T., Zhou, A., Abbeel, P., Levine, S.: Soft actor-critic: off-policy maximum entropy deep reinforcement learning with a stochastic actor. In: International Conference on Machine Learning, pp. 1861–1870. PMLR (2018)
21. Fujimoto, S., Hoof, H., Meger, D.: Addressing function approximation error in actor-critic methods. In: International Conference on Machine Learning, pp. 1587–1596. PMLR (2018)
22. Bottou, L., Curtis, F.E., Nocedal, J.: Optimization methods for large-scale machine learning. SIAM Rev. **60**(2), 223–311 (2018)
23. Lillicrap, T.P., et al.: Continuous control with deep reinforcement learning. arXiv preprint arXiv:1509.02971 (2015)
24. Wu, Y.F., Zhang, W., Xu, P., Gu, Q.: A finite-time analysis of two time-scale actor-critic methods. In: Advances in Neural Information Processing Systems, vol. 33, pp. 17617–17628 (2020)
25. Todorov, E., Erez, T., Tassa, Y.: MuJoCo: a physics engine for model-based control. In: 2012 IEEE/RSJ International Conference on Intelligent Robots and Systems, pp. 5026–5033. IEEE (2012)
26. Brockman, G., et al.: OpenAI Gym.arXiv preprint arXiv:1606.01540 (2016)
27. Kingma, D.P., Ba, J.: Adam: a method for stochastic optimization. arXiv preprint arXiv:1412.6980 (2014)
28. Goodfellow, I.J., Shlens, J., Szegedy, C.: Explaining and harnessing adversarial examples. arXiv preprint arXiv:1412.6572 (2014)

Detection of Infostealer Variants Through Graph Neural Networks

Álvaro Bustos-Tabernero(✉), Daniel López-Sánchez, and Angélica González Arrieta

University of Salamanca, Plaza de los Caídos, 37008 Salamanca, Spain
{alvarob97,lope,angelica}@usal.es

Abstract. Cybersecurity technology is capable of detecting malicious software that is recognized by signatures, heuristic rules, or that has been previously seen and stored in a database. However, threat actors try to generate new strains/variants of existing malware, by obfuscating or modifying part of the code to evade antivirus engines. One of the most common malicious programs are infostealers, which aim to obtain personal or banking information from an infected system and exfiltrate it. In this work, a pipeline is proposed that allows to analyze the infostealers through their assembler instructions, extract a feature vector associated with the functions and determine a binary classification by applying graph neural networks.

Keywords: Cybersecurity · threat hunting · threat intelligence · deep learning · graph neuronal network · infostealer

1 Introduction

Cybersecurity is an ongoing effort to maintain a balance between those seeking to exploit technology and those aiming to protect it. Attackers constantly explore new avenues for attack, whereas defenders reactively try to adapt to these new threats [20]. However, waiting until after an attack has occurred can have serious consequences for the victim. To counteract this, Cyber Threat Intelligence (CTI) was born to gather information and secure defenses. Cyber Threat Intelligence is comprehensive and evidence-based understanding of a current or potential threat to assets, which takes into account its context, mechanisms, indicators, implications, and provides practical recommendations to inform the subject's response to such a threat [10].

Many of these advanced threats are distributed across the Internet. The management and effort of the volume of data they can generate is quite high; therefore, a certain degree of automation is necessary to deal with them. Owing to the high prevalence of malware, several antivirus engines have attempted to improve their malware detection tools by employing rules [11], static code analysis, or sandboxes [16,17]. In order to solve these tasks, machine learning techniques, specifically neural networks, will be applied.

S. Ossowski et al. (Eds.): DCAI 2023, LNNS 740, pp. 65–73, 2023.
https://doi.org/10.1007/978-3-031-38333-5_7

In this study, we focus on the early detection of possible malware variants of known threats through graph neural networks [19]. Specifically, we focused on malicious software that infects computers to steal information, personal, banking, or otherwise, and exfiltrate it. One of the examples of infostealer that is evading the defense provided by some antimalware is Rhadamanthys [14], which spreads via fake advertisements or spam emails.

Among all the varieties of malware, infostealers are malicious programs that obtain information from a target computer and send it to the attacker's server. The first famous case was the ZeuS Trojan, a malware first seen in the wild in 2007 that aimed to obtain banking information from a Windows OS client computer. Once ZeuS has infected the system, it collects information and sends it to Command and Control server (C&C) [18]. Subsequently, variants of this executable file have emerged where they focus on various types of information they want to filter. Still, the objective is always the same: to infect, capture information and send this information.

This article presents a comprehensive review of state-of-art frameworks to be utilized. Subsequently, a proposal is put forth, incorporating the aforementioned algorithms and the generated model, to conduct experiments aimed at detecting infostealers within binary files.

2 Related Work

This section provides an overview of the state-of-the-art models that have informed our study and served as the basis for our proposed model. Two main concepts are highlighted: the assembly language as an initial point, and the use of graph neural networks as an architecture for analyzing the flow of a program, where the assembler functions serve as nodes.

2.1 Vector Representations of Assembly Code

Assembly language is a low-level programming language where instructions are written at the CPU level before being compiled into machine code. Although the instruction set is limited, it is possible to develop programs with the same or very similar functionality with completely different codes.

The disassembly of a binary transforms the machine code into a sequence of assembler instructions. Given the different methods of natural language processing, it is possible to extract the characteristics of these instructions to a numeric vector. One of the current existing models is the Asm2Vec [4], which is capable of extracting the features of an assembler function and mapping them into a feature vector.

Given a function from a assembly function set $f_s \in \mathbf{R}$, Asm2Vec model maps a function f_s into $\theta_{f_s} \in \mathbb{R}^{2\times d}$, where θ_{f_s} is the vector representation of the function f_s, and $2 \times d$ because of the concatenation of the operands and operators.

This model maximizes the log probability of finding a token t_c given an assembly function f_c and its neighbor:

$$\sum_{f_s}^{\mathbf{R}} \sum_{seq_i}^{\mathcal{S}(f_s)} \sum_{in_j}^{\mathcal{I}(seq_i)} \sum_{t_c}^{\mathcal{T}(in_j)} \log P(t_c|f_s, in_{j-1}, in_{j+1}) \tag{1}$$

Extending the details of Eq. 1, a function f_s of a repository $\mathbf{R}$ has several sequences $\mathcal{S}(f_s) = seq[1 : i]$. Then a sequence seq_i is represented as a list of instructions $\mathcal{I}(seq_i) = in[1 : j]$. Finally, an instruction in_j is composed of a list of operands $\mathcal{A}(in_j)$ and operators $\mathcal{P}(in_j)$, which their concatenation constitutes a list of tokens $\mathcal{T}(in_j) = \mathcal{P}(in_j)\|\mathcal{A}(in_j)$ [4].

In this work, we will focus on identifying infostealers by analyzing binary code via artificial neural networks. Concretely, we explore Graph Neural Networks (GNN).

2.2 Graph Neural Networks

A graph is defined as a mathematical structure $G = (V, E)$, where $V = \{v_1, \ldots, v_n\}$ is the set the nodes represented by this architecture and $E = \{(x, y)|(x, y) \in V^2\}$ is the set of links or edges between these vertices. In addition, each node may contain certain features, denoted as x_v for $v \in V$. We define $h_r \in \mathcal{R}^d$ as a vector representing a node $v \in V$ or the whole graph G. Given a graph structure, a Graph Neural Network (GNN) extracts the features of the nodes or the entire graph and outputs a vector h_r [2]. Once the network features have been extracted, they would be classified to determine if the binary is an infostealer.

Acquiring information and obtaining features from a network can be a heavy and computationally intensive task. However, there are models that have **improved their efficiency and effectiveness, such as GraphSAGE** [5]. Given an input graph G, GraphSAGE is an inductive embedding model of the nodes through the node's own features such as its attributes, information profile or degree, to learn an embedding function for previously unseen nodes [5]. Moreover, by incorporating this learning, it allows to know the topology of the node's neighborhood and the distribution of the features of the neighboring nodes. This model does not intended to learn an embedding vector of node features, but rather the algorithm learns to aggregate the feature information from the local neighborhood of a node [5]. In this case, if we have K aggregation functions, denoted as $AGGREGATE_k, \forall k \in \{1, \ldots, K\}$ which propagates the node information for all the layers, as well as the weight matrices W^k. Given a graph $G = (V, E)$, with $x_v, \forall v \in V$ the feature vector of node v, it is intended to extract the feature vector h_v^k from node v at step k. First, we have to know the feature vector of the node v in its neighborhood $\{h_u^{k-1}, \forall u \in \mathcal{N}(v)\}$ into the vector $h_{(v)}^{k-1}$.

$$h_{\mathcal{N}(v)}^k = AGGREGATE_k(\{h_u^{k-1}, \forall u \in \mathcal{N}(v)\}) \tag{2}$$

This aggregated vector is taken as the input to a fully connected network with a nonlinear function σ as the activation function.

$$h_v^k = \sigma\left(W^k \cdot CONCAT(h_v^{k-1}, h_{\mathcal{N}(v)}^k)\right) \tag{3}$$

The aggregation function is usually defined as the mean, LSTM, minimum or maximum, among others.

Another proposed model is Graph Convolutional Networks [7]. This algorithm propagates the convolutional idea seen in convolutional neural networks, but applied to graphs. The objective of this architecture is not to derive a comprehensive vector representation of the graph, but to generate multiple graph representations for subsequent analysis. This model is composed of several convolutional layers. Then, another layer allows the sorting of the data of the previous layer according to its structural representation. Finally, dense layers are joined for the vector representation of the input graph. With these models described above, a new approach to detect infostealers has been proposed. Its formulation is described in the following sections.

3 Proposal

Based on the work [2], the same pipeline has been followed for the creation of the dataset creation and model establishment flow. Upon acquisition of a comprehensive dataset of both infostealers and legitimate files, our pipeline undergoes several phases. Initially, we obtain the graph of an executable file wherein its nodes represent assembler functions. Subsequently, we extract feature vectors from said assembler functions. Lastly, we classify the graphs into benign or malicious files. This analytical process enables us to discern executable file behavior and assess any potential security threats. Figure 1 shows the pipeline of this model.

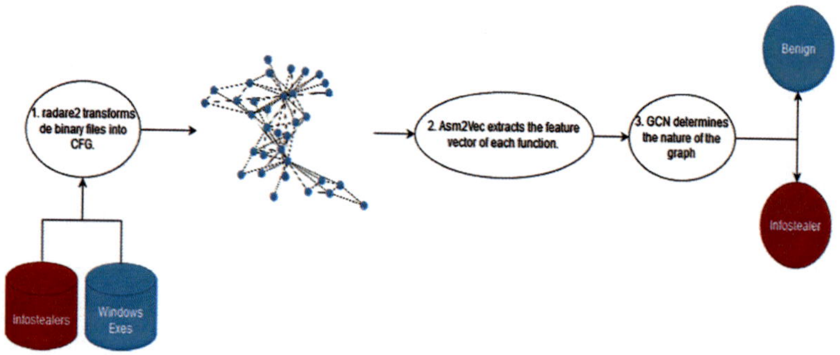

Fig. 1. Model pipeline.

Assembler functions can be represented as a Control Flow Graph (CFG). This would allow the construction of a directed graph where each node would contain assembly instructions of a sequence and the edges would represent the execution flow. The instructions stored in the nodes would be the input to the Asm2Vec model for training and to be able to extract the node's features. Figure 2 represents a CFG of an infostealer sample.

In order to disassemble a file, we used the Radare2 tool [15]. Radare2 is an open tool for manipulating and performing low-level tasks on binary objects [15]. The disassembly and CFG acquisition of a binary file is achieved through the execution of a command. Specifically, the following command is applied to the file in question: `r2 -a -w -c "adfg > file.dot" -qq file.exe`. `r2` is the Radare2 tool explained above with the following parameters: `"-a"` would allow to perform an analysis of the delivered `file.exe`. The `"-w"` flag would open the program in write mode and `"-c"` would execute the command "adfg > file.dot" which would obtain the CFG of the program in dot format and derive it in the file.dot file. Finally `"-qq"` would terminate the execution of the `r2` executable after the indicated command.

An Asm2Vec model is trained on feature extraction using all the instructions stored inside each CFG node, which represents a function. The model implementation that has been used is [8]. This model instance has been trained using as dataset a set of 200 Windows executable files (with .exe extension) that have been transformed into assembly code and relocated in their corresponding functions to provide it as input to the model. Once Asm2Vec has been trained, it is frozen and then the next process in the pipeline of our model is trained.

Once a graph has been obtained whose nodes have a feature vector, a model is needed to extract the information from the network and to classify it into benign or infostealer.

3.1 Model Configuration

Through the GNN development framework, StellarGraph [3], the following model has been built in order to perform a classification of graphs in relation to the networks created by the assembler functions provided by the aforementioned `r2` tool.

From the dot files provided by the `radare2` tool, we used a Python library for network management called NetworkX [12]. This library allowed us to read the dot files and instantiate them as directed graphs. In addition, it allowed us to manipulate the nodes containing the assembler instructions of the functions and to insert another attribute to the node consisting of the feature vector that we got from the Asm2Vec trained model given the assembler function of the node.

In order to develop a function to classify whether a binary file is an infostealer, the Deep Convolutional Graph Neural Network (DeepCNN) [21] architecture has been used. This architecture is composed of three fundamental layers: graph convolutional layers (GCN), SortPooling layers and some convolutional and fully-connected layers. The first layer allows to extract the local characteristics of the node and those of its neighborhood. On the other hand, SortPooling Layer

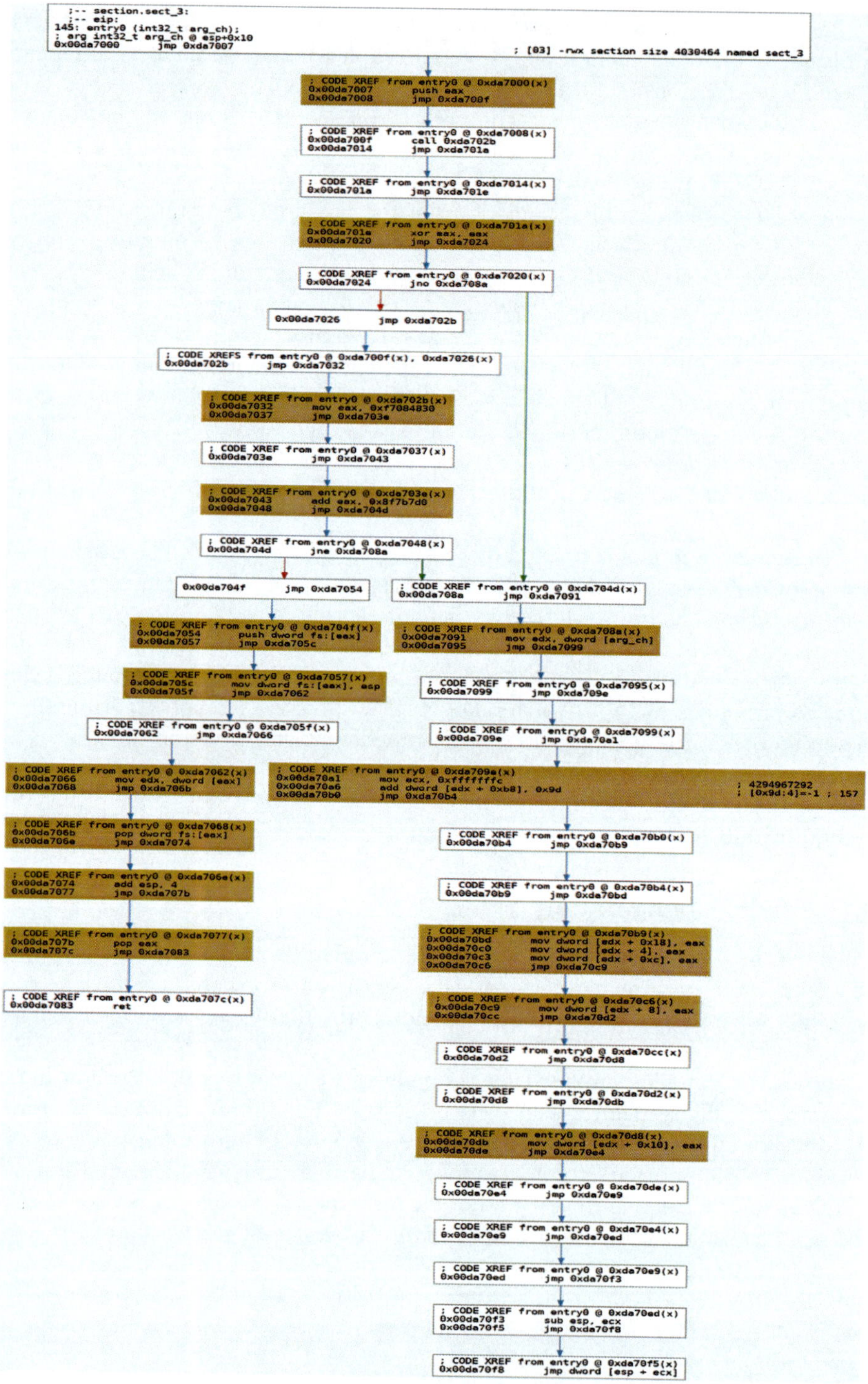

Fig. 2. CFG of an infostealer sample [1].

generates a vector sorted according to the structural roles [13] of the nodes. Finally, Conventional convolutional layers and fully connected layers process data to determine a label in the classification of an input graph.

Figure 3, show the final architecture of the proposed model. This model has been trained with Adam optimizer [6] with a learning rate $\alpha = 1 \cdot 10^{-4}$. We have used cross-entropy loss function for the network training.

During training, the use of a strategy called "Repeated Stratified K-Fold Cross Validation" has been adopted. The training set is divided into K folds with an approximate percentage of existing classes for both training and validation. This process is repeated n times for each fold with a different randomization in each repetition. In this work, we use $K = 10$ cross validation and $n = 5$ for each fold, where n is the number of repetitions.

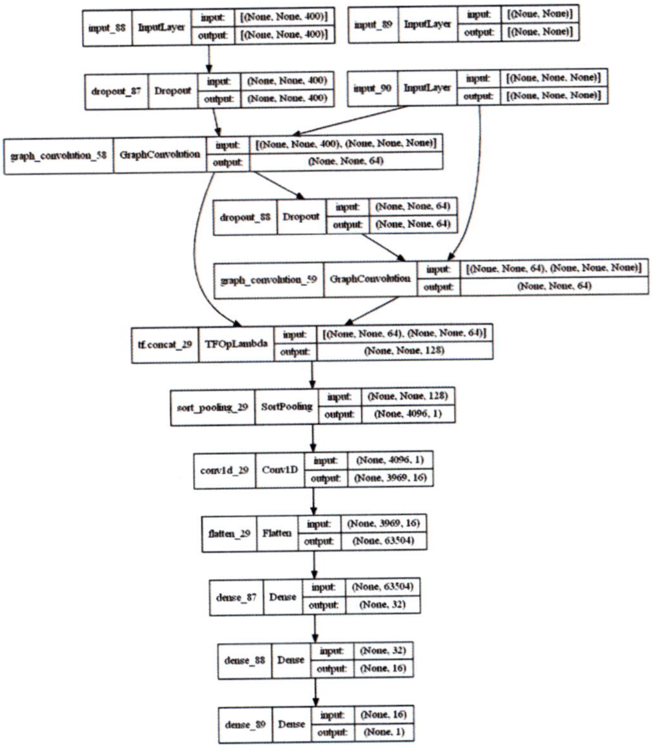

Fig. 3. Our GCN model architecture.

4 Preliminary Results

The malicious files catalogued as stealers have been obtained from the Malware-Bazaar website [9], which, through its API, allows downloading files that are in its database, which contains the 180 hashes of infostealers. On the other hand, 137 legitimate Windows operating system files have been obtained.

A random subset with 70% of the collected executables was assigned to the training set, while the remaining 30% was used for testing.

The model has undergone evaluation using a carefully selected subset of tests, the outcomes of which indicate that we have achieved an accuracy rate of 91.25%, a precision rate of 82.05%, and an outstanding recall rate of 99.98%.

We can observe the metrics table generated by the model given the test subset. According to the criteria we have chosen, label TP is an infostealer and TN is a legitimate binary file. As it can be seen, the metrics used are quite high, with recall being the most notable. The recall allows us to determine which of the true positives were correctly detected, which this model indicates, almost all of them. From a security point of view, it is much more dangerous to have a substantial false negative rate than a false positive rate. A false positive raises more alerts than necessary and increases the workload for the analyst to review. However, a false negative means that a file has been classified as good, but is actually malicious, and can compromise a computer or a company.

5 Conclusions

The present paper proposes a novel model that effectively addresses one of the most significant cybersecurity threats, which is the theft and exfiltration of sensitive data. The model presented in this study provides an efficient and reliable solution to tackle this issue. Most of the defenses that our computers can get are not updated at the time of the new strategies that can develop the infostealers. Our model, based on the DeepGCN architecture, has made it possible to analyze the assembler functions of the graphs and determine if it is an infostealer with 91.25% accuracy. This proposal would make it possible to analyze unknown binary files to antivirus engines and identify unseen infostealers.

For future research, we intend to incorporate the GraphSAGE architecture, as inspired by [2], which aims to detect malware families. In addition, it is proposed to study the graph that it generates dynamically interacting with the rest of the system and what communications it establishes. Right now, the analysis is only static, i.e., it only obtains data from the code. To move forward, and be more precise, it is proposed to study with a GNN model its behavior in a computer.

References

1. Bazaar, M.: MalwareBazaar | SHA256 (2023). https://rb.gy/6hyai
2. Chen, Y.H., Chen, J.L., Deng, R.F.: Similarity-based malware classification using graph neural networks. Appl. Sci. **12**(21) (2022). https://doi.org/10.3390/app122110837. https://www.mdpi.com/2076-3417/12/21/10837
3. CSIRO: Stellargraph - machine learning on graphs (2020). https://www.stellargraph.io/

4. Ding, S.H., Fung, B.C., Charland, P.: Asm2vec: boosting static representation robustness for binary clone search against code obfuscation and compiler optimization. In: Proceedings of the IEEE Symposium on Security and Privacy, May 2019, pp. 472–489 (2019). https://doi.org/10.1109/SP.2019.00003
5. Hamilton, W.L., Ying, R., Leskovec, J.: Inductive representation learning on large graphs. In: Advances in Neural Information Processing Systems, December 2017 (2017). https://arxiv.org/pdf/1706.02216.pdf
6. Kingma, D.P., Ba, J.L.: Adam: a method for stochastic optimization. In: 3rd International Conference on Learning Representations, ICLR 2015, Conference Track Proceedings (2015)
7. Kipf, T.N., Welling, M.: Semi-supervised classification with graph convolutional networks. In: 5th International Conference on Learning Representations, ICLR 2017, Conference Track Proceedings (2016). http://arxiv.org/abs/1609.02907
8. Lancern: Github - lancern/asm2vec: an unofficial implementation of asm2vec as a standalone python package (2020). https://github.com/Lancern/asm2vec
9. MalwareBazaar: Malwarebazaar (2023). https://bazaar.abuse.ch
10. McMillan, R.: Definition: threat intelligence (2013). https://www.gartner.com/en/documents/2487216
11. Micro, T.: Yara rules parent topic (2019). https://docs.trendmicro.com/all/ent/ddi/v5.5/en-us/ddi_5.5_olh/YARA-Rules.html
12. NetworkX: Networkx - network analysis in python (2014). https://networkx.org/
13. Niepert, M., Ahmad, M., Kutzkov, K.: Learning convolutional neural networks for graphs. In: 33rd International Conference on Machine Learning, ICML 2016 (2016)
14. Olyniychuk, D.: Rhadamanthys malware detection: new infostealer spread via google ads & spam emails to target crypto wallets and dump sensitive information. https://socprime.com/blog/rhadamanthys-malware-detection-new-infostealer-spread-via-google-ads-spam-emails-to-target-crypto-wallets-and-dump-sensitive-information/
15. Radare2: Radare2 (2023). https://rada.re/n/radare2.html
16. Run, AA: Any.run: Interactive MISC malware analysis sandbox (2023). https://app.any.run/
17. Sandbox, J.: Automated malware analysis - joe sandbox cloud basic (2023). https://joesandbox.com/
18. Sarojini, S., Asha, S.: Botnet detection on the analysis of Zeus panda financial botnet. Int. J. Eng. Adv. Technol. **8**, 1972–1976 (2019). https://doi.org/10.35940/ijeat.F7941.088619
19. Scarselli, F., Gori, M., Tsoi, A.C., Hagenbuchner, M., Monfardini, G.: The graph neural network model. IEEE Trans. Neural Netw. **20** (2009). https://doi.org/10.1109/TNN.2008.2005605
20. Schneier, B.: How changing technology affects security. IEEE Secur. Priv. **10**(2), 104–104 (2012). https://doi.org/10.1109/MSP.2012.39
21. Zhang, M., Cui, Z., Neumann, M., Chen, Y.: An end-to-end deep learning architecture for graph classification. In: The Thirty-Second AAAI Conference on Artificial Intelligence (AAAI 2018) (2018)

Activity Classification with Inertial Sensors to Perform Gait Analysis

David Martínez-Pascual[1(✉)], José. M. Catalán[1], José. V. García-Pérez[1], Mónica Sanchís[2], Francisca Arán-Ais[2], and Nicolás García-Aracil[1]

[1] Robotics and Artificial Intelligence Group of the Bioengineering Institute, Miguel Hernández University, Avenida de la Universidad, s/n. 03202 Elche, Alicante, Spain
{david.martinezp,jcatalan,j.garciap,nicolas.garcia}@umh.es
[2] INESCOP, C/ Alemania, 102., 03600 Elda, Alicante, Spain

Abstract. The human gait can be analyzed to prevent injuries or gait disorders and make diagnoses or evaluate the progress during the rehabilitation therapies. This work aims to develop a system based on wearable inertial sensors to estimate flexion/extension angles of the lower limbs. Moreover, we have trained a classifier that allows the developed system to differentiate and classify autonomously between four activities of daily living. The proposed classifier is based on a feedforward neural network and has shown an accuracy of 98.33% with users that were not involved in the model's training.

Keywords: Gait analysis · IMU · Feedforward Neural Network · Activity classification

1 Introduction

The analysis of the human gait can be employed to analyze the range of motion (ROM) of the lower limb joints, which could be reduced in patients who suffered from neuromuscular pathologies such as cerebral palsy, Parkinson's disease or hemiplegia [1,2]. The reduction of the ROM of the lower limb joints presents difficulties in the realization of activities of daily living (ADLs), such as walking, ascending or descending ramps, and climbing stairs.

The lower limbs motion analysis could be helpful to make detailed diagnoses, plan an optimal treatment, or evaluate the rehabilitation therapies outcomes [3]. In order to evaluate gait performance, visual observation of the lower limb kinematics could be used to determine gait disorders and the treatment results [4,5].

This work was supported by the Spanish Ministry of Universities through the Research and Doctorate Supporting Program FPU20/05137; by the Ministry of Universities and European Union, "financed by European Union - Next Generation EU" through Margarita Salas grant for the training of young doctors; by the Research Grants of the Miguel Hernández University of Elche through the grant 2022/PER/00002; by the Ministry of Science and Innovation through the project PID2019-108310RB-100; and by the Valencian Innovation Agency through the project GVRTE/2021/361542.

S. Ossowski et al. (Eds.): DCAI 2023, LNNS 740, pp. 74–82, 2023.
https://doi.org/10.1007/978-3-031-38333-5_8

The angle of the human lower limb joints can be estimated with artificial vision motion capture techniques [6]. However, the main disadvantage is that these techniques require several cameras to acquire motion capture data, so this method could limit the space of analysis [7,8]. The angle of the human lower limb joints can also be estimated by means of wearable inertial measurement units (IMU) [9,10]. These devices are placed on each of the main segments of the lower limbs. In this way, it is possible to estimate the angle value of the joints using the information provided by each IMU.

Knowing which gait ADLs are being performed at any given moment is necessary when analyzing lower limb movement. The development and improvement of machine learning algorithms allow the construction of robust classifiers that are able to differentiate between gait ADLs using the information provided by different types of sensors to perform gait analysis [11]. The classification of gait activities with machine learning techniques can be made by extracting temporal and frequency domain features from the IMU signals [12].

This work aims to develop a system to perform lower limb kinematics analysis during gait based on wearable inertial sensors. The system should be able to detect the onset and end of each step to analyze the flexion/extension angles of the lower limb joints during the cycle gait. Furthermore, we want to provide the system the ability to classify and separate the users' steps into four groups of gait ADLs: ground-level walking, ascending a ramp, descending a ramp, and climbing stairs. With this purpose, artificial neural network models could be employed to classify different gait ADLs by using the angle of the human lower limb joints.

2 Materials and Methods

An experimental laboratory session has been carried out with able-bodied subjects in order to develop a system to perform lower limb kinematics analysis that could autonomously differentiate between several gait ADLs. The information about the participants, the experimental protocol and setup, the acquired data, and the trained classifier are described in the following sections.

2.1 Subjects

The experimental sessions involved 12 participants, 10 male and 2 female, without motor or cognitive impairment. The users' heights were between 1.65 m and 1.87 m (176.2 ± 7.4 cm), with weights ranging between 56.1 kg and 90.2 kg (76.0 ± 12.5 kg), and ages between 23 and 52 years old (29.8 ± 7.4).

2.2 Experimental Setup

The devices used during the experimental session are shown in Fig. 1.

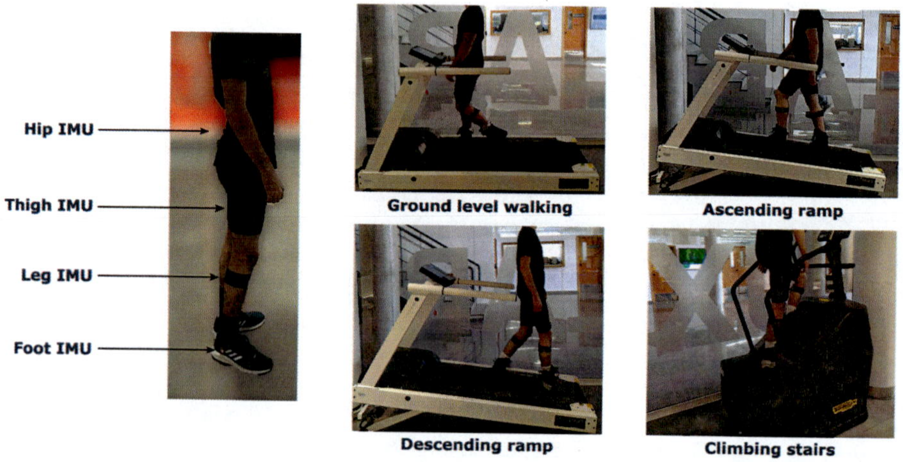

Fig. 1. Experimental setup and performed tasks. On the left image, the IMU locations to estimate the lower limb joints' angles have been indicated. During the experimental session, the users walked on a treadmill and a stairs machine simulating gait on a flat surface, gait on a 12% positive slope, gait on a 12% negative slope, and climbing stairs.

To acquire the lower limbs' motion, 4 XSens Dot IMUs were employed at a sampling rate 60 Hz. The inertial sensors were placed over the pelvis, the thigh, and the shank with elastic trips. The remaining inertial sensor was placed on the foot with stickers.

To simulate and collect data during several gait ADLs, we employed a treadmill h/p/cosmos 150/50 to simulate ground-level walking, ramp ascending, and ramp descending. To simulate climbing stairs, we employed an stair climber machine.

2.3 Study Protocol

At the beginning of the session, the IMUs were attached to the subjects in the specified locations (Fig. 1), and the height and weight of the users were measured. Then, the participants were told to perform calibration movements to estimate the flexion/extension angles: they performed arbitrary movements of the leg for 30s, and they walked on the treadmill for 1 min at a comfortable speed.

Once the calibration process was completed, the subjects realized four gait ADLs over the treadmill and stairs machine:

1. Ground-level walking. The users walked on the treadmill at 4.5 km/h for 5 min.
2. Ascending ramp with 12% slope. The users walked on the ascending treadmill at 2.5 km/h for 5 min.
3. Descending ramp with 12% slope. The users walked on the descending treadmill at 2.5 km/h for 5 min.
4. Climbing stairs. The users walked on the stairs machine at a speed of 50 stairs/minute for 3 min.

The walking speeds and slope were set according to INESCOP Footwear Technology Center [13] protocols for footwear certification, and stair climbing time was reduced to 3 min to avoid excessive participant fatigue.

2.4 Acquired Data and Processing

Acceleration and gyroscope data were acquired from the IMUs placed over the lower limbs to measure the angle of the joints, which were estimated in the sagittal plane through the method proposed by T. Seel et al. [14].

Some methods collected in the bibliography require specific and accurate calibration movements to set up the IMU reference systems to the subject joint axis [15]. By contrast, the algorithm proposed by T. Seel et al. makes use of arbitrary calibration movements, as well as joint movements during gait to determine the sensors' situation with respect to the joint they measure, exploding kinematic constraints. Therefore, this method would be ideal for biomechanical analysis due to the ease of sensor placement. In addition, the error in the measurement of joint angles does not depend on the placement of the sensors.

This method has shown a root mean squared error (RMSE) between 1 and 3° in flexion/extension movements, so it can be assumed high accuracy in the angle measurement. In addition, it should be noted that this method has already been employed in previous studies [9,16].

From the IMU located on the pelvis, we have made use of the anteroposterior acceleration to detect the foot-ground contact [17]. We have applied a forward-backward low pass filter with 2 Hz cutoff and then computed the local maxima corresponding to the left and right foot contacts. Once we detect the foot-ground contacts, we know the gait cycle. As a matter of convention, we detect the onset and end of each step and transform the data from the temporal domain to the gait cycle domain (0–100% cycle gait).

Each step has been described by employing six temporal domain features to train the classifier. The angular trajectories are described according to the minimum and maximum value and instant during the step, the root mean square (RMS), and the mean absolute deviation (MAD). Moreover, the tasks have been codified with a one-hot encoding to be introduced to the classifier for the learning process.

2.5 Classifier Architecture

A Feedforward Neural Network (FNN) to classify several gait ADLs was trained, which is based on neurons that receive an input signal and process them by their activation function to an output signal [18].

As we want to differentiate between four gait ADLs, we have trained an FNN with Keras framework [19] to solve a multinomial logistic regression problem. The proposed architecture is shown in Fig. 2. We have trained an FNN model with one input layer equal to the number of inputs (6 descriptors per joint, a total of 18 inputs), 5 hidden layers with 10 neurons, and an output layer with four neurons (1 per ADL). We have employed the Adam algorithm as the optimizer to adjust the model weights and the Categorical Cross-Entropy as the loss function. We have used the Rectified Linear Unit (ReLu) function as the activation function of the hidden layer neurons, and the Softmax function in the output layer neurons, so we have four outputs between 0 and 1 which could be interpreted as the probability of a performed task. To avoid overfitting, we have used the dropout regularization method [20]. In addition, we have imposed that the maximum norm of the weights does not exceed 4.

We have divided the data into three parts in order to evaluate the performance of the classifier. First, we used data collected from 9 users to train the classifier (training users). This data collection was randomly split into two subsets to validate the model performance by cross-validation: 80% of the data for the training and the 20% remaining for validation. The data from the three remaining users were used to evaluate the classifier performance with users that are not involved in the learning process (test users). In addition, we have scaled each feature of the training data from its minimum and maximum, so we have numerical values between 0 and 1. We have also employed the training scaler to evaluate the accuracy of the classifier with the test data.

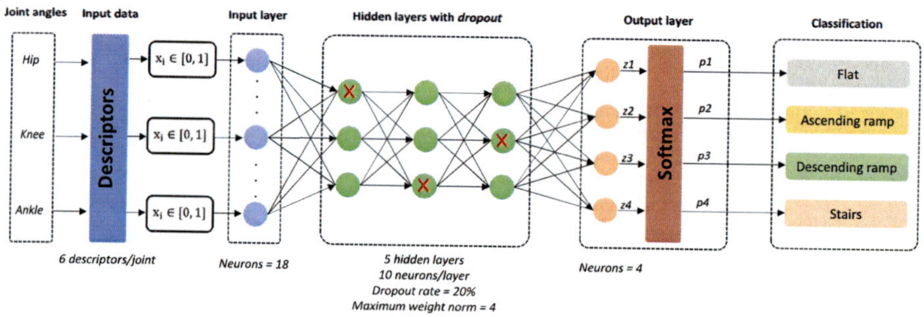

Fig. 2. Feedforward Neural Network (FNN) model architecture proposed to classify 4 gait ADLs.

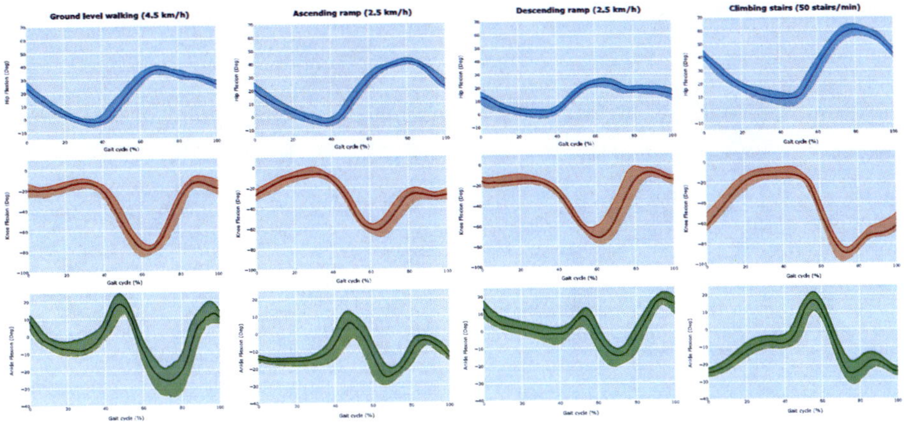

Fig. 3. Estimated lower limb flexion/extension angles during the experimental session. In the above graphs, the median angles have been represented, and the shaded areas correspond to the values between the first and third quartiles. The performed tasks are divided by columns (flat gait, positive slope gait, negative slope gait, climbing stairs), and the joints' angles are organized by rows (hip, knee, ankle).

3 Results

The measured angles of the lower limb joints of the participants have been represented in Fig. 3. We have represented the median angles during the gait cycle, and the shaded areas correspond to the values between the first and third quartiles.

The FNN has been trained for 25 epochs and a batch size of 8. The results obtained with the trained classifier are collected in Fig. 4, which shows confusion matrices with the validation and test data, as well as the achieved accuracy.

The results show that the proposed classifier achieves a validation accuracy of 99.83% and a test accuracy of 98.33%.

The confusion matrices shown in Fig. 4 represent the steps in the performed task, while each column represents the steps in the predicted task. The validation confusion matrix (Fig. 4.a) shows that the activities of walking on an ascending and descending ramp are all classified correctly. Nevertheless, the task of walking on a flat surface is incorrectly classified 0.22% of the time as walking on an ascending ramp. At the same time, the steps performed during climbing stairs are also incorrectly classified 1.2% of the time as walking on an ascending ramp.

The confusion matrix with the test users is represented in Fig. 4.b. It can be observed that the descending steps are classified 97% of the time correctly, being confused with steps on a flat surface 3% of the time. Walking on a flat surface is classified with a rate of 98%, where the 2% of ground-level walking steps are confused with descent steps. Moreover, the classifier achieves almost a 100% prediction rate walking on an ascending ramp, although 0.19% of the

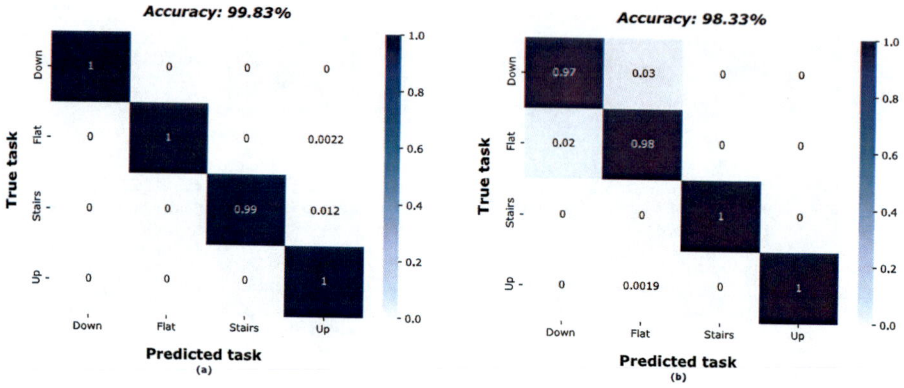

Fig. 4. Confusion matrices obtained with the FNN classifier proposed for (a) validation data and (b) test data. Each row of the matrix represent the ADL performed, and the columns represent the predicted ADL. The four gait ADLs performed during the experimental session are collected in the matrices: ground-level walking, ascending a ramp (Up), descending a ramp (Down), and climbing stairs.

steps are labeled as walking on a flat surface. It should be noted that all the performed steps while climbing stairs are correctly classified.

4 Discussion

In this work, an FNN model has been trained to classify four gait ADLs: ground-level walking, ramp ascent, ramp descent, and stair climbing. Accelerometer and gyroscope data from IMUs have been previously used to recognize different gait activities [12]. In contrast, we propose to use data from multiple IMUs to measure the joints of the lower limb in order to perform biomechanical analysis. Therefore, we explored the use of hip, knee, and ankle flexion/extension angles during gait as inputs to an FNN. This method has a major benefit: the estimation of joint angles does not depend on the location and direction of the IMUs. So, we can expect that the error in the gait activity classification will not change due to variations in sensor placement. Also, the calibration method of this algorithm is easy and does not need accurate movements, as the calibration relies on walking and doing random leg movements for a short time.

The results show that the proposed classifier achieves high accuracy with the participants in the training process, with a validation accuracy of 99.83%. Although the classifier achieves a lower accuracy with the test data, the results also suggest good accuracy since the FNN achieves a 98.33% accuracy. This percentage of accuracy suggests that by using different IMUs to measure the angles of the lower limbs together with the proposed model, similar or even better results than previous studies are obtained: in [21] a 93% accuracy is achieved using a convolutional neural network, in [22] an accuracy of 89.38–98.23% is achieved using different models, and in [23] a 92.71% accuracy is achieved using a one-dimensional convolutional network.

The results shown in the confusion matrices could be explained by using the joint angles during gait. In Fig. 3 can be observed the similarity between the activities of walking on a flat surface, ascending, and descending a ramp in terms of ROM, maxima or minima angles, and the instant of the gait cycle in which they occur. Therefore, it may happen that, due to these similarities, certain variations of the joint angles during gait may cause the classifier to interpret it as a different task. Nevertheless, the measured knee and ankle angles during climbing stairs show different trajectories concerning the rest of the performed gait ADLs. Hence, these differences would explain why the classifier can label this task with such a high prediction rate.

5 Conclusion

This work presents a system based on wearable inertial sensors to perform gait analysis when performing different gait ADLs, which allows us to estimate the flexion/extension angles of the lower limbs and detect foot-ground contact.

Moreover, we have trained a classifier based on an FNN model that employs a description of the steps to differentiate between four gait ADLs: ground-level walking, ascending a ramp, descending a ramp, and climbing stairs. The results suggest that the proposed classifier could achieve good generalization since the model shows a validation accuracy of 99.83% and a test accuracy of 98.33%.

In future works, we pretend to extend the number of situations detected, such as standing still, descending stairs, or getting up or sitting down from a chair, and we pretend to examine the classifier performance in a real environment. In addition, Deep Learning techniques will be explored to classify the proposed gait ADLs.

References

1. Hyodo, K., Masuda, T., Aizawa, J., Jinno, T., Morita, S.: Hip, knee, and ankle kinematics during activities of daily living: a cross-sectional study. Braz. J. Phys. Ther. **21**(3), 159–166 (2017)
2. Baker, R.: Gait analysis methods in rehabilitation. J. Neuroeng. Rehabil. **3**(1), 1–10 (2006)
3. Nadeau, S., Betschart, M., Bethoux, F.: Gait analysis for poststroke rehabilitation: the relevance of biomechanical analysis and the impact of gait speed. Phys. Med. Rehabil. Clin. **24**(2), 265–276 (2013)
4. Grant, A.D.: Gait analysis: normal and pathological function. JAMA **304**(8), 907 (2010)
5. Brunnekreef, J.J., Van Uden, C.J., van Moorsel, S., Kooloos, J.G.: Reliability of videotaped observational gait analysis in patients with orthopedic impairments. BMC Musculoskelet. Disord. **6**(1), 1–9 (2005)
6. Moeslund, T.B., Hilton, A., Krüger, V.: A survey of advances in vision-based human motion capture and analysis. Comput. Vis. Image Underst. **104**(2–3), 90–126 (2006)

7. Oh, S.E., Choi, A., Mun, J.H.: Prediction of ground reaction forces during gait based on kinematics and a neural network model. J. Biomech. **46**(14), 2372–2380 (2013)
8. Choi, A., Lee, J.M., Mun, J.H.: Ground reaction forces predicted by using artificial neural network during asymmetric movements. Int. J. Precis. Eng. Manuf. **14**(3), 475–483 (2013)
9. Filippeschi, A., Schmitz, N., Miezal, M., Bleser, G., Ruffaldi, E., Stricker, D.: Survey of motion tracking methods based on inertial sensors: a focus on upper limb human motion. Sensors **17**(6), 1257 (2017)
10. Hamdi, M.M., Awad, M.I., Abdelhameed, M.M., Tolbah, F.A.: Lower limb motion tracking using IMU sensor network. In: 2014 Cairo International Biomedical Engineering Conference (CIBEC), pp. 28–33. IEEE, December 2014
11. Chen, D., et al.: Bring gait lab to everyday life: gait analysis in terms of activities of daily living. IEEE Internet Things J. **7**(2), 1298–1312 (2019)
12. Mannini, A., Trojaniello, D., Cereatti, A., Sabatini, A.M.: A machine learning framework for gait classification using inertial sensors: application to elderly, post-stroke and Huntington's disease patients. Sensors **16**(1), 134 (2016)
13. INESCOP. https://inescop.es/en/
14. Seel, T., Raisch, J., Schauer, T.: IMU-based joint angle measurement for gait analysis. Sensors **14**(4), 6891–6909 (2014)
15. Cutti, A.G., Ferrari, A., Garofalo, P., Raggi, M., Cappello, A., Ferrari, A.: Outwalk: a protocol for clinical gait analysis based on inertial and magnetic sensors. Med. Biol. Eng. Comput. **48**(1), 17–25 (2010)
16. Martínez-Pascual, D., et al.: Machine learning and inertial sensors to estimate vertical ground reaction force during gait. In: Tardioli, D., Matellan, V., Heredia, G., Silva, M.F., Marques, L. (eds.) ROBOT2022: Fifth Iberian Robotics Conference. ROBOT 2022. LNNS, vol. 590, pp. 264–273. Springer, Cham (2023). https://doi.org/10.1007/978-3-031-21062-4_22
17. Zijlstra, W., Hof, A.L.: Assessment of spatio-temporal gait parameters from trunk accelerations during human walking. Gait Posture **18**(2), 1–10 (2003)
18. Svozil, D., Kvasnicka, V., Pospichal, J.: Introduction to multi-layer feed-forward neural networks. Chemom. Intell. Lab. Syst. **39**(1), 43–62 (1997)
19. Chollet, F., et al.: Keras. GitHub (2015). https://github.com/fchollet/keras
20. Srivastava, N., Hinton, G., Krizhevsky, A., Sutskever, I., Salakhutdinov, R.: Dropout: a simple way to prevent neural networks from overfitting. J. Mach. Learn. Res. **15**(1), 1929–1958 (2014)
21. Lopez-Nava, I.H., et al.: Gait activity classification on unbalanced data from inertial sensors using shallow and deep learning. Sensors **20**(17), 4756 (2020)
22. Alsheikh, M.A., et al.: Deep activity recognition models with triaxial accelerometers. arXiv preprint arXiv:1511.04664 (2015)
23. Lee, S.M., Yoon, S.M., Cho, H.: Human activity recognition from accelerometer data using convolutional neural network. In: 2017 IEEE International Conference on Big Data and Smart Computing (bigcomp). IEEE (2017)

Guided Rotational Graph Embeddings for Error Detection in Noisy Knowledge Graphs

Raghad Khalil(✉) and Ziad Kobti

School of Computer Science, University of Windsor, Windsor, ON N2B 3P4, Canada
{khalilr,kobti}@uwindsor.ca
https://www.uwindsor.ca/science/computerscience/

Abstract. Knowledge graphs (KGs) use triples to describe real-world facts. They have seen widespread use in intelligent analysis and applications. However, the automatic construction process of KGs unavoidably introduces possible noises and errors. Furthermore, KG-based tasks and applications assume that the knowledge in the KG is entirely correct, which leads to potential deviations. Error detection is critical in KGs, where errors are rare but significant. Various error detection methodologies, primarily path ranking (PR) and representation learning, have been proposed to address this issue. In this paper, we introduced the Enhanced Path Ranking Guided Embedding (EPRGE), which is an improved version of an existing model, the Path Ranking Guided Embedding (PRGE) that uses path-ranking confidence scores to guide TransE embeddings. To improve PRGE, we use a rotational-based embedding model (RotatE) instead of TransE, which uses a self-adversarial negative sampling technique to train the model efficiently and effectively. EPRGE, unlike PRGE, avoids generating meaningless false triples during training by employing the self-adversarial negative sampling method. We compare various methods on two benchmark datasets, demonstrating the potential of our approach and providing enhanced insights on graph embeddings when dealing with noisy KGs.

Keywords: Knowledge Graph · Knowledge Graph Embedding · Error Detection

1 Introduction

A knowledge graph (KG) organizes information as a multi-relational graph with entities and relations as nodes and edges, respectively [21]. KGs can be built using various approaches, such as expert curation (Cyc) [10], crowd-sourcing (Freebase [2], Wikidata [20]), or automatic extraction from semi-structured web knowledge bases (DBpedia [9], YAGO [16]). Automation introduces noise and errors, making KGs imperfect [13]. Existing error detection methods primarily fall into two categories: rule-based and embedding-based approaches [1,5–8,15,22]. Rule-based methods, while effective, often lack generalizability as they

S. Ossowski et al. (Eds.): DCAI 2023, LNNS 740, pp. 83–92, 2023.
https://doi.org/10.1007/978-3-031-38333-5_9

rely on user-defined, domain-specific rules. On the other hand, embedding-based methods utilize graph embeddings from Translational Distance Models (e.g., TransE [3], RotatE [19], HRotate [17]) or Semantic Matching models [21] for error detection. Despite being studied extensively, these methods often resort to uniform negative sampling, which can lead to low-quality embeddings and the generation of meaningless false triples.

This research aims to develop an Enhanced Path Ranking Guided Embedding (EPRGE) approach, improving the existing PRGE algorithm [4] for KG error detection. The current PRGE uses PaTyBRED [12] for confidence scores and TransE [3] for embeddings, but suffers from low-quality embeddings due to uniform sampling and generating false triples.

Our proposed EPRGE integrates RotatE's [19] self-adversarial negative sampling and PaTyBRED, addressing TransE's limitations in capturing relation patterns. Using RotatE instead of TransE, we expect improved embeddings and error detection performance.

This paper's main objective is to enhance PRGE's error detection by incorporating RotatE in the EPRGE model. We will demonstrate our approach's effectiveness through experimentation and analysis. Key contributions include:

1. Proposing EPRGE, a new hybrid model for KG error detection
2. Assessing various error detection methods on two benchmark datasets
3. Evaluating the proposed approach on the Dementia KG to demonstrate real-world effectiveness.

2 Problem Formulation

Let $G = \{E, R\}$ denote a KG that contains a large number of triples, where E and R denote the sets of entities and relations, respectively. Each triple comprises a head entity h, a relation r, and a tail entity t, represented as (h, r, t). We adopt the KG Error Detection definition presented in [22]. This definition states that given a triple in a KG (h, r, t), if there is a mismatch between head/tail entities and their relation r, then this triple (h, r, t) is an error. Albert Einstein, for example, was born in Elm, Germany. So, suppose a fact is given in a KG as *(Berlin, isBirthPlaceOf, Albert Einstein)*; this is a false triple, although Berlin and Albert Einstein are both correct entities. This example demonstrates that a mismatch between three components more frequently produces errors than by a single entity or relation. In this paper, we assume that any given G contains some noise ratio N percent, indicating that N percent of the triples in G are incorrect [4]. We also use a scoring function, which ranges from 0 to 1, to indicate the degree to which triples are correct. The scoring function for false triples should be close to 1. The goal of error detection is to figure out how to locate these errors in G. Given the assumptions and definitions above, we formalize the error detection problem on KGs as follows:

Given a KG $G = \{E, R\}$, we want to propose a method that takes the KG as input and returns a rank of all the triples based on their computed scoring functions, indicating the possibility of an error.

3 Literature Review

SDValidate [14] is a statistical approach for error detection in KGs. This method uses the distributions of entity types and relations to identify anomalies, effectively using a high-level understanding of the KG structure. However, its performance is heavily reliant on the accuracy and completeness of these distributions. The KG triple confidence assessment model proposed by Jia et al. [8] relies on a reachable paths inference algorithm. While this model provides useful insights into KG error detection, it may struggle with complex graphs where the reachable paths may become overwhelmingly numerous or intricate. The original PRGE model by Bougiatiotis et al. [4] uses a simple path ranking (PR)-based approach, PaTyBRED [12], which applies a path ranking algorithm and sub-graph feature extraction. Despite the model's simplicity, it effectively validates the utility of PR for KG error detection. Translational distance models like TransE [3] and RotatE [19] are foundational to KG embeddings (KGE). These approaches convert KG entities and relations into lower-dimensional vectors, aiding in error detection by transforming the problem into a more tractable space. However, these methods can struggle with noisy input, limiting their effectiveness in real-world KGs that often contain imperfections. Several other KGE-based error detection methods have been proposed to improve embedding quality in the presence of errors [4,8,11,18]. For instance, PTransE [11] extends TransE with paths, enabling it to capture more complex relational patterns. CKRL [18] introduces a triple confidence score to account for the uncertainty inherent in KGs. The KGTtm approach [8] generates confidence scores using a crisscrossed neural network structure, demonstrating the potential of more complex machine learning methods in this domain. However, this complexity may not always be necessary. The PRGE model [4] shows that a simple path ranking approach can effectively guide KG embeddings construction. Our proposed EPRGE model builds upon the PRGE framework, integrating a higher-performing combination of path ranking and KGE models. This approach aims to capture the strengths of these methods while mitigating their weaknesses, offering a balanced solution to KG error detection. Direct comparisons with existing models like KGTtm are difficult due to the different underlying assumptions and design choices. Nonetheless, our model contributes to the ongoing exploration of hybrid models that integrate various techniques for enhanced error detection in KGs.

4 Proposed Approach

In this section, we describe our proposed hybrid approach, EPRGE, an enhanced variant of PRGE [4]. Basically, EPRGE is a hybridization of PaTyBRED [12] and RotatE [19] models. EPRGE uses the RotatE model to generate embeddings, and it computes a confidence score too based on the inspiration drawn from PaTyBRED.

4.1 PaTyBRED

PaTyBRED [12] is a PR-based algorithm with error detection in mind. Paths are used as features in PaTyBRED, with a path defined as a sequence of relations $r_1 \rightarrow r_2 \rightarrow ... \rightarrow r_n$ connecting a head h and a tail t. The algorithm uses these paths as features to determine whether a given triple is noisy. Specific paths are kept per relation using pruning and other heuristics to indicate whether a triple is incorrect. Finally, for each triple, a confidence score with values in the range [0, 1] is determined, with low scores indicating noise.

4.2 RotatE

The basic idea behind RotatE [19] is that given a triple (h, r, t), the head h and the tail t are mapped to the complex embeddings, i.e., $h, t \in C^k$. Then the functional mapping induced by each relation r is defined as an element-wise rotation from the head entity h to the tail entity t. In other words, given a triple (h, r, t), $t = h \circ r$ where $|r_i| = 1$ and $\circ$ is the element-wise product. According to the above definition, for each triple (h, r, t), the fitness of the model is calculated through the following distance function:

$$d_r(h, t) = ||h \circ r - t|| \tag{4.1}$$

which is minimized using a self-adversarial negative sampling loss function in Eq. (4.1) for training to optimize the model effectively.

$$L_{RotatE} = -log\sigma(\gamma - d_r(h, t)) - \sum_{i=1}^{n} p(h'_i, r, t'_i)log\sigma(d_r(h'_i, t'_i) - \gamma), \tag{4.2}$$

where γ is a fixed margin, σ is the sigmoid function, $p(h'_i, r, t'_i)$ is the weight of the negative sample, n is the number of negative samples, and (h'_i, r, t'_i) is the i-th negative triplet.

4.3 Enhanced Path Ranking Guided Embeddings (EPRGE)

To calculate the confidence scores, denoted as $P(h, r, t)$, we chose the PaTyBRED method because it is the most simplified and robust of the PRA-based methods and RotatE is the KGE method. TransE employs a uniform negative sampling strategy, which means that the head or tail entity is corrupted randomly. Such an approach could be more efficient because many samples are false as training progresses, providing no meaningful information. RotatE, on the other hand, employs a self-adversarial negative sampling technique that samples negative triples based on the current embedding model. This technique is required to avoid producing meaningless false triples, which results in model inefficiency. In addition, RotatE is capable of modelling and inferring various relation patterns found in KGs, such as symmetry/antisymmetry, inversion, and composition.

To integrate the confidence score of each triple into the loss function, we introduce a novel modification to Eq. (4.2), as shown in Eq. (4.3):

$$Loss = -log\sigma(\gamma - d_r(h,t)) \cdot P(h,r,t)^{\lambda} - \sum_{i=1}^{n} p(h'_i, r, t'_i) log\sigma(d_r(h'_i, t'_i) - \gamma), \quad (4.3)$$

where λ is a scaling parameter used to adjust the significance of the confidence score. As such, EPRGE provides an enhanced modular framework for combining methodologies from the error detection and graph embedding models to tackle challenges in real-world applications where noise in KGs usually exists. Algorithm 1 demonstrates the proposed method.

Algorithm 1. Learning EPRGE

Input: Training Set $S = (h, r, t)$, PaTyBRED confidence scores $P(h, r, t)$
Initialize: Hyperparameters λ, γ, α, b, learning rate, hidden dimension, and n;
Initialize entity and relation embeddings randomly
while *terminal condition not met (i.e. maximum training steps not reached)* **do**
  $S_{batch} \leftarrow$ sample b tuples from S;
  Generate negative samples using self-adversarial generation for each (h, r, t) in S_{batch};
  for *each* (h_i, r_i, t_i) *in* S_{batch} *and* n *negative samples* (h', r, t') **do**
    $score \leftarrow d_r(h,t) = ||h \circ r - t||$
    $score \leftarrow d_r(h',t') = ||h' \circ r - t'||$
    Update embeddings w.r.t:
    $-log\sigma(\gamma - d_r(h,t)) \cdot P(h,r,t)^{\lambda} - \sum_{i=1}^{n} p(h'_i, r, t'_i) log\sigma(d_r(h'_i, t'_i) - \gamma)$
  end
end

5 Experiment Evaluation

5.1 Datasets

We evaluate the proposed method and other competitors on different tasks related to error detection. We perform experiments on two commonly used KG datasets. Table 1 shows a statistical overview of the datasets.

5.2 Evaluation Metrics

Following the same steps as [4], we compute the distance function in Eq. (4.1) for each triple in the dataset. Then, we generate a ranking for all triples based on this distance function score. The smaller the value of the distance function of the triple, the more valid the triple is. As such, the erroneous triples would have much greater value than the correct ones. To measure this we use the filtered mean rank (*fMR*) and the filtered mean reciprocal rank (*fMRR*) [12]. For (*fMR*), lower is better while for (*fMRR*), higher is better.

Table 1. Dataset Statistics

Dataset	#Relations	# Entities	# Training	# Validation	# Test
WN18	18	40,943	141,442	5,000	5,000
FB15k	1345	14,951	483,142	50,000	59,071
Dementia	64	48,008	135,000	4999	5862

5.3 Experimental Setup

To make a fair comparison, we ensured that our experimental setup is identical to the one used in PRGE [4]. The details are highlighted here:

1. **Error Imputation Protocol**: We used the new datasets created by [4] with different percentages of noise levels to simulate real-world KGs constructed automatically. For the FB15K dataset, [4] constrained the noise generation in the sense that the new head h' and tail t' should have appeared in the dataset with the same relation r. This constraint focuses on generating more confusing noise for any method. On the other hand, [4] performed negative sampling on WN18 and Dementia without constraint.
2. **Hyper-parameters** We conducted experiments on two benchmark and one real-world datasets using various baseline methods, applying the suggested settings and parameters from each method's original authors. For our proposed Enhanced Path Ranking Guided Embedding (EPRGE), which builds upon PRGE and integrates RotatE [19], we conducted hyperparameter optimization using a grid search with the following ranges: embedding dimension ($k \in 125, 250, 500, 1000$), batch size ($b \in 512, 1024, 2048$), self-adversarial sampling temperature ($\alpha \in 0.5, 1.0$), and fixed margin ($\gamma \in 3, 6, 9, 12, 18, 24, 30$). The best hyperparameter settings for EPRGE on several benchmarks are reported in Table 2.

Table 2. The best hyperparameter setting of EPRGE on several benchmarks.

Dataset	n	b	α	γ	lr	k
WN18	1024	512	0.5	12	0.00100	500
FB15k	128	2048	1.0	24	0.00010	1000
Dementia	1024	512	0.5	12	0.00100	500

5.4 Results

Tables 3, 4, and 5 demonstrate the results of all approaches for the error detection task on all datasets. We present insights and observations stemming from the results here:

1. **WN18 Dataset**: EPRGE was found to outperform other models on the WN18 dataset, confirming both hypotheses. Adding RotatE to the EPRGE model improved the quality of embeddings and enhanced its ability to detect errors and predict false information in KGs. Compared to other hybrid KGE algorithms, such as PRGE, PTransE, and CKRL, EPRGE was more robust and effective in KG error detection and completion tasks on the WN18 dataset, with AUC scores ranging from 0.6719 to 0.9894. EPRGE achieved the highest AUC scores across all error ratios, indicating its superiority in detecting errors in KGs
2. **FBK15 Dataset**: Similar to WN18, the experimental setup on FB15K showed that as the error ratio increased from 10% to 40%, the performance of all models deteriorated. However, EPRGE consistently outperformed all other models across all error ratios and evaluation metrics, confirming the effectiveness of using RotatE in the EPRGE model for KG error detection. Despite FB15K's larger number of relations and triples, EPRGE achieved better fMR, fMRR, and AUC scores than the other models. The lower AUC scores of all models on FB15K compared to WN18 may be due to the increased complexity and size of the dataset. Additionally, the FB15K dataset has a significant number of symmetry/antisymmetry relations, supporting the use of RotatE in the PRGE model. Although the performance of all models on FB15K was generally lower than on WN18, EPRGE still outperformed all other models, demonstrating its ability to handle noisy and complex KGs. Notably, the error injection protocol on FB15K was different from WN18 and the Demensia KG, generating harder and more confusing noise for all methods.
3. **Dementia Dataset**: The results on the Dementia KG dataset indicate that as the error ratio increases from 10% to 40%, all models experience a significant decrease in performance. However, EPRGE outperforms all other models in fMR, fMRR, and AUC scores, demonstrating its ability to detect errors and predict false information in the Dementia KG dataset. Notably, the Dementia KG dataset is more complex and noisy than the benchmark datasets (WN18 and FB15K), making it more challenging for KGE methods. Despite this difficulty, EPRGE still shows superior performance, indicating its potential to handle real-world KGs with complex relation patterns and high levels of noise. Overall, these results support the hypotheses of the study that using RotatE in the PRGE model results in a more efficient and effective error detection methodology for handling real-world KGs.
4. **Effect of noise**: As the noise level rises from 10% to 40%, the performance of all models starts to deteriorate in all datasets. However, both PRGE and EPRGE methods show smaller fluctuations in performance when compared to all other models.

Table 3. Error Detection Results for WN18 (Imputing Random Errors)

	WN18-10%			WN18-20%			WN18-40%		
	*f*MR	*f*MRR	AUC	*f*MR	*f*MRR	AUC	*f*MR	*f*MRR	AUC
TransE	38942	0.0002	0.7246	39338	0.0003	0.7219	44465	0.0005	0.6857
PTransE	45722	0.0007	0.6763	45393	0.0003	0.6791	46410	0.0002	0.6719
CKRL	15735	0.0009	0.8887	16966	0.0007	0.8800	39251	0.0011	0.7225
PRGE	3679	0.0009	0.9742	3868	0.0009	0.9727	3671	0.0008	0.9740
EPRGE	**3522**	**0.0016**	**0.9807**	**3702**	**0.0016**	**0.9894**	**3514**	**0.0014**	**0.9805**

Table 4. Error Detection Results for FB15K (Imputing with Same Relation Errors constrained approach).

	FB15K-10%			FB15K-20%			FB15K-40%		
	*f*MR	*f*MRR	AUC	*f*MR	*f*MRR	AUC	*f*MR	*f*MRR	AUC
TransE	127938	0.0002	0.7351	133761	0.0001	0.7230	169486	0.0000	0.6490
PTransE	166347	0.0000	0.6559	167995	0.0000	0.6525	173641	0.0000	0.6407
CKRL	96115	0.0001	0.8010	101585	0.0001	0.7896	112327	0.0001	0.7676
PRGE	73997	0.0006	0.8471	89167	0.0005	0.81553	86350	0.0002	0.8214
EPRGE	**40952**	**0.0008**	**0.8527**	**55667**	**0.0008**	**0.8211**	**54252**	**0.0004**	**0.8269**

Table 5. Error Detection Results for Dementia (Imputing Random Errors)

	Dementia-10%			Dementia-20%			Dementia-40%		
	*f*MR	*f*MRR	AUC	*f*MR	*f*MRR	AUC	*f*MR	*f*MRR	AUC
TransE	58014	0.0001	0.5702	59421	0.0001	0.5599	59835	0.0000	0.5568
PTransE	59718	0.0002	0.5576	61518	0.0001	0.5443	65533	0.0000	0.5146
CKRL	60584	0.0001	0.5512	61034	0.0001	0.5479	61089	0.0001	0.5475
PRGE	57642	0.0001	0.5730	58258	0.0001	0.5685	59314	0.0001	0.5606
EPRGE	**55336**	**0.0003**	**0.5959**	**55927**	**0.0004**	**0.5912**	**56941**	**0.0003**	**0.5830**

To reflect on these results, we can confirm that the PRGE model is, in fact, modular as claimed in [4], allowing for different combinations of techniques for the loss function and the confidence score that could enhance results for the error detection task. Furthermore, this study aims to show the impact of using the self-adversarial negative sampling method in RotatE over the uniform sampling method used in TransE. Such a uniform negative sampling suffers the problem of low performance since many samples are false as training goes on, which does not provide any meaningful information. Therefore, the self-adversarial negative sample we used, which samples negative triples according to the current embedding model, showed better results.

6 Conclusion

We proposed the Enhanced Path-ranking Graph Embedding (EPRGE) model, which uses PaTyBRED and RotatE to address the shortcomings of PRGE. Our experiments on WN18, FB15K, and Dementia datasets showed that EPRGE outperformed other models in terms of fMR, fMRR, and AUC scores and was more robust to increasing levels of noise. TransE, which is used in PRGE, uses uniform sampling, leading to model inefficiency, while RotatE proposes a more efficient self-adversarial negative sampling technique. EPRGE's superiority in error detection is attributed to the use of RotatE and the PaTyBRED method, which produce better embeddings and accurate error detection, even in the presence of noise. Our results suggest that EPRGE is a promising approach for detecting errors in KGs, particularly in noisy and real-world scenarios. However, there is still room for improvement, especially in handling extremely sparse datasets such as Dementia.

Acknowledgement. We acknowledge the support of the Natural Sciences and Engineering Research Council of Canada (NSERC) [funding reference number 03181].

References

1. Agrawal, R., Imieliński, T., Swami, A.: Mining association rules between sets of items in large databases. In: Proceedings of the 1993 ACM SIGMOD International Conference on Management of Data, pp. 207–216 (1993)
2. Bollacker, K., Evans, C., Paritosh, P., Sturge, T., Taylor, J.: Freebase: a collaboratively created graph database for structuring human knowledge. In: Proceedings of the 2008 ACM SIGMOD International Conference on Management of Data, pp. 1247–1250 (2008)
3. Bordes, A., Usunier, N., Garcia-Duran, A., Weston, J., Yakhnenko, O.: Translating embeddings for modeling multi-relational data. In: Advances in Neural Information Processing Systems, vol. 26 (2013)
4. Bougiatiotis, K., Fasoulis, R., Aisopos, F., Nentidis, A., Paliouras, G.: Guiding graph embeddings using path-ranking methods for error detection in noisy knowledge graphs. arXiv preprint arXiv:2002.08762 (2020)
5. Cheng, Y., Chen, L., Yuan, Y., Wang, G.: Rule-based graph repairing: semantic and efficient repairing methods. In: 2018 IEEE 34th International Conference on Data Engineering (ICDE), pp. 773–784. IEEE (2018)
6. Galárraga, L.A., Teflioudi, C., Hose, K., Suchanek, F.: AMIE: association rule mining under incomplete evidence in ontological knowledge bases. In: Proceedings of the 22nd International Conference on World Wide Web, pp. 413–422 (2013)
7. Guo, S., Wang, Q., Wang, L., Wang, B., Guo, L.: Knowledge graph embedding with iterative guidance from soft rules. In: Proceedings of the AAAI Conference on Artificial Intelligence, vol. 32 (2018)
8. Jia, S., Xiang, Y., Chen, X., Wang, K.: Triple trustworthiness measurement for knowledge graph. In: The World Wide Web Conference, pp. 2865–2871 (2019)
9. Lehmann, J., et al.: DBpedia-a large-scale, multilingual knowledge base extracted from Wikipedia. Semant. Web **6**(2), 167–195 (2015)

10. Lenat, D.B.: CYC: a large-scale investment in knowledge infrastructure. Commun. ACM **38**(11), 33–38 (1995)
11. Lin, Y., Liu, Z., Luan, H., Sun, M., Rao, S., Liu, S.: Modeling relation paths for representation learning of knowledge bases. arXiv preprint arXiv:1506.00379 (2015)
12. Melo, A., Paulheim, H.: Detection of relation assertion errors in knowledge graphs. In: Proceedings of the Knowledge Capture Conference, pp. 1–8 (2017)
13. Paulheim, H.: Knowledge graph refinement: a survey of approaches and evaluation methods. Semant. Web **8**(3), 489–508 (2017)
14. Paulheim, H., Bizer, C.: Improving the quality of linked data using statistical distributions. Int. J. Semant. Web Inf. Syst. (IJSWIS) **10**(2), 63–86 (2014)
15. Tanon, T.P., Stepanova, D., Razniewski, S., Mirza, P., Weikum, G.: Completeness-aware rule learning from knowledge graphs. In: d'Amato, C., et al. (eds.) ISWC 2017. LNCS, vol. 10587, pp. 507–525. Springer, Cham (2017). https://doi.org/10.1007/978-3-319-68288-4_30
16. Rebele, T., Suchanek, F., Hoffart, J., Biega, J., Kuzey, E., Weikum, G.: YAGO: a multilingual knowledge base from Wikipedia, wordnet, and geonames. In: Groth, P., et al. (eds.) ISWC 2016. LNCS, vol. 9982, pp. 177–185. Springer, Cham (2016). https://doi.org/10.1007/978-3-319-46547-0_19
17. Shah, A., Molokwu, B., Kobti, Z.: HRotatE: hybrid relational rotation embedding for knowledge graph. In: 2021 International Joint Conference on Neural Networks (IJCNN), pp. 1–8. IEEE (2021)
18. Shan, Y., Bu, C., Liu, X., Ji, S., Li, L.: Confidence-aware negative sampling method for noisy knowledge graph embedding. In: 2018 IEEE International Conference on Big Knowledge (ICBK), pp. 33–40. IEEE (2018)
19. Sun, Z., Deng, Z.H., Nie, J.Y., Tang, J.: Rotate: knowledge graph embedding by relational rotation in complex space. arXiv preprint arXiv:1902.10197 (2019)
20. Vrandečić, D., Krötzsch, M.: Wikidata: a free collaborative knowledgebase. Commun. ACM **57**(10), 78–85 (2014)
21. Wang, M., Qiu, L., Wang, X.: A survey on knowledge graph embeddings for link prediction. Symmetry **13**(3), 485 (2021)
22. Zhang, Q., Dong, J., Duan, K., Huang, X., Liu, Y., Xu, L.: Contrastive knowledge graph error detection. In: Proceedings of the 31st ACM International Conference on Information & Knowledge Management, pp. 2590–2599 (2022)

Distributed Control for Traffic Light in Smart Cities: Parameters and Algorithms

Pedro Uribe-Chavert[1], Juan-Luis Posadas-Yagüe[2], and Jose-Luis Poza-Lujan[2](✉)

[1] Doctoral School, Universitat Politècnica de València. Camino de vera, sn 46022, Valencia, Spain
pedurcha@doctor.upv.es

[2] Research Institute of Industrial Computing and Automatics. Universitat Politècnica de València. Camino de vera, sn 46022, Valencia, Spain
{jopolu,jposadas}@upv.es

Abstract. This article focuses on communicating and transmitting meaningful parameters between junctions in traffic control systems. Decisions taken at one junction can have a domino effect on subsequent junctions, and good communication can allow for better coordination and optimisation of traffic throughout the system. Traffic control algorithms must know the parameters to determine how distributed the control should be, and the distance the information will be sent to control nodes is necessary. In this article, we present an orientation to the distribution of the control algorithms that support the architecture presented in previous works. We review the consequences of having isolated control per junction, collaborative control where information is shared, and consensual control where information is known by all nodes involved at various junctions. The article also describes the proposed control methods adapted to the architecture presented in previous works and presents the tested algorithms. Finally, conclusions and possible lines of progress are presented. The goal of this article is to contribute to the optimisation and coordination of traffic control systems in smart cities.

Keywords: City Traffic Control · Distributed Intelligent Control · Smart Cities

1 Introduction

Smart cities are characterised by their ability to use information and communication technologies to improve the quality of life of their citizens. One of the important challenges cities face is traffic management. The ever-increasing population and the growing number of vehicles on the roads lead to increased traffic congestion and decreased transport efficiency. Traffic control in smart cities offers a solution to this problem by enabling more efficient traffic management by automating traffic lights and other signalling devices. These traffic control systems can also optimise public transport, reducing passenger waiting times and improving overall system efficiency [1].

S. Ossowski et al. (Eds.): DCAI 2023, LNNS 740, pp. 93–102, 2023.
https://doi.org/10.1007/978-3-031-38333-5_10

However, designing traffic control systems in smart cities is a complex task, as it must consider multiple variables and consider private vehicle drivers, public transport users, and pedestrians. In addition, the technological and infrastructural challenges in implementing such control systems must be addressed. In this context, research on distributed algorithms for traffic control can provide effective solutions for traffic management in smart cities [3].

In this context, traffic control implies having relevant information to make decisions and significantly increasing or decreasing the time for vehicles or pedestrians to pass in one direction. For example, if a traffic light knows the queue lengths of vehicles or pedestrians waiting on each street over time, it can attempt to match these queues. Nevertheless, suppose the traffic light has more information, such as traffic density at previous junctions. In that case, it can predict the load and attempt to empty the junction to prepare for the vehicle increase.

Junction control algorithms in smart cities can use a variety of parameters to manage how long each street should be green or red to reduce vehicle queues at the junction. One important parameter that can be used is the queue length on each intersecting street. If the queue on the street is detected to be too long, the algorithm can give more green time to that street to allow more vehicles to pass. Similarly, if the queue on the street is short, the algorithm may give less green time to that street to allow other streets with a long queue to move forward. Another parameter that can be used is the traffic volume on each street. If a street has a high traffic volume, the algorithm may allocate more green time to that street to allow more vehicles to pass, while streets with less traffic may have less green time.

In traffic control systems, it is essential to remember that decisions taken at one junction can have a domino effect on subsequent junctions. Therefore, it is essential to have good communication and transmission of meaningful parameters between junctions. If one junction can transmit relevant information to downstream junctions, such as queue length or traffic flow, it can allow for better coordination and optimisation of traffic throughout the system.

In addition, traffic control systems at junctions are designed to be adaptive and change in real time according to current traffic conditions. Communication between junctions allows the traffic control system to make informed decisions and change its behaviour based on current traffic conditions. Traffic control algorithms may know crossing information either because they are centralised at a node, clustered at zone control nodes, or because they are more distributed [2].

The algorithms must know the parameters to know how distributed the control should be. Considering the distance the information will be sent to control nodes is necessary. To answer the previous questions, this article presents an orientation to the distribution of the control algorithms that support the architecture presented in [5]. The article reviews the consequences of having isolated control per junction, collaborative control where information is shared, and consensual control where information is known by all nodes involved at various junctions. The article is organised as follows. The next section describes the proposed control methods adapted to the architecture presented in DCAI 2021. The parameters for measuring the effectiveness of the control are then presented. Next, the algorithms currently being tested are presented. Finally, conclusions and possible lines of progress are presented.

2 Control Proposal

2.1 Fixed-Time Control

A fixed-time traffic signal control system works by cycling through a predetermined sequence of signal timings that dictate when each direction of traffic is allowed to proceed.

One approach to designing a fixed-time traffic signal control system in this environment might be to use a coordinated system in which the signals at each intersection are synchronized to minimize delays and maximize traffic flow. This might involve using a series of detectors and timers to determine when traffic is approaching each intersection and adjust the signal timings accordingly.

For example, the system might be designed to prioritise the main through-streets at each intersection, allowing them to proceed longer than the side streets. The system might also include pedestrian crossings and left turn phases, which must be coordinated with the vehicle traffic signals.

Some advantages of fixed-time control systems include their simplicity and low cost, as they do not require sophisticated sensing technologies or real-time adjustments. Additionally, fixed-time systems can be effective in areas with consistent traffic patterns, where signal timings can be set to optimize traffic flow during peak hours.

However, fixed-time systems also have some disadvantages. They do not adjust to changes in traffic patterns, so they may not be effective in areas with fluctuating traffic volumes or congestion. Additionally, fixed-time systems can lead to delays and congestion if there is an unexpected surge in traffic or if there is an imbalance in traffic flow between different approaches.

Overall, the goal of a fixed-time traffic signal control system in this environment would be to balance the needs of all road users and minimise delays while ensuring safe and efficient traffic flow through both intersections.

This system was simulated in the article [4] to reach which sequence of signal timings is better for the environment described.

2.2 Isolated Control

A traffic signal control system with isolated control typically uses traffic detectors to measure the flow of traffic through each intersection. These detectors may be located in the roadway or mounted on the traffic signal poles, and they can provide information about the number of vehicles waiting at each approach and the length of queues [5].

Based on this information, the traffic signal control system can adjust the signal timings at each intersection to optimize traffic flow. For example, if one approach has a longer queue than others, the system may extend the green time for that approach to clear the queue more quickly. Similarly, if there is very little traffic on one approach, the system may shorten the green time to reduce delays for other approaches.

In some cases, traffic signal control systems with isolated control may also include features like adaptive signal control, which can adjust the signal timings in real-time based on changing traffic conditions. Focusing on the queue length of the red signal street and the timeout. The timeout is the maximum time that one street light signal could stay in red if at least one vehicle is in that street. While queue length is the minimum number of vehicles from which the street light signal is changed to green to allow all these vehicles to cross the crossing.

There is one more implementation, to know the size of the queue length that is emptying in the street with the green signal. All these parameters are used to optimize the crossing-

Overall, the goal of a traffic signal control system with isolated control is to maximize traffic flow and minimize delays at each intersection, while ensuring safe and efficient movement of vehicles and pedestrians through the area.

A traffic signal control system with isolated control can have a potential issue due to the lack of communication between the two intersections. When the first intersection turns green and allows a significant number of vehicles to proceed to the second intersection, it can cause congestion and potential saturation at the second intersection.

Without a coordinated system that takes into account the traffic flow between the two intersections, there is a risk of imbalanced traffic flow that can cause delays and safety concerns. For example, if the first intersection has a longer green time than the second intersection, it can result in a large number of vehicles arriving at the second intersection all at once, causing congestion and delays.

Traffic signal control systems with isolated control have several advantages over fixed-time systems. By adjusting signal timings in real-time based on traffic conditions, these systems can optimize traffic flow and reduce delays. Additionally, isolated control systems can be more effective in areas with fluctuating traffic patterns or congestion, as they can adjust signal timings to balance traffic flow between different approaches.

However, isolated control systems can also have some disadvantages. Without proper coordination between intersections, there can be an imbalance in traffic flow that can cause delays and safety concerns. Additionally, isolated control systems require sophisticated sensing technologies and real-time data processing, which can be costly and complex to implement.

2.3 Informative Control

The informative control system differs from isolated control in that the intersections are connected, but not all of them, only those within a "control distance" of each other. Intersection n informs intersection $n + 1$ that it has turned green and the number of vehicles it had in its queue.

This number of vehicles is multiplied by a variable called "weighting". The weighting depends on the characteristics of the intersection. If the intersection has only one exit street, then all vehicles will exit through that street, so the weighting based on intersection characteristics is maximum. If there are two exit streets, the weighting is based on which street has more traffic.

The weighting also depends on the control distance. The larger the control distance, the lower the weighting, since the intersection will likely receive fewer vehicles.

The use of informative control in traffic intersections has several advantages over fixed-time and isolated control methods. Some of the main advantages of informative control include: Increased efficiency: By communicating with nearby intersections, the system can better coordinate the flow of traffic, reducing congestion and wait times. Flexibility: Unlike fixed-time control, which relies on pre-set timing schedules, informative control can adapt to changing traffic conditions, making it more flexible and responsive.

However, there are also some potential drawbacks to consider, such as: Communication failures: Since the system relies on communication between intersections, any failure in the communication system could cause issues with traffic coordination. Cost: Implementing an informative control system requires significant investment in infrastructure and technology, which may not be feasible for smaller municipalities. Complexity: The system requires more complex algorithms and software to process and analyze traffic data, which can make it more difficult to maintain and troubleshoot.

In summary, while informative control offers significant benefits for managing traffic flow, it also requires careful planning, investment, and maintenance to ensure its successful implementation and operation.

In summary, the informative control system optimizes traffic flow by allowing intersections to communicate with each other and adjust their signal timings based on the number of vehicles waiting at the previous intersection and other factors such as the intersection's characteristics and its proximity to other intersections.

2.4 Feedback Control

Dynamic adjustment is a technique used in feedback control. It involves modifying control parameters based on the system's response to achieve optimal performance. In the case of traffic control, this technique can be applied in a system known as feedback control.

The informative control and the feedback control are the same control, with the difference that the latter updates the weighting parameter with the information returned by the downstream intersections.

Feedback control uses information about the system's performance to adjust weighting in traffic lights. This weighting is based on the difference between the desired state of the system (e.g., no congestion) and the actual state (e.g., queue length). The feedback loop allows the system to continuously adapt to changing conditions and maintain optimal performance.

One of the advantages of feedback control is that it can react quickly to changes in traffic conditions. As the system receives real-time feedback, it can adjust the signal sent to the traffic lights accordingly, reducing congestion and improving traffic flow.

However, feedback control requires a certain level of sophistication and computational power. The system needs to be able to collect data from various sources and process it in real-time. Additionally, the system needs to be designed and optimized for the specific traffic conditions and environment.

Overall, feedback control with dynamic adjustment has the potential to greatly improve traffic flow and reduce congestion. However, it requires significant investment in both hardware and software, as well as ongoing maintenance and optimization.

2.5 Self-learning Control

The self-learning control algorithm is a type of feedback control that uses machine learning techniques to optimize all control parameters, not just the weighting parameter as in the feedback control algorithm. These parameters are the services times T_r and T_g, the Min queue length mQ and Min queue length green mQ_g, Timeout T_{max}, Control Distance D_c, Weighting W and Previous Queue Length Q_p. The self-learning algorithm continuously gathers data on traffic patterns and adjusts the control parameters accordingly. This allows for more efficient and adaptive control of traffic flow, especially in unpredictable situations or in cases where traffic patterns change over time.

Compared to the other control algorithms, the self-learning algorithm has several advantages. Firstly, it is much more adaptable to changing traffic conditions, as it can learn from experience and adjust its control parameters accordingly. This makes it particularly useful in areas with high variability in traffic flow. Secondly, the self-learning algorithm can optimize all control parameters, rather than just one, which can lead to more efficient traffic flow and reduce congestion.

However, the self-learning algorithm also has some disadvantages. It requires a significant amount of data to be effective, so it may take time to see improvements in traffic flow. Additionally, the algorithm can be computationally expensive and may require more advanced hardware to implement. Finally, the self-learning algorithm may be less transparent than other control algorithms, as the decision-making process is based on machine-learning models that may be difficult to interpret or explain.

3 Parameters

There are two input parameters used for the previous algorithms. These input parameters are the elapsed time T and the queue length Q. Time is a variable that is reset every traffic light cycle or every time the traffic light changes state.

The queue length is obtained using [5] sensors and indicates the number of vehicles waiting to pass on each street at the intersection.

The control parameters used for the different controls are:

- Service times, both red service time T_r and green service time T_g. These times are used only for fixed-time control.
- Minimum Queue length mQ is the minimum queue length from which a street can switch from red to green.
- Minimum Queue length green mQ_g is the minimum queue length required for the street with the green light for the other street to switch from red to green.
- Timeout T_{max} is the maximum time allowed for a red intersection to remain red when at least one vehicle is waiting to cross.
- Control Distance D_c is the distance (in intersections) at which queue length information arrives.
- Weighting W is the weighting that the intersection has (based on its characteristics) multiplied by the weighting derived from Control Distance.
- Previous Queue Length Q_p is the queue length value just before the intersection changes to green that is sent to the following Control Distances crossings.

The following Table 1 refers to different algorithms. Each algorithm has two columns. Column 'Used' indicates if each algorithm uses the parameter or not. Column 'Mod' indicates if the algorithm modifies the parameter.

Table 1. Parameters in each control

	Fixed-Time		Isolated		Informative		Feedback		Self-Learning	
	Used	Mod	Used	Mod	Used	Mod	Used	Mod	Used	Mod
T_r	X		X	X	X	X	X	X	X	X
T_g	X		X	X	X	X	X	X	X	X
mQ			X		X		X		X	X
mQ_g			X		X		X		X	X
T_{max}			X		X		X		X	X
D_c					X		X		X	X
W					X		X	X	X	X
Q_p					X		X	X	X	X

4 Algorithm

The fixed-time control sets predetermined service times for each traffic light at the intersection, for both Road R_1 and Road R_2. These times are usually precalculated to optimize the traffic flow. The control follows the Algorithm 1, in which the traffic light signal changes from one road to another after a fixed time interval.

Algorithm 1. Fixed-time control system at an intersection

Set the initial signal timing plan
Start the timer
while the system is operating **do**
  Check the timer
  if the timer reaches the end of the signal timing plan **then**
    Reset the timer
    Switch the signal to the next phase in the timing plan
  end if
end while

The isolated control differs from the previous control in that in this case, the queues of both roads, R_1 and R_2, are taken into account, explained in Algorithm 2.

Algorithm 2. Isolated intersection control system with adaptive signal timings and queue length thresholds

Set the initial signal timing plan
Initialize the queue lengths Q_r and Q_g
Start the timer
while the system is operating **do**
  Check the timer T
  Update the queue lengths Q_r and Q_g
  if $(Q_r > mQ$ and $Q_g < mQ_g)$ or $(T > T_{max}$ and $Q_r > 1)$ **then**
    Change signals states
    Reset the timer
  else
    Maintain the current signal timings
    Update the signal timing plan
  end if
end while

In this control, there are three important parameters. The input parameter Queue Length, for both the road with the traffic light in red Q_r and the road with the traffic light in green Q_g. And the control parameters: Timeout (T_{max}), Min Queue Length (mQ) and Min Queue Length Green (mQ_g). In isolated control, the queues of both roads are monitored and controlled using the above parameters, unlike the fixed-time control, where the service times for each traffic light are predefined and optimized for the best flow.

The informative control differs from the isolated control in that nearby intersections communicate with each other. This control is explained in the Algorithm 3 This is why additional parameters need to be added to the isolated control parameters: Previous Queue Length (Q_p), Control Distance (D_c) and Weighting (W).

Algorithm 3. Informative intersection control system with adaptive signal timings and queue length thresholds

Set the initial signal timing plan
Initialize the queue lengths Q_r and Q_g
Start the timer
while the system is operating **do**
  Check the timer T
  Update the queue lengths Q_r and Q_g
  if $(Q_r + Q_p > mQ$ and $Q_g < mQ_g)$ or $(T > T_{max}$ and $Q_r > 1)$ **then**
    Change signals states
    Reset the timer
  else
    Maintain the current signal timings
    Update the signal timing plan
  end if
end while

The feedback control, Algorithm 4, differs from the isolated control in that the feedback control returns the information of vehicles that have passed through the intersection to the previous intersection. This information is used to improve Weighting W.

The control is similar to the isolated control, but with the difference that the previous queue length Q_p is optimized due to the improvement of W.

Algorithm 4. self-learning intersection control system with adaptive signal timings and queue length thresholds

Set the initial signal timing plan
Initialize the queue lengths Q_r and Q_g
Start the timer
while the system is operating **do**
  Check the timer T
  Optimize W and Q_p
  Update the queue lengths Q_r and Q_g
  Update Q_p with the new constant
  if $(Q_r + Q_p > mQ$ and $Q_g < mQ_g)$ or $(T > T_{max}$ and $Q_r > 1)$ **then**
    Change signals states
    Reset the timer
  else
    Maintain the current signal timings
    Update the signal timing plan
  end if
end while

Finally, the self-learning control seeks the history of the variables and queue lengths to optimize and improve all the previously calculated parameters. This allows for a live optimization of the parameters and makes the control much more effective. The algorithm is similar to the feedback control Algorithm 4, but in this case, all the control variables are optimized.

In summary, the self-learning control searches the history of variables and queue lengths to optimize and improve the previously calculated parameters, achieving real-time optimization and making the control much more effective.

5 Conclusions

This paper has reviewed how distributed control algorithms at control nodes require information from other nodes. This information can have more or less "distance" between nodes. This control distance can facilitate control optimisation. However, it can lead to noise, especially from very distant nodes. In addition, the parameters that are sent and received can be significant, such as the length of the queues at each junction or the waiting time. The impact of the waiting time on the journey is an important parameter. For example, a pedestrian (or scooter 25 km/h) taking 10 min from point A to B, waiting 1 min is 10% of their journey time. A public transport vehicle at 50km/h that goes from point A to B in 5 min implies that 1 min is 20% of its time and pollutes more. Moreover, if it is a bus, it is "slowing down"; it harms more people proportionally (it is comfort, after all). Future work aims to detect which control distance and which parameters are most significant. This future work will be implemented in a simulation environment made in Sumo and Python for the different algorithms. It should be considered that with current architectures and devices, more complex and valuable parameters are possible.

References

1. Aleksander, R., Paweł, C.: Recent advances in traffic optimisation: systematic literature review of modern models, methods and algorithms. IET Intel. Transp. Syst. **14**(13), 1740–1758 (2020)
2. Chow, A.H., Sha, R., Li, S.: Centralised and decentralised signal timing optimisation approaches for network traffic control. Transp. Res. Procedia **38**, 222–241 (2019)
3. Rabby, M.K.M., Islam, M.M., Imon, S.M.: A review of IoT application in a smart traffic management system. In: 2019 5th International Conference on Advances in Electrical Engineering (ICAEE), pp. 280–285. IEEE (2019)
4. Uribe-Chavert, P., Posadas-Yagüe, J.L., Balbastre, P., Poza-Luján, J.L.: Arquitectura distribuida modular para el control inteligente del tráfico. Rev. Iberoamericana Autom. Inform. Ind. **20**(1), 56–67 (2022)
5. Uribe-Chavert, P., Posadas-Yagüe, J.-L., Poza-Lujan, J.-L.: Proposal for a distributed intelligent control architecture based on heterogeneous modular devices. In: González, S.R., et al. (eds.) DCAI 2021. LNNS, vol. 332, pp. 198–201. Springer, Cham (2022). https://doi.org/10.1007/978-3-030-86887-1_20

Using Data Analytic for Social Media Posts to Optimise Recyclable Solid Waste Management Exemplary at the City of Valencia

Philipp Junge[1], Sturle Stavrum-Tång[2], José M. Cecilia[3], and Jose-Luis Poza-Lujan[3](✉)

[1] School of Engineering and Design, TU Munich, Boltzmannstrasse 15, 85748 Garching bei München, Bavaria, Germany
philipp.junge@tum.de
[2] School of Engineering and ICT, NTNU, Høgskoleringen 6, 7491 Trondheim, Trønderlag, Norway
[3] Universitat Politècnica de València, Camino de vera sn, 46980 Valencia, Spain
{jmcecilia,jopolu}@upv.es

Abstract. Optimising general waste management and collection has been shown to minimise the health hazards from overfilled garbage containers and pollution from heavy trucks collecting waste. Smart cities and the public would reap the great benefits of finding a solution to optimise the waste collection system and general waste management. Therefore, this paper proposes a solution to gather citizen waste critics from Twitter as a data source and analyse the outcome. Firstly, the paper presents the related work, framework and method. After that, analyse the data and discuss the result. The results show barriers in both collection and content of the data as it contains a lot of false data. The main contribution will therefore be a foundation for further research and experiment of data collection and analysis.

Keywords: Smart City · Waste Management · Route Optimisation · Twitter Data

1 Introduction

Due to the rapid growth of the human population and increasing consumption, waste has become a health challenge to urban residents [2]. The voluminous waste generation comes from the industrialisation of cities, increased household incomes, economic growth, development, and urban exponential growth [8]. Waste management has therefore become a critical challenge for engineers and city planners all over the world [3], especially as improper disposal has resulted in several health hazards [5]. Fixing the ever-growing waste problem will, therefore, not only result in a cleaner city but increase the health and happiness of the

S. Ossowski et al. (Eds.): DCAI 2023, LNNS 740, pp. 103–112, 2023.
https://doi.org/10.1007/978-3-031-38333-5_11

inhabitants [5]. Developing a general and expandable solution for waste management will, therefore, better global health from the perspective of the UN SDG (sustainable development goals) number 3, especially in "substantially reducing the number of deaths and illnesses from hazardous chemicals and air, water and soil pollution and contamination" [7].

Based on statistics extracted from Spanish Statistics Institute (INE), Spain has seen a slow population growth from 2011 until 2021. Nevertheless, Spain is following the worldwide trend of increasing urbanisation. Valencia, the focal point of this paper, has seen steady population growth since the start of the 21st century and, consequently, a growing waste problem. Considering this event, the motivation for this paper is not only to better joint health but identify the key enablers and barriers resulting in proper professionalisation and implementation of waste management [10]. SDG 11 indicates the importance of waste management, and the subject touches on numerous other SDGs, such as SDG 6 (Clean water and sanitation), SDG 9 (Industry, Innovation, and Infrastructure), SDG 12 (Responsible consumption and production) and more) [7].

In Valencia, an existing project is already tackling the waste management problem: The VLCi Impulse project [9] by "Smart City Valencia". They advertise their approach with the slogan "Garbage containers that warn before they get filled". According to their homepage, waste hauling is more critical and expensive in Valencia's northern and southern territories. Therefore they aim to monitor the containers there wireless and in real time by deploying sensors in 125 plastic and 138 glass containers. Information like container fullness levels, weight, volume, and temperature can be gathered. This solution will allow more efficient waste management and could improve the quality of life for the citizens.

But the current highly technological solution has several drawbacks. Deploying IoT sensors in over 250 containers is expensive in development, production mounting, and maintenance. Additionally, it's not that easy to implement because the fullness level of a garbage container depends not simply on weight but also on the density of the garbage. Another problem is vandalism, especially with such expensive technology in publicly accessible places.

This paper shows an alternative way of improving waste management in Valencia. Therefore, we analyse Twitter complaints regarding waste issues and perform data analysis to extract critical information about the waste situation in Valencia.

The paper is structured as follows. The introduction shows the motivation, the current technological state and our proposal. Next, the framework section gives a graphical overview of our model. In Methods, the whole data collecting and filtering process is described. After that, the results are shown and discussed in the Conclusions section.

2 Framework

All the solutions mentioned in the previous chapter have in common that they are either expensive, complex in implementation or need the active participation of the citizens. Our approach is to use existing citizen data and extract useful information from that. First, we will look at our system design, starting with the edge layer. What we want to get is information about complaints regarding garbage in Valencia. One example is if and where an overfilled container is, but also other complaints are looked into. We use Twitter posts as a data source; therefore, we discussed a soft sensor with humans as "actuators". The data we aim to extract are tweets regarding current waste problems, not those where waste has a different meaning. We do not have a fog layer because we are not working with complex sensors and IoT techniques. In the cloud layer, we store our data from Twitter posts. This data is stored in our database, and we aim to extract knowledge about critical locations, our dates for overfilled garbage bins or other complaints. If our approach is successful, Valencia can install new containers in critical areas or temporarily in critical times based on the waste information we gathered. Another possibility would be the optimisation of the garbage vehicle routes.

Figure 1 shows the model used and contains five symbols: green circles, which represent a trigger action, red circles, which represent the end of an operation; rounded white boxes, which describe actions; tilted Squares, which means questions to be assessed, layered Cylinder, which represents a database.

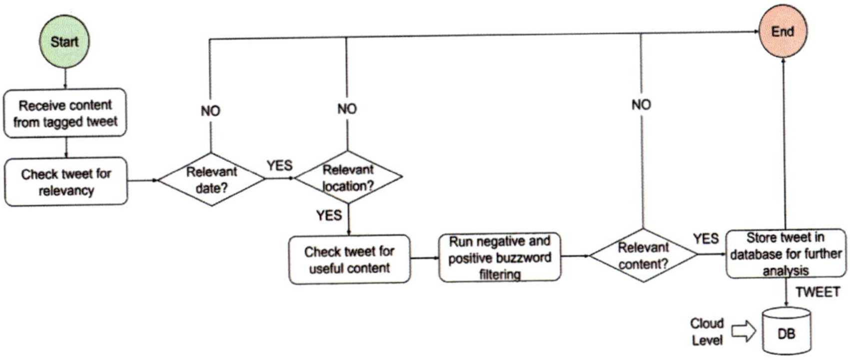

Fig. 1. Model used in the study to obtain the data.

The initiator is represented by the citizens or visitors of Valencia contributing to the smart citizens of the smart city concept [6]. The process starts when the initiator makes a tweet about the waste system in Valencia. The next operation is triggered when tags referring to Valencia's waste situation are used. For example #WasteValencia, #Waste, #Valencia, etc. Previous questions are necessary to prevent the algorithm from asserting a disproportionate number of tweets to find the most relevant tweets. A positional code or geo-tag other than the tagged position would be beneficial but not crucial in this model's state.

The Tweet Sorting Algorithm is the first stage of the Cloud level. The algorithm is triggered when the tags have been used in a tweet. Tweets can be simply achieved by integrating Twitter with some trigger operations connected to our cloud. Services like Microsoft Azure Logic App and API management could be suitable for integration like this (Azure logic apps). When the trigger is set, and the text is received, the algorithm will start assessing and sorting the message. First, the message will be tested for relevancy. The first check is for the date and location represented by the two first actions of Fig. 1. This is the coarse filtering to remove unwanted data before checking for content relevancy which is more complicated. The content relevancy runs several negative and positive filtering analyses using keywords, which will be discussed later in the paper. The more data processed through this algorithm, the more precise the filtering will become. After the assessment, the location data and fullness level are sent to the database for further data handling.

From the user's point of view and based on the information extracted from the tweets, route optimisation for the waste collectors could be done. Several experiments and projects have been done using the Travelling Salesman Problem and shown successful results [4].

Therefore, a potential end user of this service will be the waste collectors themselves, as they are the ones who physically collect the waste. The route will be optimised continuously constantly to save kilometres and time. The result will be a highly flexible, high-performance team of waste collectors to maximise the waste collected in the most current locations.

3 Method

3.1 Data Gathering and Collection

The data needed for the experiments were social media posts filled out by Valencian people regarding the waste situation in Valencia. Critical data would be Data about the overfilled streets and containers, the fullness level of the containers and general complaints about the waste situation in Valencia.

The data acquired from Twitter were fetched through a Twitter API from 2007 until March 2022. Many open-data sets of tweets with the geolocation Spain are available, but the specifics of our experiment made them useless. As Twitter was first launched in 2006, we can assume there are few relevant tweets from the first years of its operation. The relevancy of tweets from this period will also have less value just because of the relevance concerning age. Nevertheless, in this stage, the experiment treats all tweets equally to have as much data as possible so that the sorting algorithm can be tested with the maximum amount of data. The queries used to extract the information consisted, first and foremost, of the geographical location of Valencia. The location was obtained through "Valencia" or "#valencia". The next step was finding the right buzzwords to extract the tweets concerning the waste situation. Multiple queries were tested to obtain as much data as possible. In this stage, the experiment encountered several challenges that had to be solved. Some queries gave less than 50 tweets, resulting in a broader range of buzzwords. After some iterations, the final set of buzzwords contained the words in Spanish: "basura" (garbage), "derrames" (littering), "desecho" (waste), "escombrera" (dump or landfill), "resíduo" (residue), "vertido" (spill). The result of this query was 762 tweets.

3.2 State of the Data

The 762 tweets ranged from 2011 until March 2022, written in four different languages, where more than 700 of the tweets were written in Spanish, then Catalan and finally English. Some tweets consisted only of internet links and were categorised as "und". Figure 2a shows the language distribution. As expected, the initial years gave no result. Figure 2b shows the distribution of tweets within the period. The graph shows little trend, but a slight increase in volume over the past years. This correlates with the increase in "profitable daily active international Twitter users" from the first quarter of 2017 through 2021 [1]. There is a significant spike in volume in 2019. There can be many reasons behind the numbers, but as "basura" (garbage)is commonly known to describe both politicians and football teams and players, there may be some correlations between the spike and the 2019 regional election and Copa del Rey (a popular football championship). A deeper analysis regarding events like these could help optimise the filtering but has not been the focus of this paper.

Regarding the most frequent buzzwords, "basura" (garbage) stands out. The word can be found 698 times. Far behind comes "vertedero" (landfill) with 40, then "desecho" (waste) and "residuo" residue) with 13 and the rest seven and below. A final analysis of relevant tweets and key buzzwords should be done to optimise the initial filtering further.

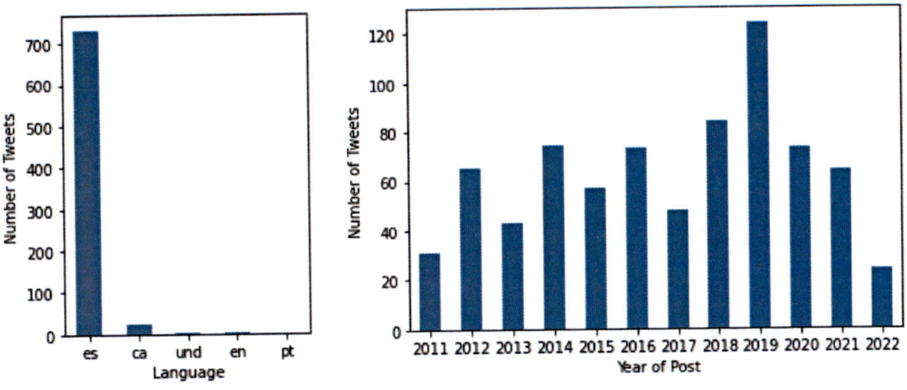

(a) (b)

Fig. 2. Languages of the tweets (a), and Year of posting of the tweets (b).

3.3 Twitter Filtering

Even though waste-related buzzwords were already used to gather only tweets related to our topic, there are still plenty of useless posts for us. The reason is that Spanish words like *basura* or *desecho*, translated to waste, have multiple meanings in Spanish. Like in many other languages - in English or German, for example - people in Spain also use *trash* to speak negatively about things or persons. Therefore a second filtering had to be applied, where the goal was to only get the waste management-related posts out of our data-set. For this, two sub-methods were developed:

- The"negative filtering" contains a set of filter words which are not at all related to waste management. If a tweet includes one of those words, it is considered useless and deleted from the data-set. After applying the negative filtering, we got a number of the remaining tweets, which ideally relate to waste management.
- The "positive filtering" contains filter words related to waste management. If a tweet includes one of those words, it is considered useful and added to a new data-set. After applying the positive filtering, we got a new data-set with several tweets, ideally all relating to waste management.

To review whether this method is suitable to filter Twitter post data-sets for waste-related topics, comparing both sub-methods is used. Therefore, the formula

$$(tweets_n - tweets_p)/tweets$$

is applied. Here, $tweets_n$ is the number of useful tweets after negative filtering, $tweets_p$ is the number of useful tweets after positive filtering and $tweets$ is the number of tweets in the original data-set. An approach like this must be used because the tweets need to be labelled, and there are too many to review each by hand. As tools for filtering, Python3 and the Pandas module were used.

Negative Filtering

To come up with suitable filter words for the negative filtering, 100 tweets were taken out and analysed by hand. If a tweet was considered unrelated to waste management, the most apparent word in that tweet was added to the filter table. After 100 tweets, we got a table divided into four categories which can be seen in Fig. 3a.

The biggest category was sports, mainly football. Many tweets also had a political topic; some were about the media. The other ones were more individual cases, so a fourth category for the rest was added.

Positive Filtering

The keywords for the positive filtering were found going through roughly 180 tweets. The main struggle here was finding tweets correlating to the actual waste situation in Valencia. As we already have discussed, there is no actual use. The next problem was finding any words besides "basura" and the other buzzwords that would be reasonable to use. Ultimately, the main keywords were words within the collection, and other complaints correlated with the waste situation, like rodents and dog owners not picking up their dog's poop. See Fig. 3b.

Sport	Media	Politics	Rest
equipo	programa	conspiracion	arte
juega	TV	político/a	profesional
deport	libro	LimGoHome	muertos/as
VFC/VCF	película	manipulada/o	Diseño
minutos		izmierda/o	bandera
físico/a		alcalde	ave
primera parte		politiquillos/as	telefonica
ocacion		dictador/a	fallecido/a
LaLiga		comunista/o	animal
ValenciaCF		genocida	
Valencia cf		asesino/a	
partido		catalan	
F1		manifa	
SpanishGP		fascismo/a	
jugador			
fútbol			
espectador			

(a)

Camión
Recogida
inmundicias
caca
suciedad
Ratas
no cabía
calle

(b)

Fig. 3. Filter-words for negative (a) and positive (b) filtering

4 Results

The following describes the results of the filtering on the Twitter data set.

Negative Filtering
As shown in Fig. 4, sports significantly reduced the number of tweets. From 760 in the beginning, it went down to 625, which means 135 were thrown out. Media (34), politics (44) and the other filter words (43) each had a minor impact, resulting in 504 remaining tweets. Looking at the months of postings in Fig. 5, from May until November, the number corresponds with the number of tourists in Valencia; only in July are there fewer posts. A low number of complaints in December also seems reasonable, as well as the higher number of postings in Fallas, March.

In Fig. 6, only one day sticks out: The 4th of June, with nine postings on this day over all years. Manual analysis of these nine posts revealed: Six were made in 2012 when the body of a boy was found in a waste dump. The other three posts also have no connection with waste management:

The example of these posts shows that negative filtering is not suitable. All three posts have nothing to do with waste management, yet no key filter word could have sorted them out. It must be assumed that there are several more posts of this type in the 504 filtered posts.

With the used filter words from the original 760 posts, only 64 are left at the end. In Fig. 7 again, the number of complaints corresponds with the Fallas in March and the number of tourists in the summer and autumn, except in July.

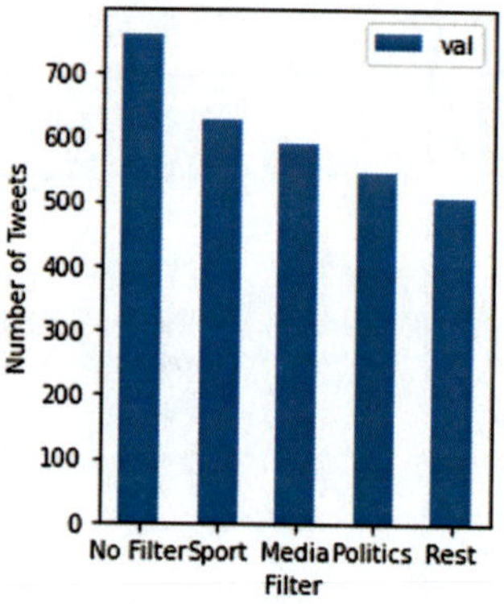

Fig. 4. Remaining tweets

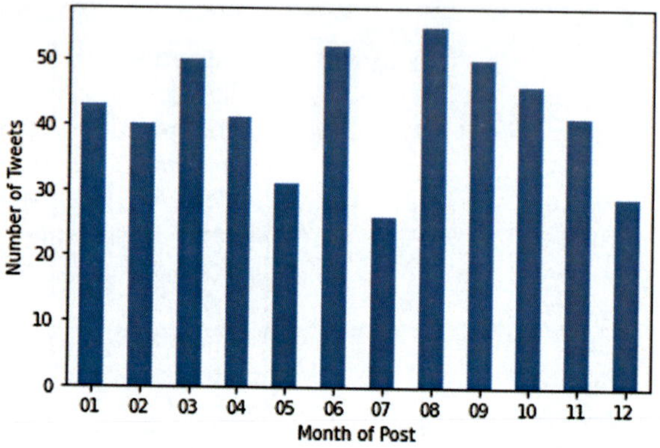

Fig. 5. Month of posting of filtered tweets

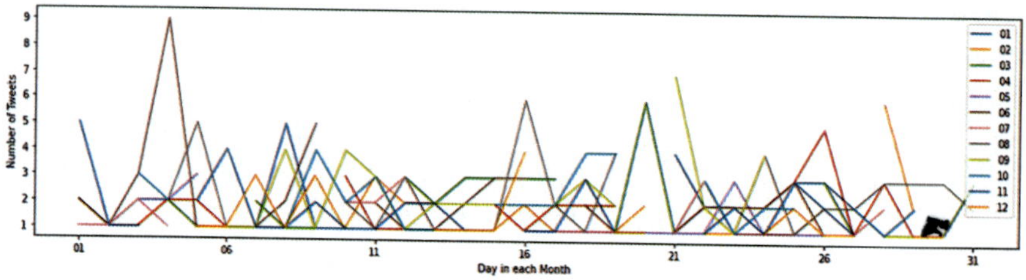

Fig. 6. Day of posting of filtered tweets

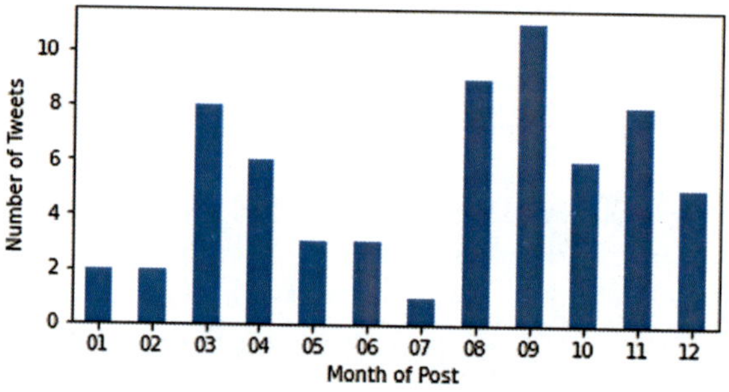

Fig. 7. Day of posting of filtered tweets

Also, there's an untypical amount of posts in September, so again, a manual analysis of posts in that month was conducted. There is no indication of an extraordinary event in September, and most complaints concern waste. Nevertheless, despite the positive filtering, some posts have nothing to do with rubbish.

In this case, the post was only selected because one mentioned person has the word *caca* in his username. But there are also posts where people used filter words, which should typically state a waste-related topic but refer to something different, like here, where the combination of *basura* and *calle* is used.

In summary, the positive filtering is not convincing either. Although it probably has a smaller percentage of false positive posts, there are still some false filtered tweets in the data set and who knows how many wrong negative posts were filtered out.

The formula for reviewing the method was already shown before. With the resulting tweets for the negative (504) and positive (64) filtering we got the formula

$$(504 - 64)/760$$

which results in 0.58. As expected after looking at the results, this is a low value; the optimal would be a minimal number. But why is the result of the social media analysis that bad? This question will be discussed in the next chapter.

5 Conclusions

The main question remaining is the usability and suitability of the filtering of the tweets. Will this method of filtering out relevant tweets about the waste situation in Valencia change how the waste management collection is done? Based on the filtering analysis, several barriers and challenges have been submerged.

The data collected through the initial queries shows us that most data can't be used. Firstly, more than the quantity of the remaining tweets is needed to draw any reasonable conclusions. Secondly, with a precise Geo-location connected to the individual tweets, there is a way of pinpointing any complaints to a specific location in Valencia. Therefore, achieving a heat map of the waste situation or gathering data for potential route optimisation solutions will be close to impossible with the state of the current data. Nevertheless, the critical outtake from the filtering experiment is what the data tell us about the missing information.

Comparing the result with the model and the tweet sorting algorithm, being able to check the tweet's content towards an actual waste situation in a specific location will require more specific tweet content and a large volume of relevant tweets. Therefore, it is reasonable to conclude that the state of the current way of tweeting about waste complaints is not suitable for the model described in this paper. More people should post about waste problems on Twitter to use rigorous filtering and still have a sufficient number of posts. Several words, especially in the waste topic, can be used for several meanings (waste, politics, sports, etc.). In the end, people don't tweet with the intention that a simple bot can read it.

As Twitter, in its current state, has yet to be an excellent tool for obtaining relevant data, there could be two options to solve the problem. Either change or develop a waste report tool or try to change the structure and content of how people report the waste situation in Valencia. An optimal tweet regarding our model would be a distinct tag immediately setting off the algorithm, with precise Geo-location and a detailed problem description.

In future work, it is planned to combine Twitter with other networks, such as Instagram, to obtain a proper information channel. Among other alternatives, developing a complaints portal or a mobile application where citizens could register their waste complaints would be particularly interesting.

Acknowledgements. This work has been supported by the projects TED2021-130890B, funded by MCIN/AEI/10.13039/501100011033 and by the European Union NextGenerationEU/PRTR, and Ramon y Cajal Grant RYC2018-025580-I, funded by MCIN/AEI/10.13039/501100011033, "FSE invest in your future" and "ERDF A way of making Europe".

References

1. monetize daily active users international. https://www.statista.com/statistics/1032751/monetizable-daily-active-twitter-users-international/. Accessed 2022
2. Abuga, D., Raghava, N.: Real-time smart garbage bin mechanism for solid waste management in smart cities. Sustain. Urban Areas **75**, 103347 (2021)
3. Das, S., Lee, S.H., Kumar, P., Kim, K.H., Lee, S.S., Bhattacharya, S.S.: Solid waste management: Scope and the challenge of sustainability. J. Clean. Prod. **228**, 658–678 (2019)
4. Das, S., Bhattacharyya, B.K.: Optimization of municipal solid waste collection and transportation routes. Waste Manag. **43**, 9–18 (2015)
5. De, S., Debnath, B.: Prevalence of health hazards associated with solid waste disposal-a case study of Kolkata, India. Procedia Environ. Sci. **35**, 201–208 (2016)
6. Marzouki, A., Mellouli, S., Daniel, S.: Towards a context-based citizen participation approach: a literature review of citizen participation issues and a conceptual framework. In: Proceedings of the 10th International Conference on Theory and Practice of Electronic Governance, pp. 204–213 (2017)
7. Sachs, J.D.: From millennium development goals to sustainable development goals. Lancet **379**(9832), 2206–2211 (2012)
8. Silva, B.N., Khan, M., Han, K.: Towards sustainable smart cities: a review of trends, architectures, components, and open challenges in smart cities. Sustain. Urban Areas **38**, 697–713 (2018)
9. de Valencia, A.: Recyclable solid waste management (2020). https://smartcity.valencia.es/vlci/recyclable-solid-waste-management/
10. Ziraba, A.K., Haregu, T.N., Mberu, B.: A review and framework for understanding the potential impact of poor solid waste management on health in developing countries. Arch. Public Health **74**(1), 1–11 (2016)

Detection of Human Falls via Computer Vision for Elderly Care – An I3D/RNN Approach

João Leal[1](✉), Hamed Moayyed[2], and Zita Vale[2]

[1] Polytechnic of Porto (ISEP/IPP), Porto, Portugal
1180723@isep.ipp.pt

[2] GECAD - Research Group on Intelligent Engineering and Computing for Advanced Innovation and Development, LASI - Intelligent Systems Associate Laboratory, Polytechnic of Porto, Porto, Portugal
{mha,zav}@isep.ipp.pt

Abstract. As the population continues to age, the number of elderly individuals is increasing at an unprecedented rate as the active age group in developed countries continues to shrink. This demographic shift has resulted in a shortage of resources and limitations in the provision of adequate elderly care. Among the many health risks faced by the elderly, falls are a common and serious problem, which can lead to injuries, hospitalizations, and even death. Despite the prevalence of falls, traditional manual detection methods are often unreliable and inadequate. In response to this, this paper proposes an automatic fall detection system based on video cameras that uses a unique combination of an inflated 3D convolutional neural network (I3D) and a recurrent neural network (RNN). This model was evaluated using multiple fall detection datasets, including a newly developed dataset of simulated falls. The results of this study demonstrate that this hybrid model is highly effective at detecting falls and is on par with other state-of-the-art systems, boasting 94% F1 score and 96% recall values. This technology has the potential to significantly improve the quality of life and safety of the elderly population by providing timely and accurate detection of falls.

Keywords: Automatic fall detection · Deep Learning · Computer vision · I3D · RNN

1 Introduction

As life expectancy increases, so does the aging population. In the First World especially, the number of seniors is starting to exceed the number of people who can provide care for them, resulting in many situations where senior citizens cannot look after themselves, nor do they have adequate access to elderly care. Seniors also suffer from an increased risk of falling. For the elderly, falls might not only cause injuries on impact, but can also incapacitate them, resulting in severe medical complications.

A timely response to falls is thus necessary to prevent serious injuries, which requires the immediate detection of a fall event. Even in nursing homes, however, falls can remain

S. Ossowski et al. (Eds.): DCAI 2023, LNNS 740, pp. 113–122, 2023.
https://doi.org/10.1007/978-3-031-38333-5_12

undetected for long periods of time, especially in secluded areas. The senior can resort to calling for help, but this is not always sufficient. Some nursing homes place manual alarms in places like seniors' rooms, such as pull cord alarms, but victims of falls cannot be expected to activate them on every occasion, and much less in a timely fashion. As it stands, traditional, manual ways of detecting falls are impractical and ineffective, with falls remaining undetected for long periods of time.

Technology can help automate elderly care, lessening the effects of an eventual medical personnel shortage and improving seniors' sense of independence. Vital signs could be monitored automatically and constantly by specialized sensors, robots could assist in social interaction or facilitate mobility. Finally, automated solutions for fall detection (FD), seeking to eliminate the human element in detecting and alerting to fall events, thereby avoiding the negative health effects falls induce, have the potential to substantially improve the quality of life for seniors by ensuring that timely and appropriate care is provided. Deep Learning (DL) models are very popular in general action recognition problems, but their specialized use in FD is underexplored. The purpose of this paper is to test one of the newer state-of-the-art models intended for action recognition and ascertain its suitability for FD.

In the next section, other existing fall detection systems (FDSs) are explored, followed by a listing of public datasets for visual FD. Next, the chosen approach is quickly explained, as is the methodology of this study. Finally, the results obtained are displayed and briefly discussed.

2 Related Work

2.1 Visual Fall Detection Systems

Existing fall detection systems (FDSs) can be categorized based on the type of sensors they use. Wearable FDSs use data from accelerometers and gyroscopes embedded in wearable devices such as bracelets, belts, smartwatches, or smartphones; Environmental FDSs use sensors such as acoustic, vibration, pressure, infrared, radar, or WiFi to detect falls; Visual FDSs, on the other hand, use computer vision algorithms to capture falls through camera feeds from RGB, thermal, or depth cameras, with methods such as pose recognition [1] or optical flow analysis [2], and these systems often use DL techniques, such as 2D Convolutional Neural Networks (CNNs) [2–4], 3D CNNs [5–7], and Recurrent Neural Networks (RNNs) [7, 8].

Wearable FDSs are cheap and accessible but are overly sensitive to noise and only work if they are worn, which is an issue as they tend to be uncomfortable to wear and elderly people are prone to forget or refuse to put them on. In contrast, environmental FDSs are not intrusive nor obstructive, and can detect falls without the need to be adjusted or turned on every day due to their "place-and-forget" nature. However, their detection range is limited, and their performance tends to suffer from the existence of "blind spots" and a hypersensitivity to noise. Visual FDSs have better accuracy due to the use of DL algorithms but are expensive, especially in the case of thermal - and, to a lesser extent, depth-based FDSs. Furthermore, obtaining training data is difficult since it cannot be "scrapped" or requested from nursing homes and hospitals, also due to privacy issues.

If RGB cameras are used in visual FDSs, privacy concerns arise, and data protection methods must be employed.

2.2 Public Visual Fall Detection Datasets

Vision-based FD DL models must rely on publicly available datasets specifically designed for fall detection. Unfortunately, such datasets are few and their quality is sub-par compared to general action recognition datasets. The largest public FD dataset, UP-Fall [9], has only 1,122 videos, while a relatively small general action recognition dataset like UCF101 contains 13,320 videos. Since FD datasets can only feature simulated and acted falls, it is difficult to create a diverse enough dataset. Moreover, most datasets lack variety in backgrounds, number of actors, and motions simulating physically impaired individuals, which can lead to model bias and limit its ability to adapt to new, real-life data.

The following table presents all the major publicly available FD datasets that consist of either video segments or video frames (Table 1):

Table 1. Major public fall detection datasets

Dataset	Falls	ADLs	Actors	Backgrounds	Type of data
UP-Fall	510	612	17	1	RGB, wearable
eHomeSeniors [10]	448	0	6	1	Thermal
HQFSD [11]	275	85	10	1	RGB
Multicam [12]	176	192	1	1	RGB
FDDBg [13]	146	79	9	5	RGB
TST V2 [14]	132	132	11	1	Depth
EDF & OCCU [15]	70	110	5	1	Depth
UR-Fall [16]	60	40	5	7	RGB, depth

3 Chosen Approach

The proposed approach for FD relies on implementing DL algorithms and techniques, specifically computer vision, to develop a model that can accurately differentiate falls from Activities of Daily Life (ADLs) present in RGB footage. This approach was chosen because of the possibility of using computer vision models, which means greater accessibility and accuracy than other types of FDSs, and despite the privacy issues, which can be ameliorated by data protection measures. It is meant to improve upon conventional FD methods by being quicker, more consistent, and by eschewing the need for the intervention of other people.

This is achieved using real-time data streams from standard RGB cameras, which are more accessible and can be placed inside a room to provide adequate visual coverage.

Depending on the size and layout of the room, more than one camera may be needed to eliminate blind spots. The proposed final FDS consists of three main components: the cameras; the computer that accesses the camera's output, sends video segments to the model, receives its classification, and activates an alarm if needed; lastly, the model, which receives processed video segments for classification and sends its output to the computer.

To train the model effectively, the construction of a comprehensive dataset is required. Since no single public dataset is large or varied enough to fully train a DL model, several public datasets were combined and curated, taking into account imbalances and variations in data quality respective to each dataset. Additionally, a private dataset was built from recordings of the author, which includes details such as obscured falls, object interactions, different lightings, simulated physical impairments, and the presence of walking aids.

Recent studies have shown that models like 3D CNNs and RNNs are particularly useful for detecting certain motion sequences of humans, as they can capture spatio-temporal features that traditional 2D CNNs cannot. By using a 3D CNN for feature extraction, low-level spatio-temporal features can be captured, while RNNs can extract high-level features from the CNN's output to increase detection accuracy.

While some visual FDSs do use a hybrid 3D CNN/RNN model, which has proved to be very effective, no FDS has yet utilized Carreira and Zisserman's [17] two-stream Inflated 3D CNN (I3D) as an alternative to regular 3D CNNs. The I3D has improved performance and transfer learning capabilities, making it a promising choice for action recognition. This paper advocates for using an I3D in conjunction with an RNN to form a hybrid model that is best suited to recognize the characteristic movements that occur in the event of a fall.

4 Methodology

4.1 Dataset Collection, Preprocessing, and Splitting

Datasets for FD are significantly smaller, more homogeneous, and exhibit drastically fewer variations than those for general action recognition. Despite this discrepancy in size and diversity, researchers in this field often fail to acknowledge this and treat their data with the same attitude as those in other fields.

The issue with small and homogeneous datasets is that the performance of models trained on them cannot be accurately measured using traditional evaluation and testing methods, even if many datasets are combined into one. Methods such as holdout, cross-validation, and other techniques that split the data into parts for the training, evaluation, and testing phases can only provide realistic performance measures if the data parts are effectively distinct from each other. Unfortunately, in FD datasets, the data is so small and homogenous that it is impossible to have a meaningful distinction between the parts.

This lack of dataset variety results in overfitting, leading to metrics that seem to reflect high performance, but are not entirely precise. One simple remedy to this issue is to reserve an entire dataset for validation and another for testing, ensuring that the data is entirely "unseen" between splits. However, this approach runs the risk of reducing the training process's effectiveness by not including all datasets in the training split. Additionally, the model's performance might change significantly according to the dataset chosen for testing. Nonetheless, for this project, all existing major RGB FD datasets were compiled. Efforts were made to ensure that the number of ADL videos extracted from each segment perfectly matched the number of falls to balance the data. Each dataset was used in their entirety except for UP-Fall to avoid overrepresentation. The datasets used for this compilation and their composition after editing are as follows: UR-Fall: 120 segments (falls + ADLs); HQFSD: 504 segments; FDDBg: 302 segments; Multicam: 352 segments; UP-Fall: 600 segments.

In order to augment the size and variety of the datasets, an original, private dataset was created by the author of this paper. The dataset comprises 1,028 videos, making it the second largest FD dataset. The videos were recorded over a period of two weeks, with the author falling and performing ADLs in six different rooms of his house, each recorded from at least two angles and with different levels of natural and artificial lighting. The dataset also includes scenarios with semi-obscured falls, object manipulation, walking aids such as crutches and canes, physical impairments, and even the presence of a dog. However, the dataset features only one actor, albeit wearing different sets of clothing.

To ensure that the dataset was suitable for training models, each video was edited to be at most three seconds long, with shorter videos being left unedited or having their last frame repeated. This was found to be a good compromise between the amount of information available in each video segment and storage size amount. Additionally, image augmentation was performed on each frame, altering the hue, contrast, saturation, and brightness values randomly, with a 50% chance of vertical mirroring. This step was crucial in increasing the dataset's variety.

The resulting entire dataset comprises 2,906 video segments, with a total length of 2 h, 24 min, and 44 s of footage, taking up 4.14 GB of storage. Once the datasets were collected, recorded, and processed, a decision had to be made regarding which datasets would serve as the validation set and the test set. The original dataset, UP-Fall, and Multicam were considered too large to be excluded from the train set, while HQFSD was deemed to have quality scenarios too important to be left out. Therefore, with two remaining datasets, FDDBg was chosen as the validation set, and UR-Fall was selected as the test set. It should be noted that the datasets used in the study still do not allow for a complete image of the model's performance - swapping the validation and test splits would result in an entirely different set of performance scores. Nevertheless, this approach was seen as the least one can do without access to sufficient amounts of data.

4.2 Chosen Models and Architectures

Action recognition in video is a challenging task, but recent developments in deep learning have shown promising results. One such approach involves the use of an I3D architecture, which is a 3D CNN pre-trained on a large-scale action recognition dataset, coupled with a RNN for feature extraction and classification. Liu et al. [18] successfully demonstrated that the combination of a pre-trained I3D and a LSTM achieved higher accuracy scores at classifying actions in video compared to other state-of-the-art models. This is due to the ability of RNNs to model higher-level temporal features, which are critical for recognizing complex actions.

To test this approach in the context of FD, four case studies were conducted using various model architectures, all of which feature a pre-trained I3D. In the first three case studies, the I3D was used without its final softmax layer, allowing it to output the features extracted before classification in RGB frames, optical flow frames, or both. Since the I3D is pre-trained on a general action recognition dataset, the extracted features tend to primarily capture the position, movements, and figures of humans. Raw video files are transformed into an equidistant sequence of image frames, which are cropped and resized, and inputted to the I3D. If one stream mode is chosen, the output consists of a tensor made up of n arrays, each containing 1,024 numerical values, with n being the supposed number of input frames. In the case that both streams are chosen, the I3D outputs two tensors for each feature type. If a video is shorter than three seconds, the last frame in the sequence might have to be repeated to maintain length n. In the last case study, an intact I3D was fine-tuned to be capable of classifying videos in isolation.

The first two case studies were conducted to assess the suitability of two popular RNN models, LSTM networks and GRU networks, for classification. Both models were trained to determine if there were any notable differences in performance. The third case study saw a CNN being used for classification. Since CNNs lack the ability to analyze features sequentially, it was theorized that this approach would result in worse results than the first two. To avoid overfitting, the structures of the LSTM, GRU, and CNN models were kept very simple. Dropout layers were included if needed, and various parameters such as optimizer, learning rate, learning rate decay, dropout rate, and class weights were adjusted in an effort to increase training/validation metric scores. Callback functions such as model checkpoints and early stopping were also applied to the training function.

All models but the I3D were tested with RGB features, optical flow features, and both simultaneously, while the I3D was only tested with both types. The RNN models were fitted with a Flatten-imbued TimeDistributed layer, as the extra dimension created by both feature types is incompatible with its possible input shapes, unlike with the CNN and I3D.

In summary, the case studies were devised as follows:

- 1st case study: I3D (feature extraction)/LSTM (classification)
- 2nd case study: I3D (feature extraction)/GRU (classification)
- 3rd case study: I3D (feature extraction)/CNN (classification)
- 4th case study: I3D (feature extraction and classification)

Steps were taken to improve validation performance metrics as much as possible and the best model would then be tested on UR-Fall. Recall was a prioritized metric, since, in the case of FD, false negatives are much more serious errors than false positives[1]. Nonetheless, accuracy and F1 score were also taken into account as more balanced representations of the model's performance.

[1] In this context, a positive result signifies the detection of a fall.

5 Results

Tables 2 and 3 show the results of each model's validation and testing processes, respectively. All metrics were taken from the same model, chosen by the greatest overall validation performance belonging to each model/feature type combination. The validation dataset used was FDDBg and the test dataset was UR-Fall. The best-performing feature type for each model combination (F1 score) is highlighted in both tables.

Table 2. Validation results

Model	Feature type	Accuracy	Recall	Specificity	Precision	F1 score
I3D/LSTM	RGB features	93.4%	92.1%	94.7%	94.6%	93.3%
	Optical flow	94.4%	94.7%	94.0%	94.1%	94.4%
	Both	**95.7%**	**94.0%**	**97.4%**	**97.3%**	**95.6%**
I3D/GRU	RGB features	92.1%	94.7%	89.4%	89.9%	92.3%
	Optical flow	97.0%	96.0%	98.0%	98.0%	97.0%
	Both	**98.0%**	**97.4%**	**98.7%**	**98.7%**	**98.0%**
I3D/CNN	RGB features	92.4%	90.7%	94.0%	93.8%	92.3%
	Optical flow	**94.7%**	**94.7%**	**94.7%**	**94.7%**	**94.7%**
	Both	94.4%	92.7%	96.0%	95.9%	94.3%
I3D	Both	91.4%	86.8%	96.0%	95.6%	91.0%

Table 3. Test results

Model	Feature type	Accuracy	Recall	Specificity	Precision	F1 score
I3D/LSTM	**RGB features**	**95.8%**	**91.2%**	**100%**	**100%**	**95.6%**
	Optical flow	90.0%	93.3%	86.7%	87.5%	90.3%
	Both	92.5%	95.0%	90.0%	90.5%	92.7%
I3D/GRU	**RGB features**	**94.2%**	**96.7%**	**91.7%**	**92.1%**	**94.3%**
	Optical flow	82.5%	80.0%	85.0%	84.2%	82.1%
	Both	89.0%	90.0%	88.3%	88.5%	89.2%
I3D/CNN	**RGB features**	**92.5%**	**90.0%**	**95.0%**	**94.7%**	**92.3%**
	Optical flow	85.0%	76.7%	98.3%	92.0%	83.6%
	Both	85.0%	81.7%	88.3%	87.5%	84.5%
I3D	Both	90.0%	95.0%	85.0%	86.4%	90.5%

6 Discussion

While these results are, in general, evidence that the I3D/RNN hybrid model excels at detecting falls, some observations that can be made should be highlighted.

As theorized, the RNN models did outperform the CNN, as the latter does not have the same capacity to capture temporal patterns. The videos that the CNN does tend to misclassify are those that feature getting up and lying down actions, suggesting that the model cannot see the difference between those and falls. Although the I3D/LSTM-RGB obtained better results overall, the I3D/GRU-RGB has a 5.5% better recall value, which makes up for the 1.3% difference in F1 score, making it the model that should be considered the best.

Generally, the test results tend to be worse than the validation results, however, the best validation results do not entirely correlate to the best test results. It seems that the model's test accuracy is lower when introduced to optical flow features, the exact opposite of validation accuracy. This goes to show that using different datasets results in extremely different results. It is expected that if the datasets were reversed, the results would too.

The relatively poor results of the fine-tuned I3D can be explained by its complexity, since there is simply not enough data to properly tune it, leading to worse performances than those of the simpler RNN and CNN models.

Finally, Table 4 pits the results of this approach's best model against those stated by the authors of some of the more similar visual state-of-the-art FDSs.

Table 4. Comparison of state-of-the-art performance results

FDS	Model	Accuracy	Recall	Specificity	Precision	F1 score
This approach	**I3D/GRU**	**94.2%**	**96.7%**	**91.7%**	**92.1%**	**94.3%**
Adhikari et al. [3]	CNN	74%	-	-	-	-
Fan et al. [8]	Filters/LSTM	-	91%	-	92%	91%
Ma et al. [5]	3D CNN/ Autoencoder	-	93.3%	92.8%	-	-
Rahnemoonfar and Alkittaw [6]	3D CNN	97.6%/ 93.2%	-	-	-	-
Lu et al. [7]	3D CNN/ LSTM	99.1%	-	-	-	-
Khraief et al. [2]	Optical flow/ CNN	99.7%	-	-	-	-

Overall, the I3D/RNN fares well. However, all the other FDSs in Table 4 were trained on a single dataset or did not take the steps necessary to ensure a proper separation of training and testing splits, usually settling with the holdout method or others like it. Those few who did, like Fan et al. [4], largely reflect this paper's findings that performance metrics vary wildly between datasets – even though they observed an excellent 98%

accuracy when testing with FDDBg, they also observed 74% with HQFSD and 63% with their own dataset. It is important to note that the apparent performance of other fall detection systems may, indeed, vary depending on how the data is handled, as well as the data itself, in training, validation, and testing.

7 Conclusion

The results obtained show clearly the potential of the I3D/RNN hybrid for visual FD, reflecting the outstanding results it has had in general action recognition. A FDS using this model would have a distinct, positive impact on the elderly's quality of life and safety.

However, image or object recognition and general action recognition technologies allow researchers access to not just huge datasets, but datasets capable of setting an 'industry' standard. These datasets – UCF-101, Kinetics, ImageNet, to name a few – offer the opportunity for accurate and transparent comparison. They are good enough to train and test all kinds of models and allow for an accurate portrayal of the model's capabilities in the real world. The ImageNet Large Scale Visual Recognition Challenge (ILSVRC)[2], for example, pits models trained on ImageNet (alone) against one another to see which is best at object detection and to give researchers a chance to visualize scientific progress in that field.

There is no ImageNet equivalent for visual fall detection and thus the field cannot hope to replicate this any time soon, or at least with what is available now. It is admittedly difficult to build a fall dataset: falls must be simulated, which necessitates planning, equipment, manpower, time, even creativity. To create a dataset of the same level of quality as Kinetics or UFC-101 requires a years-long project with actual funding. It is a herculean task, but one that could be a gigantic step forward in progressing visual FD. Additionally, perhaps creating a multimodal dataset would further the fall detection as a whole, not just the visual subfield. But until this is done, it is unlikely that visual FD systems, even if they show apparent outstanding results, will find a good enough justification for practical use in the real world.

Acknowledgement. This work received funding from FCT under the project UIDP/00760/2020. The authors acknowledge the work facilities and equipment provided by GECAD research center (UIDB/00760/2020) to the project team.

References

1. Hasan, M., Islam, M., Abdullah, S.: Robust pose-based human fall detection using recurrent neural network. In: 2019 IEEE International Conference on Robotics, Automation, Artificial-Intelligence and Internet-of-Things, RAAICON 2019, pp. 48–51 (2019)
2. Khraief, C., Benzarti, F., Amiri, H.: Elderly fall detection based on multi-stream deep convolutional networks. Multimedia Tools Appl. **79**(27–28), 19537–19560 (2020). https://doi.org/10.1007/s11042-020-08812-x

[2] https://www.image-net.org/challenges/LSVRC/.

3. Adhikari, K., Bouchachia, H., Nait-Charif, H.: Activity recognition for indoor fall detection using convolutional neural network. In: Proceedings of the 15th IAPR International Conference on Machine Vision Applications, MVA 2017, Institute of Electrical and Electronics Engineers Inc., pp. 81–84 (2017)
4. Fan, Y., Levine, M., Wen, G., Qiu, S.: A deep neural network for real-time detection of falling humans in naturally occurring scenes. Neurocomputing **260**, 43–58 (2017)
5. Ma, C., Shimada, A., Uchiyama, H., Nagahara, H., Taniguchi, R.: Fall detection using optical level anonymous image sensing system. Opt. Laser Technol. **110**, 44–61 (2019)
6. Rahnemoonfar, M., Alkittawi, H.: Spatio-temporal convolutional neural network for elderly fall detection in depth Video cameras. In: Proceedings - 2018 IEEE International Conference on Big Data, Big Data 2018, pp. 2868–2873 (2019)
7. Lu, N., Wu, Y., Feng, L., Song, J.: Deep learning for fall detection: three-dimensional CNN Combined with LSTM on video kinematic data. IEEE J. Biomed. Health Inform. **23**(1), 314–323 (2019)
8. Fan, X., Zhang, H., Leung, C., Shen, Z.: Robust unobtrusive fall detection using infrared array sensors. In: IEEE International Conference on Multisensor Fusion and Integration for Intelligent Systems, vol. 2017, pp. 194–199 (2017)
9. Martínez-Villaseñor, L., Ponce, H., Brieva, J., Moya-Albor, E., Núñez-Martínez, J., Peñafort-Asturiano, C.: Up-fall detection dataset: a multimodal approach. Sensors (Switzerland) **19**(9), 1988 (2019)
10. Riquelme, F., Espinoza, C., Rodenas, T., Minonzio, J., Taramasco, C.: eHomeSeniors dataset: an infrared thermal sensor dataset for automatic fall detection research. Sensors (Basel) **19**(20) (2019)
11. Baldewijns, G., Debard, G., Mertes, G., Vanrumste, B., Croonenborghs, T.: Bridging the gap between real-life data and simulated data by providing a highly realistic fall dataset for evaluating camera-based fall detection algorithms. Healthc. Technol. Lett. **3**(1), 6–11 (2016)
12. Auvinet, E., Reveret, L., St-Arnaud, A., Rousseau, J., Meunier, J.: Fall detection using multiple cameras. In: Proceedings of the 30th Annual International Conference of the IEEE Engineering in Medicine and Biology Society, EMBS 2008 - "Personalized Healthcare Through Technology". IEEE Computer Society, pp. 2554–2557 (2008)
13. Charfi, I., Miteran, J., Dubois, J., Atri, M., Tourki, R.: Optimized spatio-temporal descriptors for real-time fall detection: comparison of support vector machine and adaboost-based classification. J. Electron. Imaging **22**(4) (2013)
14. Gasparrini, S., et al.: Proposal and experimental evaluation of fall detection solution based on wearable and depth data fusion. In: Loshkovska, S., Koceski, S. (eds.) ICT Innovations 2015. AISC, vol. 399, pp. 99–108. Springer, Cham (2016). https://doi.org/10.1007/978-3-319-25733-4_11
15. Zhang, Z., Conly, C., Athitsos, V.: Evaluating depth-based computer vision methods for fall detection under occlusions. In: Lecture Notes in Computer Science. Lecture Notes in Artificial Intelligence and Lecture Notes in Bioinformatics. Springer, Heidelberg, vol. 8888, pp. 196–207, (2014). https://doi.org/10.1007/978-3-319-14364-4_19
16. Kwolek, B., Kepski, M.: Human fall detection on embedded platform using depth maps and wireless accelerometer (2014)
17. Carreira, J., Zisserman, A.: Quo Vadis, action recognition? A new model and the kinetics dataset. In: Proceedings - 30th IEEE Conference on Computer Vision and Pattern Recognition, CVPR 2017, vol. 2017, pp. 4724–4733 (2017)
18. Liu, W., et al..: I3D-LSTM: a new model for human action recognition. IOP Conf. Ser. Mater. Sci. Eng. **569**(3), 032035 (2019)

Generic Architecture for Multisource Physiological Signal Acquisition, Processing and Classification Based on Microservices

Roberto Sánchez-Reolid[1,2](✉), Daniel Sánchez-Reolid[2], Clara Ayora[1,2], José Luis de la Vara[1,2], António Pereira[3,4], and Antonio Fernández-Caballero[1,2,5](✉)

[1] Departamento de Sistemas Informáticos, Universidad de Castilla-La Mancha, Albacete, Spain
{roberto.sanchez,clara.ayora,joseluis.delavara,antonio.fdez}@uclm.es

[2] Neurocognition and Emotion Unit, Instituto de Investigación en Informática de Albacete, Albacete, Spain
daniel.sanchez@uclm.es

[3] Computer Science and Communications Research Centre, School of Technology and Management, Polytechnic Institute of Leiria, Leiria, Portugal
apereira@ipleiria.pt

[4] INOV INESC INOVAÇÃO, Institute of New Technologies—Leiria Office, Leiria, Portugal

[5] CIBERSAM-ISCIII (Biomedical Research Networking Center in Mental Health), 28016 Madrid, Spain

Abstract. The use of IoT devices is increasing and their integration into healthcare is growing. Therefore, there is a need to develop microservice-oriented hardware-software architectures that integrate all the stages from the acquisition of physiological signals to their processing and classification. In addition, the integration of physiological signals from different sources is a must in order to increase the knowledge of the monitored person's condition. In this context, the focus of this work has been to identify all the necessary workflow phases in this type of architecture, focusing mainly on scalability, replication and redundancy of the different services. This work proposes an architecture generic in terms of the number of sensors to be included and their acquisition requirements (sampling frequency and latency). We have chosen to include network protocols such as the Laboratory Stream Layer for data synchronisation and streaming. To this end, infrastructure as a service and as a machine have been included.

Keywords: Microservices · Multisource signal acquisition · Physiological signals · Signal processing · Signal classification

S. Ossowski et al. (Eds.): DCAI 2023, LNNS 740, pp. 123–133, 2023.
https://doi.org/10.1007/978-3-031-38333-5_13

1 Introduction

Internet of Things (IoT) sensors are small, specialised devices designed to collect data and transmit it over the Internet to other devices or systems. IoT sensors are increasingly being used in healthcare and clinical settings to monitor patients and track a variety of health metrics [11]. A wide range of health indicators such as heart rate, blood pressure, temperature and oxygen saturation can be measured using IoT sensors [8,10,23]. They can also be used to track medication adherence and other patient behaviours that can affect health outcomes [18]. One of the key benefits of IoT sensors in healthcare is their ability to provide real-time data. This can be used to identify potential health problems before they become more serious. This can help healthcare providers intervene earlier and provide more targeted care to patients [21].

Overall, IoT sensors are rapidly transforming the healthcare industry by enabling a more personalised and data-driven approach to patient care. They are likely to become even more prevalent and influential in shaping the future of healthcare as these technologies continue to evolve. To ensure the accuracy and reliability of the data, it is critical to create a well-designed architecture for the acquisition, processing and synchronisation of IoT signals [17]. Such an architecture can help minimise delays and errors in data collection, reduce the risk of data loss, and enable more effective analysis of the data to provide accurate information about patient health.

Fortunately, managing the collection, processing and synchronisation of physiological signals can be achieved through two models of software architecture: microservices and event-based architectures [15,19]. Microservices are software modules that can perform specific tasks efficiently and quickly on the server side. Event-driven architecture, on the other hand, is based on internal or external events that trigger workflows within the existing infrastructure. By using a message broker, event producers and consumers are decoupled, allowing asynchronous processing of different events and optimising the throughput capacity of each service [20]. Specifically, the aim of this paper is to describe a hardware/software architecture microservice oriented that streamlines this processing, allowing researchers to focus on research rather than signal acquisition, processing and classification [17].

2 Common Architectures for Physiological Signal Processing and Classification

IoT-based applications for physiological signal monitoring, typically using wireless sensor networks, are expected to have a significant impact on healthcare and daily life in the near future. For these applications to enable real-time remote monitoring of health-related biosignals, they must meet requirements such as reliability, connectivity, user interaction and moderate cost. However, the scalability of these applications is becoming increasingly important as the number of users continues to grow.

Wearable devices have been developed with sensor-based architectures and cloud systems using wireless technologies such as WiFi and Bluetooth. However, current approaches often overlook important aspects such as fault tolerance, high availability and scalability. Only some research has been done to partially implement architectures that meet these requirements. Such approaches are being used in applications where environmental monitoring is critical, with a focus on defining boundary conditions for both indoor and outdoor environments, often in conjunction with other technologies such as the Internet or 4G/5G networks. In this area, researchers have also explored the integration of heterogeneous sensor architectures using low-speed wireless personal area networks (LR-WPAN) [1], such as wireless body area networks (WBAN) [3].

Typically, architectures dedicated to physiological signal processing and classification consist of different software elements placed in different stages (see Fig. 1). These phases are assembled to form a workflow that allows us to perform the tasks of acquisition, storage, processing and classification of the signals monitored [6]. These stages are described below.

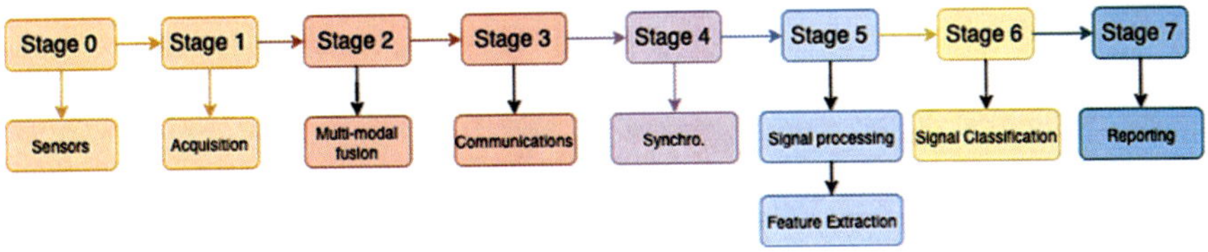

Fig. 1. Different stages in the architecture for physiological signal processing.

Stage 0: Devices and Sensors. In situations where data is derived from the real world or the environment, the hardware elements must be at the lowest level. Stage 0 represents the level at which hardware components are responsible for detecting various physiological signals.

Stage 1: Acquisition. This software layer manages the communication with the sensors. For each sensor there is a module that can interpret the incoming data. As this level is the lowest in terms of processing, it does not receive data from higher levels. Instead, it takes raw, low-level data from the sensors and passes it on to higher levels.

Stage 2: Multi-modal Fusion. One of the challenges of this type of architecture is to synchronise the different data sources (**multi-modal fusion**). Although there are numerous topologies that can be used for synchronisation; It is generally accepted that the closer the sensors are to each other, the fewer the problems of data fusion due to latencies in the network layer. Therefore, the use of a communication and synchronisation protocol will be essential in this task [13]. This network protocol can be used to add, configure and synchronise the different devices universally, independently of the type of sensor.

Stage 3: Communications. This stage consists of the various communication layers, from the device synchronisation layer to communication with the rest of the microservices infrastructure. Local network protocols such as the Laboratory Stream Layer (LSL) [5] and the MQTT protocol are often used to perform this communication process [4]. MQTT stands for Message Queuing Telemetry Transport and is a lightweight, publish-subscribe protocol designed for use in low-bandwidth, high-latency, unreliable networks. These protocols are used to retrieve and send the various sensor data to our infrastructures. Allowing an easy integration with a wide range of devices and platforms.

Stage 4: Synchronisation Strategy. Given that we are dealing with microservices architectures, data synchronisation is a critical issue that needs to be addressed. Although there are many ways to synchronise data, the most common approach is to synchronise data as close to the source as possible, thus avoiding the introduction of latencies that make this process more costly in subsequent stages. In general, two synchronisation and storage environments need to be established. The first, the local environment, will be responsible for collecting the data and sending it via various communication protocols to the second environment, the cloud environment [2]. This environment will be responsible for storing the signals and making them available for subsequent processing. This requires a Message Broker (MB) to act as an intermediary between applications, devices and services. It receives messages from different sources and routes them to their intended destinations. One of the most popular message broker protocols is the MQTT protocol.

Stage 5: Signal Processing and Feature Extraction. To deal with the variety of signals that need to be processed, it is important to consider the most commonly used signal processing methods. In terms of prototype signals, they typically go through the same processing stages, including pre-processing, processing and feature extraction. During pre-processing, various techniques are used for filtering and artefact removal. The processing phase applies specific methods for each type of signal, and finally the extracted features are used in subsequent phases. Feature extraction includes time dependent variables, frequency dependent variables, statistical parameters or signal morphology [16].

Stage 6: Signal Classification. This stage is carried out after processing the different signals obtained in the previous stages (stages 1, 2, 3 and 4). Due to the diversity of the signals and the experimental variation, the establishment of a fixed classification is a complex task. It is important to determine the most common classifiers for each type of signal and to have pre-trained models or trained models that can distinguish between different situations, such as stress and emotional states.

Stage 7: Reporting. In order to be aware of the results and to be able to plan the data collection experiment well, it is necessary to use a user interface (UI) that allows us to carry out this process. In general, these UIs are designed to be as possible.

3 Generic Architecture for Signal Processing and Classification

3.1 Description of the Architecture

Figure 2 shows the layout of the microservice architecture. Looking at the diagram from left to right, there are physiological sensors in a star node, followed by a gateway (or edge) that connects to a back-end service (called Infrastructure).

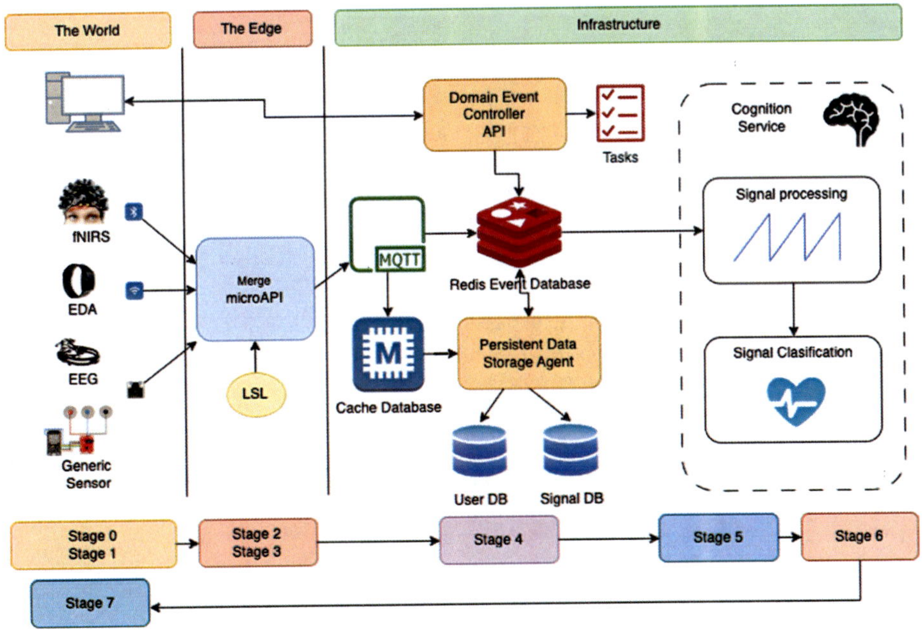

Fig. 2. Architecture diagram.

In the acquisition zone (The World) there are several sensors connected via different network standards such as Bluetooth and WiFi (corresponding to stages 0 and 1). These sensors are responsible for acquiring and sending the signals to the next step for synchronisation and forwarding. To perform this synchronisation, the LSL network protocol embedded in a micro-API (application programming interface) was chosen to route, synchronise and forward the network traffic to the next part of the infrastructure, namely The Edge (see stages 2 and 3).

LSL is a software library used to synchronise different types of data streams in real time. LSL is specifically designed to enable researchers to record and analyse multiple data streams, such as EEG, ECG, EDA, TMP and motion data, and to synchronise these streams with high precision. LSL achieves synchronisation by using timestamps and buffer management techniques to ensure that all data streams are aligned in time. This allows researchers to analyse data from different sources simultaneously, which can provide deeper insights into the relationship between different types of physiological signals [5,7]. LSL is widely used in neuroscience research and has the potential to be transferred to all types of sensors to enable more comprehensive and accurate analysis of physiological data. This external hub, which is software and hardware using a Linux-based distribution called Openwrt, is meant to provide secure connections to the rest of the infrastructure via a virtual private network (VPN) connection.

At the core of the application you will see a number of services working together, coordinated by the domain controller. The controller is responsible for managing communications, events and the general state of the application. It will handle the communications coming from the MQTT-based broker and the state management within the key-value database (Redis). On the other hand, given the acquisition speed of the various devices, it is necessary to have a mechanism for writing this data quickly. For this purpose, a cache mechanism has been implemented that allows writing to the volatile memory (RAM), so that this data is then passed to a persistent memory (see stage 4).

Our cognitive services are on the right-hand side of the diagram. These services are responsible for performing different operations on the signals. In general, the operations of feature processing and extraction (see stage 5) and signal classification (see stage 6) are within these microservices.

Finally, a web-based monitoring and reporting interface is provided, known as the Control and Reporting Web service (see stage 7). The result is a multiplatform system that can be accessed regardless of the hardware being used, such as PCs, tablets and mobile phones. In other words, the control panel used by the administrator to activate the various components and manage the components of the architecture (see stage 7).

3.2 Key Aspects of the Architecture

Monolithic Architecture vs Microservices Architecture. The first aspect addressed was choosing the right topology for the architecture. In general, you can opt for a monolithic architecture, i.e. all-in-one, or, on the contrary, an architecture that is segmented into several services or microservices. The advantages of choosing the second option are quite simple. From an implementation point of view, a monolithic architecture requires knowledge of all services in advance.

A microservices-based architecture, on the other hand, takes the opposite approach. Each of the services can be segmented and implemented in components as the architecture grows. In this way, it is possible to implement and add different types of services without compromising the integrity and functionality of the others. In our case, due to the diversity of sensors and signals to be processed, a microservices-based architecture helps us to achieve the goal of creating a path for processing and classifying each of the physiological signals.

Horizontal Auto-Scaling. The proposed architecture could be compromised by the number of requests and jobs to be processed for each of the physiological signals acquired. In terms of resources, it is necessary to provide different mechanisms that allow us to optimise the different microservices according to the demand or requests. Fortunately, the microservice orientation allows us to solve this problem.

In our approach, the use of Kubernetes [9], a technology widely used in this type of architecture, allows you to scale and model the size of the infrastructure based on the metrics of usage or capacity of your current system. Strictly speaking, it manages the number of services and is responsible for replicating them on different servers to achieve the most optimal configuration of resources at any given time. The metrics collected from the services are used to assess whether they are running optimally or inefficiently, in order to allocate more resources if necessary, or to replicate the services on the available servers. On the other hand, the metrics collected from the server allow us to establish a relationship between the workload and the available resources. The number of machines (servers) can be effectively set for each case.

Infrastructure as a Service (IaaS) and Machine as a Service (MaaS). Another key aspect to consider when developing such architectures is the use of two types of configurations for services. The first configuration consists of the layout of the different services of the architecture and their scaling through software that allows us to model them. That is, the infrastructure is scaled within different machines, but without the user or administrator noticing the change or addition of services. On the other hand, the second configuration is to deploy machines that host the services. This approach allows services to be hosted on specific machines with specific characteristics for the development of the task at hand.

In our approach we have chosen a hybrid version of both configurations. On the one hand, we have considered IaaS [12] for everything related to the first 5 stages (stages 0,1,2,3,4) and the last stage (stage 7), while for anything related to feature extraction, model training and classification (stages 5 and 6) we have chosen an approach based on MaaS [14,22]. This approach gives us the versatility and speed that IaaS brings to data acquisition, communication, data synchronisation and storage, as well as the ability to use different machines with GPU processing and training systems. This is particularly important as these processes are very demanding in terms of the hardware required to do the job optimally.

Data Persistence. Another aspect to consider in a signal processing and classification architecture is the storage and persistence of the data. In this respect, the typology of data acquired must be taken into account. Two types of data can be defined: physiological signals and user and application data. In this sense, several considerations must be taken into account. The first consideration is the replication of databases and their integrity. All databases in our architecture use sharding mechanisms. This allows for replication and high fault tolerance. The second consideration is the data typology. For physiological signals, write speed and data loss prevention must be considered. In this case, a redundant cache mechanism and persistent write is used as the data is buffered. This provides high availability and very high speed. Structured databases and key-value databases were chosen for user data and business logic.

Finally, the confidentiality of data and physiological data must be maintained. According to European regulations, the data must be decoupled and stored in different databases, using a transnational database and an agent to ensure the anonymity of the data. In the event of a security breach, the data cannot be linked to the participant.

Message Broker vs Direct Connection. Finally, with regard to the use of a message broker, in this type of architecture it is necessary to use them to provide robustness and centralise communications, with the minimum external exposure of the architecture, providing maximum availability. In this case, its use is essential. As already mentioned, for security reasons, the infrastructure must be exposed to the outside world (Internet) as little as possible, so the use of a message broker and an API facilitates its security compared to a direct connection between services.

However, in our proposal we have chosen to use distributed brokers (RabbitMQ) which allow us to mitigate this shortcoming as much as possible. Although latencies may be slightly higher, intermediate caching mechanisms (in-memory and transactional databases) work efficiently to mitigate this fact, allowing fast and continuous data streaming.

4 Conclusions

This paper has presented a generic architecture for physiological signal processing based on microservices. The different aspects to be taken into account for its development have been addressed. The different phases from the sensors to the classification of the different signals and the display of the results have been shown. Emphasis was placed on communication, architecture and orchestration of the different services. The different current technologies in the field of signal processing have been brought together to make the process as automated as possible. Finally, the advantages and disadvantages of the different technologies or configurations used and chosen in our approach were discussed. In this case, our focus was on distribution, redundancy and scalability in terms of communication, data persistence for the IaaS and configuration and deployment of different machines for training models using the MaaS methodology.

Thus, a goal of this research was to improve fault tolerance and consistency while maintaining high performance and minimal resource consumption on the server side, achieved by implementing a microservices and event-driven-based architecture. Unlike previous studies, the architecture aims to achieve all improvements within a single framework. The use of event-driven architectures and horizontal scaling enabled by microservices simplifies and solves scalability and performance issues. Technologies such as Kubernetes can be used to replicate services, allowing the architecture to scale to meet growing demand. This approach adds an extra layer that improves the fault tolerance and consistency of the system, leading to improved system quality.

The proposed architecture prioritises fault tolerance, high availability, scalability and cost-effectiveness. It uses an event-driven approach for real-time data processing and integration of heterogeneous sensor networks. The architecture's microservices allow services to be flexibly added or removed without affecting the overall system. The architecture aims to improve the performance, reliability and efficiency of IoT-based well-being applications, particularly in remote monitoring of health-related biosignals.

Acknowledgements. The work leading to this paper has received funding from the following sources: REBECCA project funded by HORIZON-KDT with reference 101097224; Grant PCI2022-135043-2 funded by MCIN/AEI/10.13039/501100011033 and by "Next Generation EU/PRTR"; Grants PID2020-115220RB-C21 and EQC2019-006063-P funded by MCIN/AEI/10.13039/501100011033 and by "ERDF A way to make Europe"; Grant 2022-GRIN-34436 funded by Universidad de Castilla-La Mancha and by "ERDF A way of making Europe"; Grants PTA2019-016876-I and RYC-2017-22836 funded by MCIN/AEI/10.13039/501100011033 and by "ESF Investing in your future". This research was also supported by CIBERSAM, Instituto de Salud Carlos III, Ministerio de Ciencia e Innovación.

References

1. Akram, H., Gokhale, A.: Rethinking the design of LR-WPAN IoT systems with software-defined networking. In: 2016 International Conference on Distributed Computing in Sensor Systems (DCOSS), pp. 238–243. IEEE (2016)
2. Al-Majeed, S.S., Al-Mejibli, I.S., Karam, J.: Home telehealth by Internet of Things (IoT). In: 2015 IEEE 28th Canadian Conference on Electrical and Computer Engineering (CCECE), pp. 609–613. IEEE (2015)
3. Alkhayyat, A., Thabit, A.A., Al-Mayali, F.A., Abbasi, Q.H.: WBSN in IoT health-based application: toward delay and energy consumption minimization. J. Sens. **2019** (2019)
4. Azzawi, M.A., Hassan, R., Bakar, K.A.A.: A review on internet of things (IoT) in healthcare. Int. J. Appl. Eng. Res. **11**(20), 10216–10221 (2016)

5. Blum, S., Hölle, D., Bleichner, M.G., Debener, S.: Pocketable labs for everyone: synchronized multi-sensor data streaming and recording on smartphones with the lab streaming layer. Sensors **21**(23), 8135 (2021)
6. Fernández-Caballero, A., Castillo, J.C., López, M.T., Serrano-Cuerda, J., Sokolova, M.V.: INT3-Horus framework for multispectrum activity interpretation in intelligent environments. Expert Syst. Appl. **40**(17), 6715–6727 (2013)
7. Lee, M.H.: Lab streaming layer enabled myo data collection software user manual. Technical report, US Army Research Laboratory Adelphi United States (2017)
8. Lozano-Monasor, E., López, M.T., Fernández-Caballero, A., Vigo-Bustos, F.: Facial expression recognition from webcam based on active shape models and support vector machines. In: Pecchia, L., Chen, L.L., Nugent, C., Bravo, J. (eds.) IWAAL 2014. LNCS, vol. 8868, pp. 147–154. Springer, Cham (2014). https://doi.org/10.1007/978-3-319-13105-4_23
9. Luksa, M.: Kubernetes in Action. Simon and Schuster (2017)
10. Martínez-Rodrigo, A., García-Martínez, B., Alcaraz, R., González, P., Fernández-Caballero, A.: Multiscale entropy analysis for recognition of visually elicited negative stress from EEG recordings. Int. J. Neural Syst. **29**(2), 1850038 (2019)
11. Mishra, S.S., Rasool, A.: IoT health care monitoring and tracking: a survey. In: 2019 3rd International Conference on Trends in Electronics and Informatics (ICOEI), pp. 1052–1057. IEEE (2019)
12. Prasad, V.K., Bhavsar, M.D.: Monitoring IaaS cloud for healthcare systems: healthcare information management and cloud resources utilization. Int. J. E-Health Med. Commun. (IJEHMC) **11**(3), 54–70 (2020)
13. Razavi, M., Janfaza, V., Yamauchi, T., Leontyev, A., Longmire-Monford, S., Orr, J.: OpenSync: an open-source platform for synchronizing multiple measures in neuroscience experiments. J. Neurosci. Methods **369**, 109458 (2022)
14. Reyes García, J.R., Lenz, G., Haveman, S.P., Bonnema, G.M.: State of the art of mobility as a service (MaaS) ecosystems and architectures-an overview of, and a definition, ecosystem and system architecture for electric mobility as a service (eMaaS). World Electr. Veh. J. **11**(1), 7 (2019)
15. Roda-Sanchez, L., Garrido-Hidalgo, C., Royo, F., Maté-Gómez, J.L., Olivares, T., Fernández-Caballero, A.: Cloud-edge microservices architecture and service orchestration: an integral solution for a real-world deployment experience. Internet of Things **22**, 100777 (2023)
16. Sanchez-Reolid, R., et al.: Emotion classification from EEG with a low-cost BCI versus a high-end equipment. Int. J. Neural Syst. **32**(10), 2250041 (2022)
17. Sánchez-Reolid, R., Sánchez-Reolid, D., Pereira, A., Fernández-Caballero, A.: Acquisition and synchronisation of multi-source physiological data using microservices and event-driven architecture. In: Julián, V., Carneiro, J., Alonso, R.S., Chamoso, P., Novais, P. (eds.) ISAmI 2022. LNNS, vol. 603, pp. 13–23. Springer, Cham (2023). https://doi.org/10.1007/978-3-031-22356-3_2
18. Selvaraj, S., Sundaravaradhan, S.: Challenges and opportunities in IoT healthcare systems: a systematic review. SN Appl. Sci. **2**(1), 139 (2020)
19. Shabani, I., Biba, T., Çiço, B.: Design of a cattle-health-monitoring system using microservices and IoT devices. Computers **11**(5), 79 (2022)
20. Surantha, N., Utomo, O.K., Lionel, E.M., Gozali, I.D., Isa, S.M.: Intelligent sleep monitoring system based on microservices and event-driven architecture. IEEE Access **10**, 42069–42080 (2022)
21. Tekeste Habte, T., et al.: IoT for healthcare. In: Ultra Low Power ECG Processing System for IoT Devices, pp. 7–12 (2019)

22. Wu, J., Zhou, L., Cai, C., Shen, J., Lau, S.K., Yong, J.: Data fusion for MaaS: opportunities and challenges. In: 2018 IEEE 22nd International Conference on Computer Supported Cooperative Work in Design (CSCWD), pp. 642–647. IEEE (2018)
23. Zangróniz, R., Martínez-Rodrigo, A., López, M.T., Pastor, J.M., Fernández-Caballero, A.: Estimation of mental distress from photoplethysmography. Appl. Sci. **8**(1), 69 (2018)

A Novel System Architecture for Anomaly Detection for Loan Defaults

Rayhaan Pirani(✉) and Ziad Kobti

School of Computer Science, University of Windsor, Windsor, ON N9B 3P4, Canada
{piranir,kobti}@uwindsor.ca
https://www.uwindsor.ca/science/computerscience

Abstract. Given the rise in loan defaults, especially after the onset of the COVID-19 pandemic, it is necessary to predict if customers might default on a loan for risk management. This paper proposes an early warning system architecture using anomaly detection based on the unbalanced nature of loan default data in the real world. Most customers do not default on their loans; only a tiny percentage do, resulting in an unbalanced dataset. We aim to evaluate potential anomaly detection methods for their suitability in handling unbalanced datasets. We conduct a comparative study on different classification and anomaly detection approaches on a balanced and an unbalanced dataset. The classification algorithms compared are logistic regression and stochastic gradient descent classification. The anomaly detection methods are isolation forest and angle-based outlier detection (ABOD). We compare them using standard evaluation metrics such as accuracy, precision, recall, F1 score, training and prediction time, and area under the receiver operating characteristic (ROC) curve. The results show that these anomaly detection methods, particularly isolation forest, perform significantly better on unbalanced loan default data and are more suitable for real-world applications.

Keywords: Anomaly detection · Unbalanced dataset · Early warning system · Loan default

1 Introduction

Recently, there has been a spike in loan defaults, especially after the COVID-19 pandemic. Fitch Ratings, a finance company providing credit ratings and research for global capital markets, increased the US institutional leveraged loan default rate forecast to 2.0%–3.0% for 2023 and 3.0%–4.0% for 2024 [1]. Nigmonov and Shams [13], in their study to understand how the COVID-19 pandemic has affected lending markets, objectively demonstrate an increase in loan default levels after the pandemic. The probability of default shown in this study before the pandemic is 0.056 and 0.079 in the post-pandemic period. Canada's largest banks have also allocated a significantly large budget in anticipation of

S. Ossowski et al. (Eds.): DCAI 2023, LNNS 740, pp. 134–144, 2023.
https://doi.org/10.1007/978-3-031-38333-5_14

more loan defaults [2]. For Q1 2022, the big six banks in Canada set aside $373 million; for Q1 2023, the allocation was almost $2.5 billion, an increase of over 6.5 times.

To mitigate such loan defaults, it is crucial to have an early warning system to warn banks of potential loan defaults, allowing them to act before the default occurs. An early warning system is a set of capabilities designed to generate and disperse timely and meaningful warning information to enable individuals and organizations to prepare and act appropriately and immediately to reduce harm or loss [12].

In finance, such a system can help financial institutions receive alerts if it is determined that a customer might default on a loan. The institutions can use these alerts to take action to mitigate any potential loan default. Such actions may include offering customers lower interest rates, different payment plans, and something as simple as checking in with the customer to determine if anything can be done to mitigate potential delinquency.

The objective of this paper is to evaluate existing anomaly detection methods to determine if they are suitable to predict loan defaults on unbalanced datasets over existing classification methods. Since loan defaults appear in low probability in the datasets, the resulting unbalanced dataset present challenges when using predictive methods, resulting in low accuracy on default prediction. We propose an early warning system architecture to detect loan defaults by first identifying an optimal anomaly detection method for the particular dataset under study. We test the system on two different datasets using an adaptation of each of logistic regression, stochastic gradient descent, isolation forest and angle-based outlier detection. The hypothesis is that the anomaly detection system can select the optimal classification or anomaly detection method to produce a more accurate prediction on unbalanced datasets. The resulting evaluation shows that the proposed anomaly detection system is more helpful in building early warning systems against potential loan defaults over supervised classification methods.

This paper first discusses the related work in loan default prediction and early warning systems and their limitations. We then discuss the reasoning for the proposed work and our approaches. We follow with a discussion on the implementation of this proposal and the results. Finally, we present our conclusion and future work.

2 Related Work

Existing literature for loan default prediction is extensive and primarily focused on balanced datasets. Most research focuses on resampling the training data to balance it and then using the resampled balanced data to train supervised classification methods. Research on anomaly detection and early warning systems on large unbalanced datasets is less extensive. In this section, we shall review related research contributions and explain their limitations.

In 2018, Qiu, Tu et al. [14] built an early warning system using anomaly detection to detect problems in users' power consumption patterns. The paper provides an application for anomaly detection in the context of early warning systems. However, the researchers only evaluate computation time and the area under the ROC curve to evaluate the metrics for only the Anomaly Detection Algorithm Based on Log Analy-SIS (ADLA) method. This paper also does not focus on large datasets or loan defaults.

Mukherjee and Badr [11] in 2022 compare four unsupervised anomaly detection methods on a large and realistic P2P loan dataset, overcoming some limitations of the previous research. They use precision and recall as evaluation metrics. This paper is one of the few to use anomaly detection in the context of loan risk evaluation. However, the paper does not evaluate more evaluation metrics, such as accuracy, the area under the ROC curve, and running time. It only evaluates the approaches on one dataset and, therefore, only on one type of loan (P2P loan). It also evaluates it in the context of binary classification alone.

Rao, Liu, et al. [15], in the same year, implement a novel approach by proposing a PSO-XGBoost model to predict loan defaults. Unlike the previous research, this paper compare existing methods using multiple metrics, including execution time. However, they only evaluate the method against one type of loan (automobile loans), compare only with three other methods, and resample the training data to balance the number of defaulters instead of using the data as is.

In 2023, Zhu, Ding, et al. [18] improved over the PSO-XGBoost model. They proposed a novel state-of-the-art approach using CNNs for feature selection and LightGBM for prediction and demonstrated higher prediction performance. However, the evaluation is performed only on one dataset and for one loan class. Also, like the previous research, the dataset is resampled to avoid imbalance instead of directly using it. The results are also binary classes and not probabilistic.

The above two papers resampled the data and converted the training data to a balanced dataset from an unbalanced dataset. Song, Wang, et al. [17] overcame this limitation by developing a novel rating-specific and multi-objective ensemble method to classify imbalanced class distribution in the case of predicting loan defaults. The methodology also focuses on maximizing sensitivity to correctly classify the minority class, i.e., customers who default on loans. However, the method does not compare execution times, only evaluates one dataset type (P2P loans) and only deals with binary classification.

From the review on existing related work, we discover that the idea of loan default prediction and credit risk monitoring is well-researched. We also observe different results demonstrated on different datasets. In addition, we observe that most approaches view loan default prediction as a classification problem with two equal classes, and the goal is to predict one class over the other. However, in designing an early warning system, because most customers in real-time data do not default, there is a need to view loan defaults as an anomaly detection problem rather than a classification problem.

The literature review [11,14,15,17,18] shows that the current research needs specific crucial contexts.

1. Most current research only tests methods against one type of loan dataset. Banks offer various kinds of loans, and testing against multiple types of loans is essential to evaluate a method holistically.

2. There needs to be more focus on the speed and performance of the algorithm. According to the CIBC Annual Report 2022 [3], CIBC has around 13 million clients; according to RBC Annual Report 2022 [5], they have 17 million. Banks usually have many clients and hence a large amount of data. Therefore, any predictive system must be fast and scalable.
3. Current research focuses on predicting loan defaults but needs to focus specifically on the probabilistic nature of such predictions. An early warning system should have the flexibility to adjust the probabilistic threshold of the predictions depending on the financial institution's risk appetite. Some institutions may be more risk-averse and want alerts even if the possibility of default is less, and vice versa. An excellent early warning system should allow for such flexibility.
4. Most current loan default prediction methods use a classification algorithm to predict defaults. However, real-world data shows that most loan customers do not default. A loan default can thus be treated as a deviation from the norm or an anomaly and not a class of a customer type. Anomaly detection methods can use real-world data directly instead of processing data to balance the number of default and non-default customers, saving much preprocessing and retraining time.
5. It is necessary to compare different approaches and different kinds of approaches to solve a problem. In most papers in the context of loan default prediction, comparisons are made with only a few approaches and a few evaluation metrics. The evaluation needs to be more comprehensive.

Considering the above factors, this paper aims to evaluate different approaches in different contexts and suggests an early warning system architecture to warn against potential loan defaults using anomaly detection. For this purpose, we selected four of the most common classification and anomaly detection methods described in Sect. 4.

3 Proposed Work

3.1 Reasoning

An excellent early warning system against potential loan defaults should have the following characteristics:

- **Versatility:** Banks offer various kinds of loans. An early warning system should work the same for any loan default prediction.
- **Fast and scalable:** Banks deal with a large amount of customer data. An early warning system should be fast enough to handle large data and offer scalability to accommodate rapid data volume changes.
- **Function in real-time:** For every change or refresh in the data, the early warning system should update its predictions and be able to notify if a customer might default.

- **Probabilistic:** Every financial company has a unique risk appetite. The early warning system should function probabilistically based on the risk threshold of a financial institution rather than providing fixed predictions.
- **Work on imbalanced datasets:** Most people do not default on their loans, and default data would thus be imbalanced. The system should work on these data.

Consequently, this paper aims to determine a machine learning approach that is fast and scalable to classify if a customer might default. This is proposed by using classification and anomaly detection methods to decide the best approach on different loan default datasets. The ideal system would have high accuracy, low prediction times, a high true positive rate (the number of correct loan default predictions) and a low false negative rate (the number of incorrect loan non-default predictions). The proposed system architecture aims to detect and report warning signs and not actual defaults. Therefore, it is imperative to note that the prediction threshold for such a classification method would be lower than for a method that would predict certain defaults.

3.2 Proposed System Architecture

Table 1. Evaluation metrics used in the evaluation

Metric	Description
Accuracy	Describes how often the model predicted a correct outcome in general; the fraction of correct outcomes among all outcomes
Precision	The percentage of relevant results
Recall	The percentage of relevant results that were predicted correctly
F1 Score	The harmonic mean of precision and recall
ROC AUC	The area under the receiver operating characteristic curve; provides an aggregate measure of performance across all classification thresholds.
Execution time	The processing time (in ms or s) taken for a computer process to finish from start to end

The initial step is to evaluate each of the selected methods using Algorithm 1. This algorithm is suitable for evaluating any classification or anomaly detection method that we might like to test. In Fig. 1 we describe the proposed early warning system architecture. The optimal algorithm obtained from Algorithm 1 would be used in this architecture. The metrics [4] for the evaluation search to obtain the most optimal approach are shown in Table 1.

Algorithm 1. Proposed experimental setup for evaluation

```
From a set of loan datasets D, and a set of anomaly detection approaches A' for a
  set of risk threshold values T
Generate a clean set of data D' by preprocessing and feature selection
for each dataset in D' do
    for each risk threshold value in T do
        Execute each algorithm in A'
        For each algorithm in A', note down the evaluation metrics
    end for
end for
Analyze the evaluation metrics
Obtain the best algorithm A from A' based on evaluation metrics
```

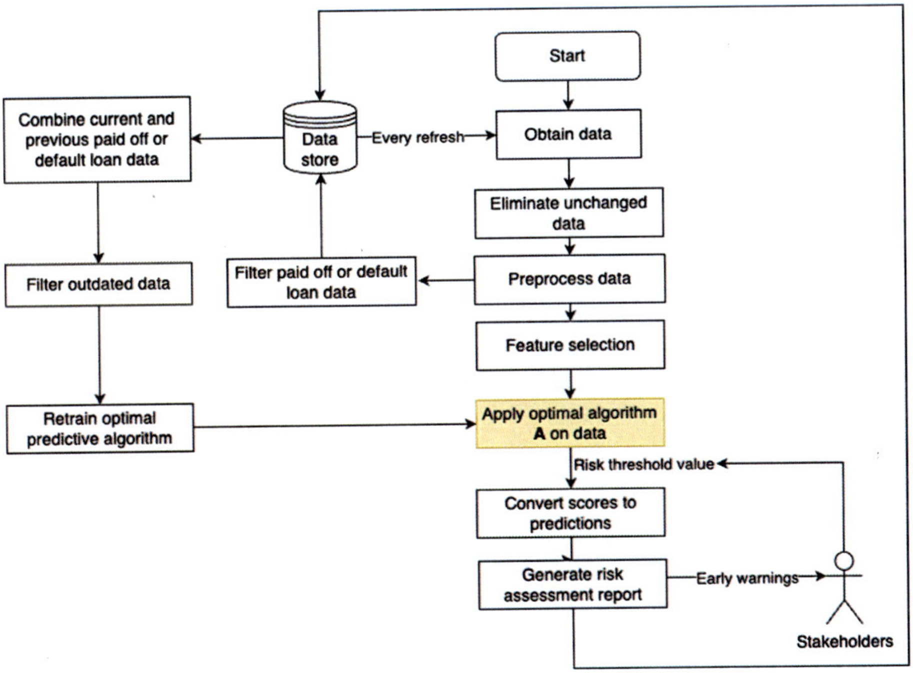

Fig. 1. Proposed early warning system architecture

4 Implementation and Results

The proposed work was implemented on Python 3.9.7 on Windows 11. For the classification algorithms, dataset split, hyperparameter tuning, and calculating metrics, the library scikit-learn v1.2.2 was used. For the anomaly detection algorithms, the library PyOD v1.0.2 was used. The libraries NumPy v1.22.3 and Pandas v1.4.3 were used for data processing.

4.1 Datasets

We use the balanced Bondora Peer-to-Peer Lending Dataset [16] and the unbalanced L&T Vehicle Loan Default Dataset [6] to compare classification and anomaly detection methods and their performance concerning balanced datasets.

The Bondora Peer-to-Peer Lending Dataset is from a reputable source and is mostly preprocessed. The dataset is also balanced, with 59% loan defaulters and the rest non-defaulters. The dataset has 48 attributes and 77,394 records.

The L&T Vehicle Loan Default Dataset is unbalanced and has around 22% loan defaulters and the rest non-defaulters. It provides a perspective on automobile loans and secured loans. It has 41 attributes and 233,154 records. It is well-cited and used in many research papers, including the one by Rao, Liu, et al. [15] discussed in the literature review.

4.2 Preprocessing

We performed the following preprocessing steps on the Bondora Peer-to-Peer Lending Dataset: We removed irrelevant columns (language, country, county, and city) since these attributes do not impact the loan default ability of customers. We converted all the binary columns to ones and zeroes from True and False. We converted all the date columns to the number of months from the date until the present date. We encoded the ordinal classes to scores (verification, education, rating, employment type, employment duration, and home ownership status). We encoded the nominal classes using one-hot encoding. We removed columns containing too many blanks. Any blank last payments were imputed with their first payment information. The reasoning is that if no previous payments were made except for the first payment, the first payment itself would be the latest. The remaining rows containing any blanks were removed (less than 10% of data).

After following the above steps, the resultant Bondora data had 85 attributes and 70,512 records left. The preprocessing steps for the L&T Vehicle Loan Default Dataset were applied as per Rao, Liu, et al. [15]. The resultant L&T Vehicle Loan data had 46 attributes and 120,165 records left.

4.3 Experimentation

Four approaches were selected to perform the experiments. Two of them were classification, and two were anomaly detection approaches. The classification approaches were logistic regression (LR) [10] and stochastic gradient descent (SGD) [7]. The anomaly detection approaches were isolation forest (IF) [9] and angle-based outlier detection (ABOD) [8].

Dataset Split. The loan default variable was considered the response variable and the remaining predictors for both datasets. The dataset was split into 80% training and 20% test data. The predictor variables were scaled so that the mean of each predictor was removed, and the variance was 1.

Hyperparameter Tuning. For both datasets and all approaches, hyperparameter tuning was performed by training a standard version of the algorithm and performing 5-fold randomized cross-validation on the split dataset over different hyperparameter values. Table 2 shows the best parameters obtained after the hyperparameter tuning processes for each dataset and each approach.

Table 2. Optimal hyperparameters obtained after tuning

	Bondora P2P Lending Dataset	L&T Vehicle Loan Default Dataset
LR	Penalty = '12', C = 100	Penalty = '12', C = 10
SGD	Penalty = 'elasticnet', Loss = 'log_loss', Learning Rate: 'optimal', alpha = 0.01	Penalty = l1, Loss = 'modified_huber', Learning Rate: 'optimal', alpha = 1
IF	Number of estimators = 1000	Number of estimators = 100
ABOD	Number of neighbors = 2	Number of neighbors = 5

Calculating Metrics. We performed another randomized train-test split of 80% training data and 20% test data for each dataset and each approach. We then used the training data to train a model using the optimal hyperparameters. We then performed predictions on the test dataset and compared these predictions with the actual response variable in the test dataset. Based on this comparison, we computed the metrics accuracy, precision, recall, F1 score, the area under the ROC curve, training time and prediction time. We performed this process ten times for each dataset on each approach, and the mean of each metric was noted. Table 3 and Table 4 show the mean values of each metric obtained during experimentation for each dataset. Table 5 and Table 6 show the same values for 30% and 70% prediction probability thresholds.

4.4 Results

Table 3. Results for Bondora P2P Dataset (balanced)

	Accuracy	Precision	Recall	F1	ROC AUC	Training time	Prediction time
LR	**0.9988**	**1.000**	**0.9981**	**0.9991**	**0.9704**	**18.97 s**	**0.01 s**
SGD	0.9709	0.9823	0.9720	0.9771	0.9950	0.23 s	0.01 s
IF	0.4438	0.5848	0.4554	0.5120	0.4990	42.09 s	12.08 s
ABOD	0.3592	0.0000	0.0000	0.0000	0.5000	5.98 s	126.23 s

Table 4. Results for L&T Auto Loan Dataset (unbalanced)

	Accuracy	Precision	Recall	F1	ROC AUC	Training time	Prediction time
LR	0.7671	0.2000	0.0005	0.0011	0.5000	0.84 s	0.01 s
SGD	0.7675	0.0000	0.0000	0.0000	0.5000	0.23 s	0.01 s
IF	**0.6613**	**0.2289**	**0.1929**	**0.2094**	**0.5354**	**5.11 s**	**1.38 s**
ABOD	0.6522	0.2092	0.1784	0.1926	0.4871	17.32 s	128.64 s

Table 3 shows the results of our experiments for the balanced Bondora P2P Lending Dataset, and Table 4 for the unbalanced L&T Vehicle Loan Default Dataset. The unbalanced dataset is more reflective of the real world.

From Table 3, it is evident that supervised classification methods performed significantly well on balanced datasets. Lgistic regression gave almost perfect accuracy in predicting loan defaults. Stochastic gradient descent was a close second. However, both anomaly detection methods performed poorly on this balanced dataset. Table 4 shows that the anomaly detection performed better than supervised classification. The isolation forest approach performed the best, followed by ABOD. Although the accuracy of both classification methods was higher by about 10–11%, other metrics like precision, recall, and F1 score were poor, highlighting that these methods failed to predict the less populated response variable, and therefore has not satisfactorily solved the problem in question. Table 5 and Table 6 reflect these findings for 30% and 70% threshold values.

Table 5. Results with various thresholds for Bondora P2P Dataset (balanced)

	Accuracy		Precision		Recall		F1		ROC AUC	
	30%	70%	30%	70%	30%	70%	30%	70%	30%	70%
LR	**0.9989**	**0.9987**	**0.9999**	**1.0000**	**0.9983**	**0.9979**	**0.9991**	**0.9989**	**0.9991**	**0.9989**
SGD	0.9177	0.9363	0.8922	0.9985	0.9914	0.9020	0.9392	0.9478	0.8889	0.9498
IF	0.4203	0.3670	0.5734	0.6305	0.3724	0.0142	0.4515	0.0277	0.4391	0.4997
ABOD	0.3592	0.3592	0.0000	0.0000	0.0000	0.0000	0.0000	0.0000	0.5000	0.5000

Table 6. Results with various thresholds for L&T Auto Loan Dataset (unbalanced)

	Accuracy		Precision		Recall		F1		ROC AUC	
	30%	70%	30%	70%	30%	70%	30%	70%	30%	70%
LR	0.7247	0.7675	0.3299	0.0000	0.1786	0.0000	0.2317	0.0000	0.5344	0.5000
SGD	0.7675	0.7675	0.0000	0.0000	0.0000	0.0000	0.0000	0.0000	0.5000	0.5000
IF	**0.6725**	**0.7589**	**0.2196**	**0.1672**	**0.1600**	**0.0093**	**0.1851**	**0.0018**	**0.4939**	**0.4976**
ABOD	0.2326	0.2326	0.2325	0.2325	0.9996	0.9996	0.3772	0.3772	0.4999	0.4999

5 Conclusion and Future Work

In this paper, we introduced an experimental setup for evaluation and an anomaly detection system architecture against loan defaults. Using this setup, we determined the best method for two datasets. The experiments show that anomaly detection methods performed better on the unbalanced dataset against classification methods typically used to solve problems like predicting loan defaults. Such datasets are more likely to be a true representative of real-world loan defaults.

By evaluating two diverse loan datasets, emphasizing speed, performance, and the probabilistic nature of predictions, and using a wide range of evaluation metrics, we conclude that anomaly detection methods are useful for building an optimal early warning system against potential loan defaults. These findings ultimately benefit financial institutions and contribute to future development in risk management systems.

In the future, we shall employ anomaly detection methods and compare them against more datasets representing different types of loans to determine the most optimal approach to solve this problem with different prediction probability thresholds.

References

1. 2023 U.S. Lev Loan Default Forecast Raised to 2.0%–3.0%; 2024 Projected at 3.0%–4.0%. https://www.fitchratings.com/site/pr/10213716
2. Canada's biggest banks set aside $2.5 billion to cover an expected wave of loan defaults. https://www.thestar.com/business/2023/03/03/canadas-big-six-banks-set-aside-25-billion-as-they-prepare-for-credit-losses.html
3. CIBC - Annual Report 2022. https://www.cibc.com/content/dam/cibc-public-assets/about-cibc/investor-relations/pdfs/quarterly-results/2022/ar-22-en.pdf
4. Metrics to Evaluate your Machine Learning Algorithm. https://towardsdatascience.com/metrics-to-evaluate-your-machine-learning-algorithm-f10ba6e38234
5. Royal Bank of Canada - Annual Report 2022. https://www.rbc.com/investor-relations/_assets-custom/pdf/ar_2022_e.pdf
6. Dhaker, M.: L&T Vehicle Loan Default Prediction Data. Kaggle (2019). https://www.kaggle.com/datasets/mamtadhaker/lt-vehicle-loan-default-prediction
7. Kiefer, J., Wolfowitz, J.: Stochastic estimation of the maximum of a regression function. Ann. Math. Stat. **23**(3), 462–466 (1952)
8. Kriegel, H.P., Schubert, M., Zimek, A.: Angle-based outlier detection in high-dimensional data. In: 14th ACM SIGKDD International Conference on Knowledge Discovery and Data Mining, pp. 444–452. ACM (2008). https://doi.org/10.1145/1401890.1401946
9. Liu, F.T., Ting, K.M., Zhou, Z.H.: Isolation forest. In: 2008 Eighth IEEE International Conference on Data Mining, pp. 413–422 (2008). https://doi.org/10.1109/ICDM.2008.17
10. McCullagh, P., Nelder, J.A.: Generalized Linear Models. Chapman and Hall (1983)

11. Mukherjee, P., Badr, Y.: Detection of defaulters in P2P lending platforms using unsupervised learning. In: 2022 IEEE International Conference on Omni-layer Intelligent Systems (COINS), pp. 1–5 (2022). https://doi.org/10.1109/COINS54846.2022.9854964
12. Mulero Chaves, J., De Cola, T.: Public warning applications: requirements and examples. In: Câmara, D., Nikaein, N. (eds.) Wireless Public Safety Networks 3, pp. 1–18. Elsevier (2017). https://doi.org/10.1016/B978-1-78548-053-9.50001-9
13. Nigmonov, A., Shams, S.: COVID-19 pandemic risk and probability of loan default: evidence from marketplace lending market. Financ. Innov. **7**(1), 1–28 (2021). https://doi.org/10.1186/s40854-021-00300-x
14. Qiu, H., Tu, Y., Zhang, Y.: Anomaly detection for power consumption patterns in electricity early warning system. In: 2018 Tenth International Conference on Advanced Computational Intelligence (ICACI), pp. 867–873 (2018). https://doi.org/10.1109/ICACI.2018.8377577
15. Rao, C., Liu, Y., Goh, M.: Credit risk assessment mechanism of personal auto loan based on PSO-XGBoost model. In: Complex & Intelligent Systems. Springer Science and Business Media LLC (2022). https://doi.org/10.1007/s40747-022-00854-y
16. Siddhartha, M.: Bondora peer-to-peer lending data. IEEE Dataport (2020). https://doi.org/10.21227/33kz-0s65
17. Song, Y., Wang, Y., Ye, X., Zaretzki, R., Liu, C.: Loan default prediction using a credit rating-specific and multi-objective ensemble learning scheme. Inf. Sci. **629**, 599–617 (2023). https://doi.org/10.1016/j.ins.2023.02.014
18. Zhu, Q., Ding, W., Xiang, M., Hu, M., Zhang, N.: Loan default prediction based on convolutional neural network and LightGBM. In: International Journal of Data Warehousing and Mining (IJDWM), vol. 19, pp. 1–16. IGI Global (2023)

Enabling Distributed Inference of Large Neural Networks on Resource Constrained Edge Devices using Ad Hoc Networks

Torsten Ohlenforst(✉), Moritz Schreiber(✉), Felix Kreyß, and Manuel Schrauth

Communication Systems, Fraunhofer Institute for Integrated Circuits (IIS), Am Wolfsmantel 33, 91058 Erlangen, Germany
{torsten.ohlenforst,moritz.schreiber}@iis.fraunhofer.de
https://iis.fraunhofer.de/en/dsai

Abstract. Processing neural network inferences on edge devices, such as smartphones, IoT devices and smart sensors can provide substantial advantages compared to traditional cloud-based computation. These include technical aspects such as low latency or high data throughput – but also data sovereignty can be a concern in many applications. Even though general approaches of distributed inference have been developed recently, a transfer of these principles to the edge of the network is still missing. In an extreme edge setting, computations typically are severely constraint by available energy resources and communication limitations among individual devices, which makes the distribution and execution of large-scale neural networks particularly challenging. Moreover, since mobile devices are volatile, existing static networks are unsuited and instead a highly dynamic network architecture needs to be orchestrated and managed. We present a novel, multi-stage concept which tackles all associated tasks in one framework. Specifically, distributed inference approaches are complemented with necessary resource management and network orchestration in order to enable distributed inference in the field – hence paving the way towards a broad number of possible applications, such as autonomous driving, traffic optimization, medical applications, agriculture, and industry 4.0.

Keywords: Distributed Inference · Edge Computing · MANET

1 Introduction

Given the steadily increasing impact of artificial intelligence (AI) and machine learning (ML) in many fields of science and technology, it can be expected that these methods will also be applied to volatile networks and Internet of Things (IoT) devices [2] in the near future. Highly dynamical and heterogeneous local

S. Ossowski et al. (Eds.): DCAI 2023, LNNS 740, pp. 145–154, 2023.
https://doi.org/10.1007/978-3-031-38333-5_15

area networks are required to provide low latency, high reliability and extremely low energy consumption simultaneously [8]. In order to make this possible, one trend in modern innovative communication concepts is to steadily shift cloud computing approaches towards the (extreme) edge of the network [3,12] and to execute AI applications as locally as possible in the user environment [10].

Common approaches to process inference of pre-trained neural network (NN) models on extreme edge IoT nodes are either based on usage of existing single devices, which limits the model size, leading to limitations in performance [14], or additional deployment of specialized ML hardware, resulting in high deployment costs. In use cases with strict latency requirements, privacy concerns, lack of stable connectivity or limited available data rate, both approaches can become infeasible. It therefore can prove beneficial to conceptually *combine* ad hoc network processes and AI/ML methods already from the start. This way, an additional hardware deployment step can be avoided. Moreover, this approach allows to distribute an inference task in a way that individual workloads are adapted to the specific capabilities of the edge devices available. For example, as permanent storage access operations tend to be particularly expensive in terms of energy consumption, one would prefer a *local* inference task not to exceed the available RAM or CPU cache on its designated device.

Ad hoc networks have the potential to greatly facilitate distributed AI models, as devices within the network are able to communicate without the need of a centralized infrastructure. This becomes particularly important in real-world scenarios where an adequate cell coverage is not ensured. In the absence of suitable network technologies, we present a solution called *Routed ad hoc Networks for dynamic and low power Environments (randle)*. The ad hoc network paradigm allows devices to form a self-organized, autonomous network, which is capable of performing complex tasks. One prime example are distributed inferences of large NN models, which will be of particular interest in this article.

In general, there are several ways to distribute a machine learning model across multiple mobile nodes. One way is to split the NNs layer-wise and distribute the parts among available nodes. During an inference, the input data is fed into the first part of the model and its output is passed to the node carrying the subsequent part of the model. It is worth mentioning that in this approach the ML model does not suffer from any accuracy loss, since no pruning [9] or other manipulation steps which go beyond the actual splitting have been performed. In addition, it has been shown that single device inference can be outperformed in terms of data throughput and energy consumption [9,13].

Another way to run a ML model on multiple nodes is to create multiple smaller models that can only work on a subset of the classes of the original, full model [5]. The particular way in which the smaller models are downsized is known as class-dependent pruning. Obviously, given an original NN with n classes, it can be distributed among up to n nodes, (ideally) providing a speed up factor of up to n times when executed fully parallel. Hence the overall inference latency is drastically reduced compared to the sequential splitting approach [4].

Having discussed recent approaches to split NN inference among distributed compute nodes, we emphasize that in this work we expand the scope of the problem significantly in that we also consider the networking of the devices, the management of the resources, the automated splitting and distribution, and the subsequent inference. Hence, in summary, we are drawing the concept for the construction of an overall system framework incorporating all these individual tasks. We even move away from traditional, static settings (in terms of wired devices) and instead consider volatile mobile edge nodes with highly restricted resources. In particular, we present a novel multi-stage approach of how inference of large ML models on a network of mobile, volatile and heavily resource constraint edge nodes can be realized without requiring a centralized network infrastructure, using only device-to-device (D2D) communication. This approach will be discussed in detail in the following.

2 A Concept to Enable Inference in the Field

We present a three-stage Network Management and Orchestration (NMO) method that enables spatially close IoT nodes to process a service request such as, for instance, a NN inference, locally. Specific to our approach is that we consider special requirements for mobile, volatile and even heterogeneous IoT nodes equally important as the network management system itself.

2.1 Ad Hoc Cluster Initialization with randle

In the first stage, a computing cluster is initialized based on the pool of available nodes. For this purpose, the nodes connect to each other by sharing routing information as soon as they are able to successfully establish a radio connection. This is a mandatory background process that is carried out continuously, independent of the actual computation task. To enable this network connection, a robust but also lightweight Mobile ad hoc network (MANET) [6] is required as the nodes run on restricted resources. We use an ad hoc network concept which is adapted to several special boundary conditions that cannot be adequately covered by existing technology. Reliability is a very important factor. Due to the constant addition and removal of subscribers, a solution for consistent routing information as well as responsive adjustments is required. Equally important is a simple and robust implementation that efficiently uses the limited energy and processing power of the microcontrollers to ensure a long run time.

randle is based on the idea of simple and lightweight advertisement packets flooding over the entire network. These are called Lookup Table Advertisements (LUT-ADs) and are sent periodically in the background by each node that requests to become part of the network. The MANET concept is based on table driven broadcast routing and can be seen as a lightweight version of Wireless Routing Protocol since link distances and costs are unified in randle. Disseminating routing information about themselves solves several problems with dynamically changing participants and their position in the ad hoc network [1,11].

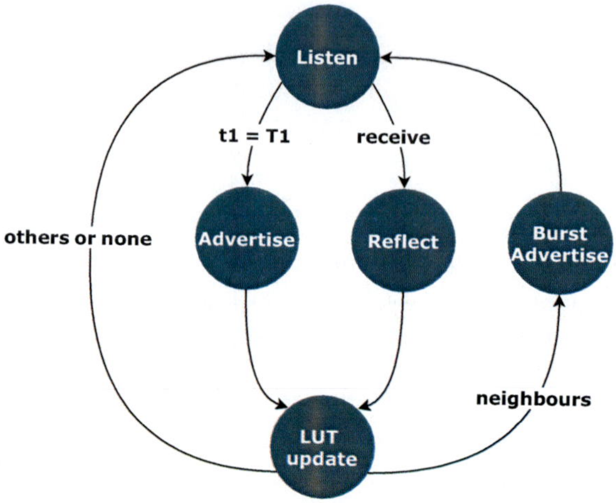

Fig. 1. *randle* state machine (simplified)

Node State Machine. Every connected node is running identical background processes for distributing routing information across the ad hoc network. Figure 1 shows the simplified state machine implemented on *randle* devices. Starting in the default state *Listen*, the node is ready for receiving LUT-ADs from others. If a LUT-AD arrives, it is reflected to forward this packet to all connected nodes. This step includes the detection of cyclic and redundant information to keep the channel load low which is not further discussed here. Then, the received LUT-AD is parsed and merged with the existing routing table. In any case, the *last-seen* value is updated and if there are any changes concerning direct neighbor routing, a *burst-advertise* sequence is initiated. This means that the node does not only send one LUT-AD as a routing update but multiple ones in short intervals in order to maximize the chances every node in range is able to gather this changed information.

In multiple predefined timing intervals, each node is providing updated routing information by transmitting LUT-ADdc dfdfds. These intervals differ in order to maximize energy and channel efficiency. Thus, without any changes, only so-called *alive-messages* are sent. This ensures that even in static networks there is permanent connectivity through updated routing tables.

Routing Table. Network members should be able to exchange data with any other node, even if they are not reachable via direct link. Therefore it is necessary to route packets over multiple nodes which are able to establish a connection to the next hop. For storing this routing information, every node is building its own routing table, filled by information received via LUT-ADs. This results in each node knowing all other network participants and the associated next hop for the best route to reach them. In fact, an important design goal for *randle* is the simple and lightweight distribution of routing information. Since we rely on low energy mobile communication standards like Bluetooth 5, data rates are very limited and collisions handling is not guaranteed [1].

Creating and Joining the Network. As described by Fig. 1, any Node wanting to establish a *randle* connection starts sending LUT-ADs and listens for foreign LUT-ADs on the same radio channel. Nodes receiving these will update their routing table which enables them to send data to this new network members. In the same domain the new routing information is distributed through sending an own LUT-AD update. This way listening nodes are able to update their routing table respectively. By repeating this process, a full network can be established.

Figure 2 explains the process of joining *randle* in four steps. At the beginning, the pre-existing network contains three nodes with balanced routing table. Each nodes LUT contains routing information to all other participants with three parameters. *Destination* node is reachable over *Next Hop* with calculated *Weight* as priority key. In this example *node 1* is able to send packets to *Node 2* over different routes, but the direct one is preferred since the *Weight* of this route is higher (0.8 instead of 0.54). These priority factor is defined by multiplication of weights per sub-route through LUT-ADs. All routed weights are defined between zero and one, higher values are better.

In the second step, a new *Node 4* (brown) moves within radio range of the existing *randle* with an empty LUT. Therefore, this node is receiving and sending LUT-ADs from and to *Node 3* which leads to an party filled LUT for the existing network participants. Like shown in the example Fig. 2, only *Node 3* is able to communicate with *Node 4* since *Node 1* and *2* are not aware of the existence of the new *randle* member.

Finally in the forth step, *Node 3* sends its updates routing information to all other connected nodes via LUT-ADs. Consequently these nodes can update their LUT. In the end, all *randle* participants are able to communicate with each other and the network state is balanced until the next event is occurring.

Leaving the Network. Nodes that leave *randle* will be signed out of the respective LUTs by the absence of the periodically sent LUT-ADs from the missing node. This leads to a changed routing table, which is made known to the other network members via corresponding LUT-ADs. Due the simple and universal algorithm described earlier, the leaving node detects the missing network connection to its former network members running identical processes. A single node without entries in its LUT can enable power saving states in multiple ways. It is possible to extend the advertising interval. Disabling advertising in general is potential not preferred as this could lead to very long recognition rates and even to a collapse of the network, if multiple nodes are not sending LUT-ADs anymore.

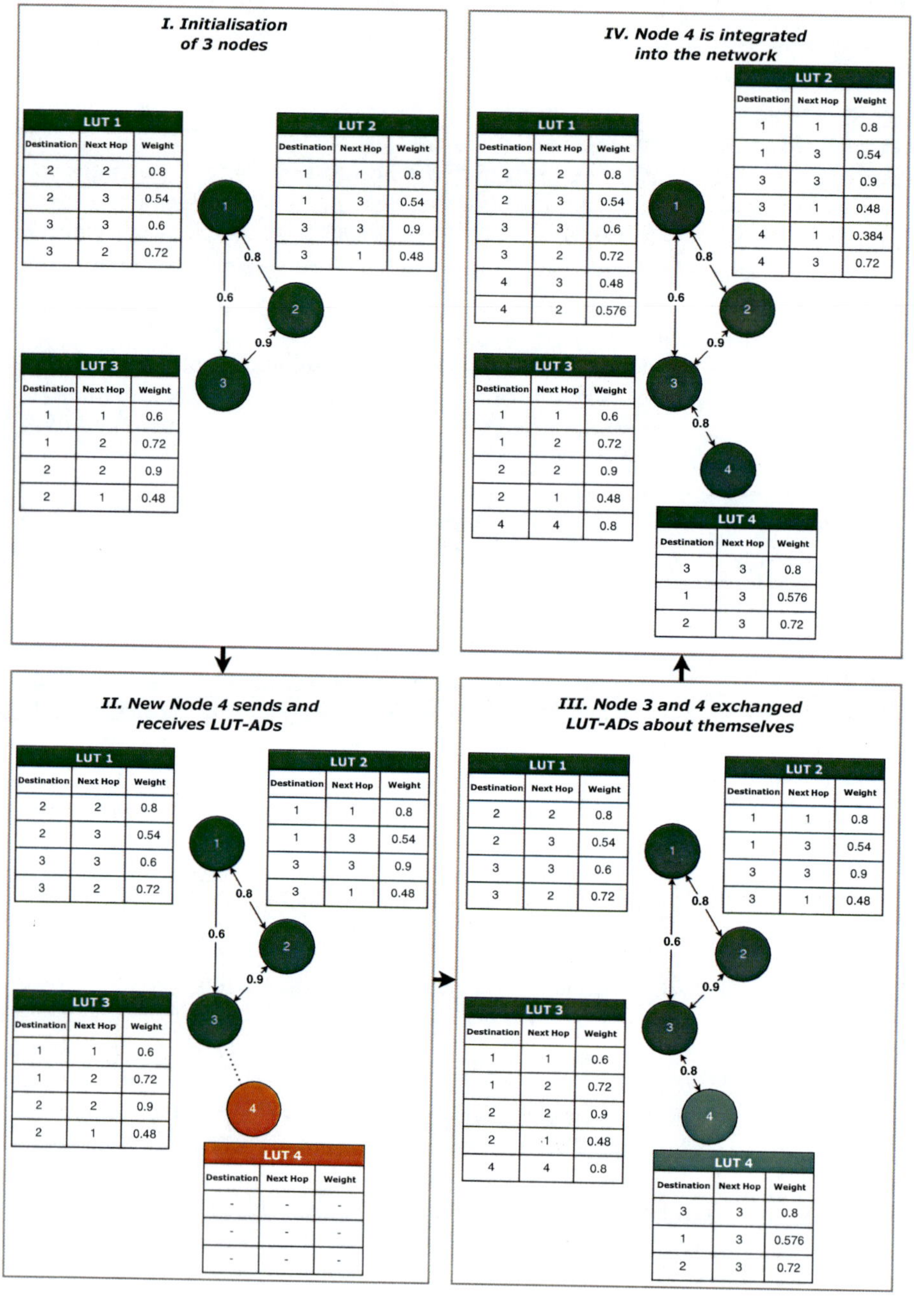

Fig. 2. Example of adding a new node to an existing *randle* (simplified)

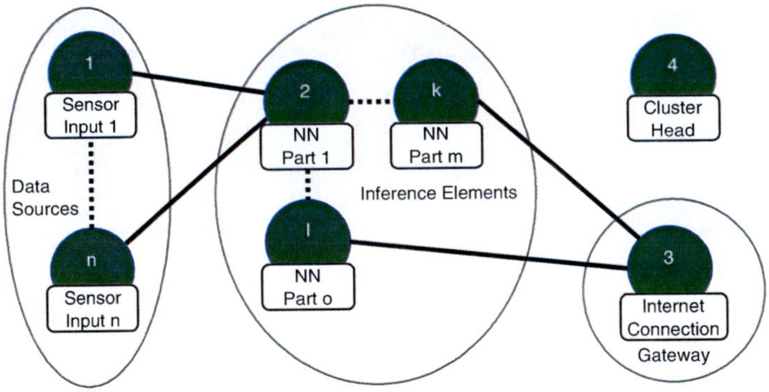

Fig. 3. System design for distributed inference on resource restricted mobile edge nodes

2.2 Service Tailored Network Orchestration

In the second stage, the network is orchestrated on demand to process requested AI inferences which exceed the computing capabilities of a single node. In order to execute AI inferences effectively, we form a processing cluster from all available nodes which is specifically tailored to match the required optimization goal, as schematically depicted in Fig. 3. Given a use case where latency optimization is to be achieved, we resort to a parallel partitioning of the NN, as outlined in References [4] and [14], whereas in case a high data throughput is required, a sequential pipeline architecture (compare, e.g., Ref. [9]) is preferred.

Designation of a Cluster Head. The management and orchestration of the network to execute the requested AI inference is performed by a cluster head. Furthermore, the cluster head must ensure that nodes that leave the network are replaced and no data loss occurs. By choosing a cluster head, the previously equal network hierarchy is broken up and divided into two levels, server and client. The cluster head fulfills the role of a server and is jointly selected by a voting of the nodes based on an almanac that consists of performance parameters such as the number of connections to other nodes, processing power and remaining battery level. An exchange of these parameters is triggered once an AI inference is requested. Each node broadcasts its own performance parameters and creates a local almanac containing the received performance parameters of the neighboring nodes. Once this process is complete, the nodes vote to elect a cluster head.

Allocation of Different Functional Roles. The cluster head determines which functional roles are required to execute the requested AI inference and to transmit the result to its destination. It assigns these to suitable nodes in the network based on the entries listed in the almanac. The following roles are at the cluster head's disposal:

- **inference element (#1, #2, ... ,#n):** The inference elements are nodes that deploy a part of the NN and perform the inference. The destination of the processed data is communicated to the node by the cluster head.
- **data source:** The data source usually provides sensor data that should be processed in raw form. It is possible to preprocess acquired sensor data, to reduce the amount of data to be transmitted via Device-to-Device (D2D) communication. The request for an AI inference can, for example, come from a sensor node that has collected data that is now available for evaluation. Furthermore it is possible to have several nodes as data sources if the NN has the necessary inputs.
- **gateway:** The gateway transmits the results of the AI inference to the target location e.g. via terrestrial or satellite networks. The background for this role is that the node requesting the inference is not always the one who needs the result. A classic case for this would be when IoT sensor nodes make the request and the results are to be collected centrally.
- **not in use:** Nodes which are not used.

Routing. In this process step, the cluster head identifies available network routing options and selects a suitable one. For this purpose, it informs the respective nodes which D2D communication paths (TX and RX) are relevant. Single-hop methods are preferred for calculating the most efficient communication possible, but multi-hop methods can also be used depending on the given framework conditions.

2.3 Service Execution

In the third stage we execute the requested AI inference service using the distributed architecture we initialized in the previous stages. The inference is executed on the network architecture created in the previous step. The built up setup can be used for inferences as long as required and as long as the nodes stay connected. The distributed NN on the edge devices produces the same result as the original NN in the data center, but with some advantages. Depending on the optimization it benefits in terms of latency, data throughput or power consumption. The local processing of the data allows to maintain data sovereignty and also does not require an internet connection.

If processing is successful and the results are transmitted, the original service request is answered positively. If an error occurred during processing or transmission of the results and cause the service to be aborted, or if the required architecture could not be built with the available resources, the service request is answered negatively.

3 Conclusion and Outlook

In this research article, we lay the foundation of how complex computations, such as large-scale AI inference tasks can be accomplished *in the field*, using the capabilities of modern distributed computing. At the heart of our concept, a three-stage NMO architecture is designed to handle a network of mobile, volatile, resource constraint, heterogeneous edge nodes and to process individual task according to specific limitations of available devices in the network. Our robust ad hoc network concept is able to establish and maintain connects between heterogeneous hardware without any preexisting cell communication in low power environments. Given this foundation we group the participating nodes into suitable clusters to deliver optimal performance and reliability judged by multiple criteria. Execution data and instructions are transferred to the appropriate nodes. Finally, the execution of the distributed inference is coordinated and ensured.

Since common techniques of AI size reduction are not applied here, our concept does not result in any loss of predictive accuracy. Furthermore, we enable the possibility of additional data transfer with the developed ad hoc network concept *randle*. This feature can also be used to greatly improve the accuracy and reliability of the AI with significantly larger input vectors.

Regarding an actual implementation of our concept we like to mention that connecting many moving mobile nodes in an ad hoc network can be a complex and challenging task. One of the most significant challenges is dealing with sudden disconnections and changes in routing information that can occur as nodes move around, leading to inconsistent network situations. This inconsistency of network and cluster members also complicates the decision of assigning sub-tasks to nodes. Therefore, not only the choice of the most reliable cluster head is of great importance, but also the redundant distribution of data and execution. Another critical issue is the energy consumption of the nodes in the network. Mobile nodes are typically battery-powered, so energy consumption is a considerable issue. An efficient algorithm must minimize the energy consumption of the nodes while maintaining a reliable connection.

Our future goal is to bring the technology presented in this article into the field. Before this is possible, as the next step, we plan to implement and simulate concrete network scenarios using the OMNeT++ platform [7]. One particular advantage of this approach is that once the algorithms outlined above have been implemented in OMNeT++, they can be adapted to actual micro controllers with only very little effort.

Acknowledgements. This work was supported by German Aerospace Center (DLR), by Bayerisches Staatsministerium für Wirtschaft, Landesentwicklung und Energie (StMWi) and Bundesministerium für Wirtschaft und Klimaschutz (BMWK).

References

1. Badihi, B., Ghavimi, F., Jäntti, R.: On the system-level performance evaluation of bluetooth 5 in IoT: Open office case study. In: 2019 16th International Symposium on Wireless Communication Systems (ISWCS), pp. 485–489 (2019). https://doi.org/10.1109/ISWCS.2019.8877223
2. Bughin, J., Seong, J., Manyika, J., Chui, M., Joshi, R.: Notes from the AI frontier: modeling the impact of AI on the world economy. McKinsey Glob. Inst. 4 (2018)
3. Giordani, M., Polese, M., Mezzavilla, M., Rangan, S., Zorzi, M.: Toward 6G networks: use cases and technologies. IEEE Commun. Mag. **58**(3), 55–61 (2020). https://doi.org/10.1109/MCOM.001.1900411
4. Hemmat, M., Davoodi, A., Hu, Y.H.: EdgenAI: distributed inference with local edge devices and minimal latency. In: Proceedings of the Asia and South Pacific Design Automation Conference, ASP-DAC, vol. 2022-January, pp. 544–549. Institute of Electrical and Electronics Engineers Inc. (2022). https://doi.org/10.1109/ASP-DAC52403.2022.9712496
5. Hemmat, M., Miguel, J.S., Davoodi, A.: CAP'NN: class-aware personalized neural network inference. In: 2020 57th ACM/IEEE Design Automation Conference (DAC), pp. 1–6. IEEE (2020). https://doi.org/10.1109/DAC18072.2020.9218741
6. Hinds, A., Ngulube, M., Zhu, S., Al-Aqrabi, H.: A review of routing protocols for mobile ad-hoc networks (manet). Int. J. Inf. Educ. Technol. **3**(1), 1 (2013)
7. OpenSim Ltd.: Omnet++ discrete event simulator (2023). https://omnetpp.org
8. Nakamura, T.: 5G evolution and 6G. In: 2020 IEEE Symposium on VLSI Technology, pp. 1–5 (2020). https://doi.org/10.1109/VLSITechnology18217.2020.9265094
9. Parthasarathy, A., Krishnamachari, B.: DEFER: distributed edge inference for deep neural networks. In: 2022 14th International Conference on COMmunication Systems and NETworkS, COMSNETS 2022, pp. 749–753. Institute of Electrical and Electronics Engineers Inc. (2022). https://doi.org/10.1109/COMSNETS53615.2022.9668515, arXiv: 2201.06769
10. Peltonen, E., et al.: 6G white paper on edge intelligence. arXiv preprint arXiv:2004.14850 (2020)
11. Royer, E., Toh, C.K.: A review of current routing protocols for ad hoc mobile wireless networks. IEEE Pers. Commun. **6**(2), 46–55 (1999). https://doi.org/10.1109/98.760423
12. Shahraki, A., Ohlenforst, T., Kreyß, F.: When machine learning meets Network Management and Orchestration in edge-based networking paradigms. J. Netw. Comput. Appl. **212**, 103558 (2023). https://doi.org/10.1016/j.jnca.2022.103558
13. Stahl, R., Hoffman, A., Mueller-Gritschneder, D., Gerstlauer, A., Schlichtmann, U.: DeeperThings: fully distributed CNN inference on resource-constrained edge devices. Int. J. Parallel Prog. **49**(4), 600–624 (2021). https://doi.org/10.1007/s10766-021-00712-3
14. Zhao, Z., Barijough, K.M., Gerstlauer, A.: DeepThings: Distributed adaptive deep learning inference on resource-constrained IoT edge clusters. In: IEEE Transactions on Computer-Aided Design of Integrated Circuits and Systems, vol. 37, no. 11, pp. 2348–2359. Institute of Electrical and Electronics Engineers Inc. (2018). https://doi.org/10.1109/TCAD.2018.2858384. ISSN: 02780070

Using Neural Network to Optimize Bin-Picking in the SME Manufacturing Digital Transformation

Philippe Juhe[1] and Paul-Eric Dossou[2,3](✉)

[1] Icam, Site of Toulouse, 31300 Toulouse, France
philippe.juhel@icam.fr
[2] Icam, Site of Grand Paris Sud, 77127 Lieusaint, France
paul-eric.dossou@icam.fr
[3] SPLOTT/AME, University of Gustave Eiffel, 77420 Champs-sur-Marne, France

Abstract. The recent increase of logistics cost, due to the political and economic situation in the world, and the Covid crisis have accelerated the relocation of industrial companies in developed countries. Industry 4.0 concepts contribute to improve these company's performance through the management of added value and non-added values in their manufacturing processes. Despite their success in large companies, they are not sufficiently exploited by SMEs. This paper presents a sustainable methodology to digitally transform the SMEs processes by exploiting lean manufacturing, SMED method and DMAIC method. A focus is specially made on the elimination of non-added values in the processes. To achieve this goal, a method, based on artificial intelligence tools such as deep learning, is developed that allows a cobot to learn by itself how to grasp objects, without human needs.

Keywords: Industry 4.0 concepts · Lean manufacturing · Grasping and Bin picking

1 Introduction

The concepts of industry 4.0 have demonstrated their relevance through the transformation of large companies [1]. Nevertheless, their exploitation in small and medium enterprises (SMEs) stays confidential [2]. Their efficiency lies in the use of concepts [3] such as advanced robotics, artificial intelligence (AI), Internet of Things (IoT), to increase the company's global performance. Frameworks have been elaborated for the company digital transformation such as in [4] through the vertical and horizontal dimensions integration, and the end-to-end cycle optimization with the value chain improvement. This framework has been completed by organizing the company digital transformation with sustainability as the kernel of the transformation [5]. The objective is to exploit new technologies as tools for aiding employees in the realization of their tasks. Three axes have been defined (physical, informational and decisional) to efficiently and digitally transform the company. The advantage of this framework is that it takes into account the

S. Ossowski et al. (Eds.): DCAI 2023, LNNS 740, pp. 155–164, 2023.
https://doi.org/10.1007/978-3-031-38333-5_16

brakes of the industry 4.0 concepts implementation in SMEs for proposing an adapted transformation solution. Indeed, the concepts make SMEs more flexible and faithful to their values integrating social, societal and environmental aspects in the company transformation. They are involved in industrial applications and are used to solve many industrial technical or organization problems such as defect (object) detection [6], wasting time reduction, or non-added value cost reduction. Lean manufacturing methodology integrating tools [7] such as Value Stream Mapping (VSM), Kanban, and SMED (Single Minute Exchange of Die) can be combined with scientific techniques as vision, form recognition, eye tracking, voice picking exploiting artificial intelligence techniques (neural network, deep learning, machine learning) [8] to solve the industrial problems. Then, this paper presents the human machine collaboration [9] on a production post by defining non-value tasks and added values tasks and attributing only the added-value one to the operator, letting the cobot (Collaborative robot) manage the rest of added value tasks and the non-added values. Artificial intelligence techniques have been used for defining the best way to solve technical and organizational problems in this post.

A focus is made on the optimization of grasping and bin-picking by a cobot to reduce the wasting time issued from the change of series in a manufacturing process. Grasping objects in bulk with CAD object model-based methods requires to perceive each of the objects present, their orientation and their position, as well as their relative position to each other (above or below), and this, despite possible occlusions. This paper presents a Deep Learning approach using RGB-D images for unloading object instances piled up in bulk one by one using a robotic arm [10]. The article is organized as follows: firstly, a literature review is presented for describing concepts that will be exploited for solving the problem; then, the concepts and methodologies elaborated are exposed; finally, a focus is made on the system architecture and an experimental illustration is detailed.

2 Related Work

The literature review is focused on the industry 4.0 concepts and their impacts on the company performance. Then, this section contains a part on organizational methods, new technologies. The methods and tools about grasping and bin-picking will also be discussed for showing what will be the contribution of this work.

2.1 Organizational Methods

Many organizational methods are already used to improve the company performance [5]. GRAI methodology describes the company through five models (functional, physical, decisional, processes and informational) during an enterprise modeling and can be combined with other operational methodologies [11]. Lean manufacturing methodology and its tools are predestined to raise the efficiency of the production department by focusing on added-values processes and reducing non-added values processes [12]. This methodology has become an admired conglomeration [12] of efficient tools such as Value Stream Mapping, Kanban, Kaizen, and SMED (Single Minute Exchange of Die). The SMED method is a setup of techniques that aims to reduce the setup time of a machine [13]. The combination of new technologies with the methodologies presented above will increase the efficiency of the company digital transformation. The next sub-section presents the new technologies to exploit.

2.2 New Technologies

The improvement of company performance involves the definition of a structured methodology to analyze the actual organization and find points to improve. The operational tasks associated with the manufacturing processes allow the implementation of efficient solutions. Industry 4.0 context creates the possibility to use new technologies to increase the company's global performance and to contribute to the company digital transformation [14]. The digital transformation involves the obtention of smart manufacturing [15]. But barriers of their implementation in SMEs concern as well cost (not sufficient money in SMEs) as the fear of employees about their job, or their difficulties to manage the technical knowledge required for these technologies' utilization. The sustainable exploitation of these technologies corresponds to an efficient response to SMEs [5]. To reduce the manufacturing processes waste and to increase the processes added value, new technologies can contribute to the non-added values elimination. IoT, advanced robotics and artificial intelligence [16] can be exploited to solve operational problems such as defect detection, manufacturing orders optimization and product realization. AI is defined as computer algorithms that imitates biotic mental processes or activities [17] such as learning, estimating, problem solving, suggestion or decision making. AI techniques are divided into classical methods (used for engineering design) and learning methods (used for defect detection) [18]. For the wasting time reduction through digital transformation, learning methods are adapted and will be exploited in this paper for the SME manufacturing processes optimization. The next section focuses on the use of these AI tools for grasping and bin picking methods to realize this wasting time reduction.

2.3 Grasping and Bin-Picking Methods by Using AI Tools

This section focuses on AI techniques utilization in grasping and bin-picking methods. There are mainly four types of robotic grippers: pneumatic, hydraulic, vacuum and servo-electric grippers [19], the two last ones being the most popular. Two categories (gripper oriented and object oriented) are proposed to classify the different methods used for bin-picking [10, 19]. **Gripper oriented methods** consist of detecting grasp opportunities for the best parallel grippers locations, or the best locally planar areas for vacuum cups. Indeed, Neural Networks training is realized through the Cornell Grasping Dataset [20]. Each image is hand-annotated with several ground-truth positive and negative grasping rectangles. The selection and ranking of potential grasping candidates is done through two deep networks. From RGB images, a two branch deep neural network is used to predict grasps and semantic segmentation [21]. Then, the system can predict the best pose for each object (98,2% of accuracy on the Cornell Dataset). **Object oriented methods** consist of detecting objects to grasp, independently of the gripper model. Two sub-classes (model based and model free methods) are defined in this category. **Model-based methods** rely on an object model like a CAD model or a previously scanned model to identify a suitable grasp pose [22]. Sensor data (from depth cameras) are used to match a 3D CAD model, to identify objects and their pose, then to compute a grasp pose and finally to generate a path to reach it. A regression task is used from a single depth image to solve the 6D object pose estimation for same type objects in bulk [22]. Iterative Closest Point (ICP) algorithms are also used [23] to match a cloud point with 3D CAD model to detect objects and to estimate object pose [24]. **Model-free methods** are mainly based on machine learning and require only labeled data to train a system. These labeled data consist of success or fail grasp poses. Their acquisition is obtained by using demonstration of grasping [19], by a human [25], and exploiting reinforcement learning [26], heuristics, human labeling (like Cornell Dataset) or automatic labeling, from synthetic data or a real robot. This present work belongs to model-free methods and uses a real robot acquired dataset. The use of synthetic images requires qualitative and well-treated images. This could be done with domain adaptation [27] exploiting techniques such as GAN (Generative Adversarial Network) to improve synthetic source images or with domain randomization [28] applying randomized transformations to the simulated images. Indeed, the prediction of picking regions can be done with suction gripper, without recognition and pose estimation and learning this picking skill from scratch [29]. For instance, an RGB-D image and a Convolutional Neural Network (CNN) can be used to predict a grasp success probability map [29]. The method used is an object-oriented model free method. The training method is inspired by Reinforcement Learning [30]. It needs 2000 attempts for training and has a success rate of 79% for new configurations (more objects with different colors). All is done in simulation with Coppelia-V-REP [31].

This paper exploits the concepts and methods presented above to define a sustainable methodology for increasing SME performance through the optimization of the grasping and bin-picking of cobots.

3 Concepts and Methodology

This section presents the concepts and methods that have been elaborated to ensure the SME manufacturing performance. A global methodology and a structured framework have been developed. Then, specific concepts for optimizing grasping and bin-picking in the frame of advanced robotics have been presented.

3.1 The Sustainable Methodology

The sustainable hybrid methodology elaborated to increase the SME performance through its digital transformation is structured for responding to the company expectations by finding levers to eliminate barriers and accelerate the industry 4.0 concepts implementation in SMEs [15]. Indeed, sustainability through its three pillars (economic, social and environmental) is used as the core of the company's digital transformation. Then, the specificity of each SME can be considered. Three axes (physical, decisional and informational) are defined to realize the transformation. The decisional axis is based on the exploitation of GRAI methodology for making the company decisions consistent. The informational axis allows to integrate computational new technologies in the company transformation. The physical axis exploits lean manufacturing (specially SMED tool) combined with DMAIC (Define, Measure, Analyze, Innovate and Control) to increase the company manufacturing system performance [5]. This paper focuses on this physical transformation axis. It aims to integrate cobots in the manufacturing process and to reduce the wasting time associated with this process. Indeed, a specific focus is made on wasting time related to grasping and bin-picking. The next chapter presents the global approach to optimize processes through the SMED tool exploitation.

3.2 The General Lean-SMED and DMAIC Optimization

This approach (see Fig. 1) involves the integration of advanced robotics in the manufacturing process to eliminate waste and optimize added values.

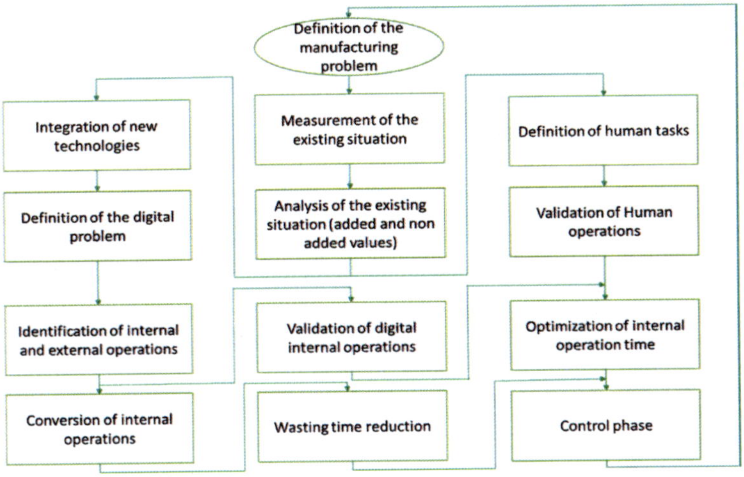

Fig. 1. The digital Lean-SMED and DMAIC optimization approach

The physical problem to solve with digitalization is defined. The existing situation measure is realized according to both DMAIC and Lean-SMED methods. Added value and non-added value tasks are analyzed in the process according to the analysis phase of DMAIC. New technologies such as mobile robots and cobots, and algorithms (AI tools) are integrated to manage both added value and non-added value tasks and used as an aid for operators in the optimization process. Humans are focused on added-value tasks. Then, for the digitized tasks, internal and external operations are identified. Some of the internal operations are converted into externals. Indeed, in the innovation phase, external operations are solved to reduce wasting time. Internal operations are optimized.

Human machine interaction tools are created to manage the collaboration. The control phase corresponds to the test of the solution, the measure of the results, the adjustment of the optimized solution and the validation of the sustainable digitization process.

The following section explains the specific utilization of the digital Lean-SMED and DMAIC approach on the grasping and bin-picking.

3.3 The Specific Approach for Grasping and Bin-Picking Optimization

The specific approach that is proposed, for grasping and bin-picking optimization, aims at emptying a bin full of similar objects with a suction gripper (see Fig. 2).

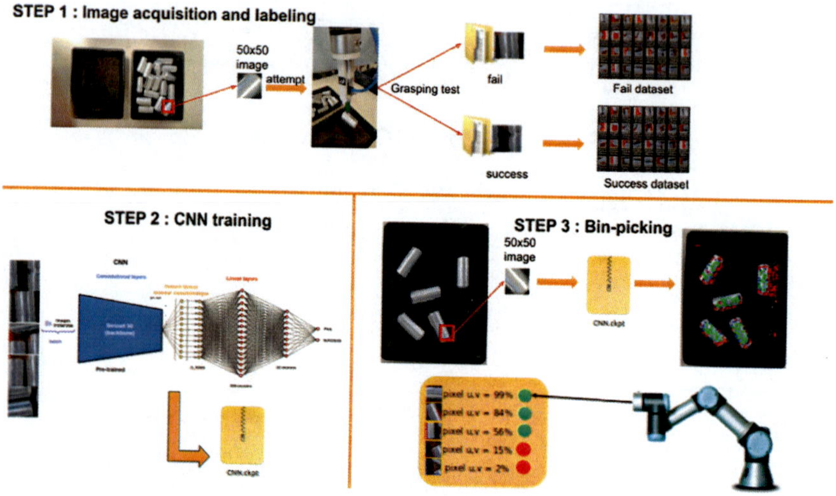

Fig. 2. The three steps: image acquisition and labelling, CNN model training and bin-picking

It consists of the three following steps:

- Image acquisition and labeling: First, the robot attempts to grasp objects at random locations in the parts container. For each attempt, an RGB-D image of the future contact area between the robot's suction cup and the targeted area is taken. This contact either results in a successful grasp or a failure. This information is used to store the acquired image in the 'success' or 'failure' directories. Thus, the system builds up its own dataset of labeled images.
- CNN training: When enough images have been acquired, the content of these two directories is used to train a CNN model in order to predict, for a given RGB-D image, the probability of success of the capture if the suction cup comes into contact with the area in the center of the image.
- Exploitation: Next, after CNN model training, we can use it to empty the bin full of objects. Two programs run in parallel. One gets many random small images, centered on a future grasping point, and computes for each their probability of successful grasping. During this time, the other program drives the robot arm to the location of the previous best probability grasping point. After this grasping, a new RGB-D image is acquired and processed in the same way until the parts container is emptied.

4 Experimental Setup

The details of the experimental setup are explained in this section. A Universal Robot UR3 cobot with a home-made suction gripper is exploited. This gripper is composed of a Venturi vacuum generator and a sensor to detect grasping. It is controlled by an

Arduino. An Intel Realsense D415 gets RGB-D 640x480 pixels images. Two computers (under Ubuntu 20.04) are used and linked to Arduino through ROS Noetic. The first uses a NVIDIA GTX 2080 Ti to perform CNN training and inferences, the second runs programs and exploits the hardware data. The CNN training is realized by a pretrained Resnet 18 (Pytorch Lightning framework). (i) For each random point inside the container, an 100x100 cropped image is captured. From the 100x100 cropped image, 36 rotations of 10° increase the image dataset (36 new experiences). This program works autonomously and improves the quality of the CNN. (ii) With 15 epochs, the pretrained Resnet 18 exploits the 224 × 224 RGB image and outputs a vector of size 512. Two hidden layers (128 and 32 neurons) realize the classification success/fail (learning rate: 0.00211, batch size: 4, training: 15 epochs). The optimization of these 4 hyperparameters is made by Ray tune. (iii) Then bin-picking is realized to empty as objects container. Two programs are exploited with an iterative process for generating a lot of random points, compute their prediction and drive the robot to the best one.

5 Results and Discussions

5.1 RGB Images Case

The objects to grasp are steel cylinders (5 cm length). 36 images are generated by each grasp attempt (39 attempts) during 25 s and produced 540 fail and 864 success images. The robot lead time is 16 min. Three sub-datasets (equal num of fail and success image) have been produced (540 (big), 432 (medium) and 324 (small) images. The dataset is composed of training (70%), validation (15%) and test (15%). The accuracy and loss Tensorboard plots are measured for the 3 datasets (Fig. 3). There is no overfitting, even for the small dataset, the accuracy is high (>98%) and the ROC curve is close to perfect which means good training. During the exploitation (bin picking) step, the number of attempts to empty the objects container is measured. The ratio #success/#attempts is obtained (big dataset: 37/38 (97%) and small dataset: 49/51 (96%)). The easy-to-grasp shape of the cylinder object explains these good results.

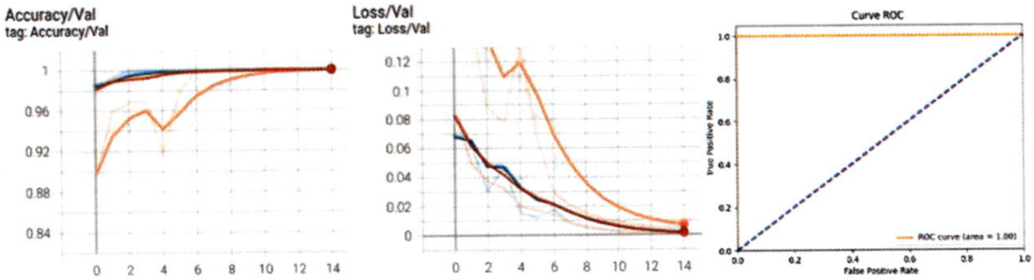

Fig. 3. Accuracy, loss and ROC curve. Orange: small, blue: medium, red: big

5.2 Generalization

This experiment uses a much more difficult-to-grasp objects (see Fig. 3) such as a plastic pump (only 3 possible grasping positions). 252 success and 720 fail images are in the dataset. The number of fail images is larger because of the difficulty of grasping. The ratio #success/#attempts is 40/57 (70%) due to this grasping difficulty (Fig. 4).

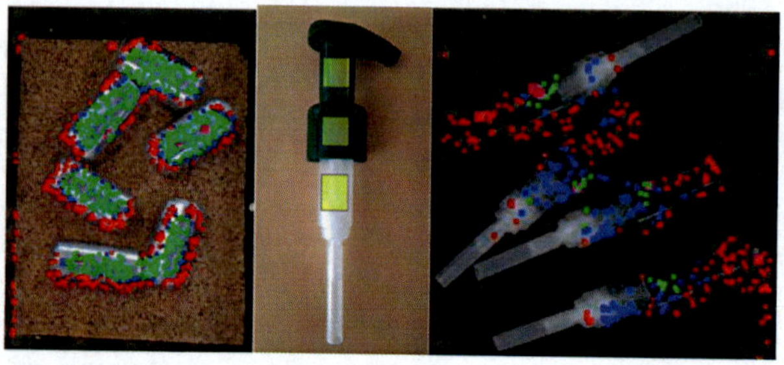

Fig. 4. Probability of success (low: red, medium: blue, high: green) for cylinder (left) and pump (right). Middle: the 3 positions to grasp the pump.

6 Limitations and Future Work

The generalization results have to be improved. One possibility is to use the Depth image with a double CNN [32].

Indeed, in the case of few images, a better management of the CNN training step is required. Then, the probability of success can be more managed. For instance, sending the cobot to the median prediction grasping point for performing an attempt, will involve new information and improve the CNN model performances. The use of Reinforcement Learning techniques represents a possibility. Simulation (with Coppelia or Gazebo) to get images for the dataset could also be exploited for reducing the training time. A part of the training will be done before the real exploitation. The SMED method aims to reduce non-added value operations and to transform internal operations into external operations. This simulation solution does not require human intervention to train a new model when one changes object shape and reduces the non-added value operations performed by the operator. A new object images acquisition requires the use of the robot (internal operation). This operation can become an external operation, with the use of simulation as much as possible for the main part of the training and the exploitation of the real robot just to finish the training.

7 Conclusion

This paper presents concepts and method of SME performance to reduce non-added values through the exploitation of human robot collaboration in the manufacturing process. A focus is made on grasping and bin-picking optimization. A system based on CNN network, has been elaborated to use a real cobot to automatically build a dataset of random images of grasping positions, to exploit this dataset to train a simple CNN for RGB images and to employ this CNN model to predict the success score for many random points and then attempt to grasp the point with the best prediction. With this system, a cobot can automatically empty a container full of objects. Even if the object to grasp changes with new series, the operations to realize will automatically be optimized. This bin-picking optimization increases the company manufacturing flexibility and contributes to its sustainable digital transformation. The next step of the research corresponds to the generalization of the solution and the experimentation in a SME.

References

1. Büyüközkan, G., Göçer, F.: Digital supply chain: literature review and a proposed framework for future research. Comput. Ind. **97**, 157–177 (2018)
2. Elhusseiny, H.M., Crispim, J.: SMEs, barriers and opportunities on adopting Industry 4.0: a review. Procedia Comput. Sci. **196**, 864–871 (2022)
3. Martell, F., López, J.M., Sánchez, I.Y., Paredes, C.A., Pisano, E.: Evaluation of the degree of automation and digitalization using a diagnostic and analysis tool for a methodological implementation of Industry 4.0. Comput. Ind. Eng. **177**, 109097 (2023)
4. Stock T., Seliger, G.: Opportunities of Sustainable Manufacturing in Industry 4.0. Procedia CIRP **40**, 536–541 (2016)
5. Koumas, M., Dossou, P.-E., Didier, J.-Y.: Digital Transformation of Small and Medium Sized Enterprises Production Manufacturing. J. Softw. Eng. Appl. **14**(12), Article no 12 (2021)
6. Ahmad, H.M., Rahimi, A.: Deep learning methods for object detection in smart manufacturing: a survey. J. Manuf. Syst. **64**, 181–196 (2022)
7. Habib, M.A., Rizvan, R., Ahmed, S.: Implementing lean manufacturing for improvement of operational performance in a labeling and packaging plant: a case study in Bangladesh. In: Results Eng. **17**, 100818 (2023)
8. Liu, Y., Ping, Y., Zhang, L., Wang, L., Xu, X.: Scheduling of decentralized robot services in cloud manufacturing with deep reinforcement learning. Robot. Comput.-Integr. Manuf. **80** (2023)
9. Baratta, A., Cimino, A., Gnoni, M.G., Longo, F.: Human robot collaboration in Industry 4.0: a literature review. Procedia Comput. Sci. **217**, 1887–1895 (2023)
10. Grard, M.: Generic instance segmentation for object-oriented bin-picking. Ph.D. thesis. Université de Lyon (2019)
11. Doumeingts, G., Ducq, Y., Vallespir, B., Kleinhans, S.: Production management and enterprise modelling. Comput. Ind. **42**, 245–263 (2000)
12. Wagner, T., Herrmann, C., Thiede, S.: Industry 4.0 impacts on lean production systems. Procedia CIRP **63**, 125–131 (2017)
13. Godina, R., Pimentel, C., Silva, F.J.G., Matias, J.C.O.: A structural literaturre review of the single minute exchange of die: the latest trends. Procedia Manuf. **17**, 783–790 (2018)
14. Monshizadeh, F., Sadeghi Moghadam, M.R., Mansouri, T., Kumar, M.: Developing an Industry 4.0 readiness model using fuzzy cognitive maps approach. Int. J. Prod. Econ. **255** (2023)
15. Jan, Z., et al.: Artificial intelligence for industry 4.0: Systematic review of applications, challenges, and opportunities. Expert Syst. Appl. **216** (2023)
16. O'Brien, K., Humphries, J.: Object detection using convolutional neural networks for smart manufacturing vision systems in the medical devices sector. Procedia Manuf. **38**, 142–147 (2019)
17. Russel, S., Norvig, P..: Artificial Intelligence: A Modern Approach, pp. 1–2, 3rd edn. Prentice Hall Press, USA (2009)
18. Yüksel, N., Börklü, H.R., Sezer, H.K., Canyurt, O.E.: Review of artificial intelligence applications in engineering design perspective. Eng. Appl. Artif. Intell. **118**, 105697 (2023)

19. Cordeiro, A., Rocha, L.F., Costa, C., Costa, P., Silva, M.F.: Bin picking approaches based on deep learning techniques: a state-of-the-art survey. In: 2022 IEEE International Conference on Autonomous Robot Systems and Competitions (ICARSC), pp. 110–117 (2022)
20. Lenz, I., Lee, H., Saxena, A.: Deep learning for detecting robotic grasps 2014 (2023). http://arxiv.org/abs/1301.3592. Accessed 21 Aug 2014
21. Ainetter, S., Fraundorfer, F.: End-to-end trainable deep neural network for robotic grasp detection and semantic segmentation from RGB. In: 2021 IEEE International Conference on Robotics and Automation (ICRA), pp. 13452–13458 (2021)
22. Kleeberger, K., Huber, M.F.: Single shot 6D object pose estimation. In: 2020 IEEE International Conference on Robotics and Automation (ICRA), pp. 6239–6245 (2020)
23. Besl, P.J., McKay, N.D.: A method for registration of 3-D shapes. IEEE Trans. Pattern Anal. Mach. Intell. **14**(2), 239–256 (1992)
24. Lee, J., Kang, S., Park, S.-Y.: 3D pose estimation of bin picking object using deep learning and 3D matching. In: Proceedings of the 15th International Conference on Informatics in Control, Automation and Robotics, pp. 328–334, Porto, Portugal (2018)
25. Song, S., Zeng, A., Lee, J., Funkhouser, T.: Grasping in the wild: learning 6dof closed-loop grasping from low-cost demonstrations. IEEE Robot. Autom. Lett. **5**(3), 4978–4985 (2020)
26. Quillen, D., Jang, E., Nachum, O., Finn, C., Ibarz, J., Levine, S.: Deep reinforcement learning for vision-based robotic grasping: a simulated comparative evaluation of off-policy methods. In: 2018 IEEE International Conference on Robotics and Automation (ICRA), pp. 6284–6291 (2018)
27. Bousmalis, K., et al.: Using simulation and domain adaptation to improve efficiency of deep robotic grasping. In: 2018 IEEE International Conference on Robotics and Automation (ICRA), pp. 4243–4250 (2018)
28. Tobin, J., Fong, R., Ray, A., Schneider, J., Zaremba, W., Abbeel, P.: Domain randomization for transferring deep neural networks from simulation to the real world. In: 2017 IEEE/RSJ International Conference on Intelligent Robots and Systems (IROS), pp. 23–30 (2017)
29. Shao Q., et al.: Suction grasp region prediction using self-supervised learning for object picking in dense clutter. In: 2019 IEEE 5th International Conference on Mechatronics System and Robots (ICMSR), pp. 7–12 (2019)
30. Sutton, R.S., Barto, A.G.: Reinforcement Learning: An Introduction. MIT Press (2018)
31. Rohmer, E., Singh, S.P.N., Freese, M.: V-REP: a versatile and scalable robot simulation framework. In: 2013 IEEE/RSJ International Conference on Intelligent Robots and Systems, pp. 1321–1326 (2013)
32. Eitel, A., Springenberg, J.T., Spinello, L., Riedmiller, M., Burgard, W.: Multimodal deep learning for robust RGB-D object recognition. In: 2015 IEEE/RSJ International Conference on Intelligent Robots and Systems (IROS), pp. 681–687 (2015)

Neural Architecture Search: Practical Key Considerations

María Alonso-García(✉) and Juan M. Corchado

AIR Institute, PCUVA Building, Paseo de Belén, 9A, 47011 Valladolid, Spain
{marialonsogar,corchado}@air-institute.com

Abstract. The rapid development of deep neural networks has highlighted the importance of research in this domain. Neural Architecture Search (NAS) has emerged as a pivotal technique for automating and optimizing neural network designs. However, due to the complex and evolving nature of this field, staying up to date with the latest research, trends, and best practices is challenging. This article addresses the need for practical considerations, best practices, and open frameworks to guide practitioners in NAS endeavors. It discusses key considerations, challenges, opportunities, and open problems, along with a compilation of best practices and open frameworks. Readers will gain a practical guide for developing, testing, or applying NAS techniques.

Keywords: Neural Architecture Search · Artificial Intelligence · Deep Learning

1 Introduction

Neural Architecture Search (NAS) comprises a set of techniques for automating the process of designing and optimizing neural network architectures [51]. The history of this field dates back to the 1990s [56], where genetic algorithms were first used to optimize neural network architectures. Other important approaches such as NeuroEvolution of Augmenting Topologies (NEAT) [42] emerged as the development of neural networks progressed. Recently, research in NAS has experienced a resurgence with the development of more sophisticated search algorithms such as reinforcement learning [39,60] and with advances in hardware [2]. These algorithms allow for a more efficient exploration of the search space. Although NAS algorithms have been developed for different data structures (tabular, images, text, graphs, etc.), applications to real case uses are not yet widespread, but some successful cases can be found in medical applications [34].

Key practical concepts in NAS include defining and choosing proper search space, search algorithm, and evaluation strategy [38]. These are explored in this article along with other practical considerations such as the need of computational resources, baselines, and the importance of human expertise. The use of powerful computing resources is necessary for efficient exploration of the search space [2]. Baselines are important for benchmarking the performance of new

S. Ossowski et al. (Eds.): DCAI 2023, LNNS 740, pp. 165–174, 2023.
https://doi.org/10.1007/978-3-031-38333-5_17

methods against existing ones [22]. Additionally, human expertise is crucial in determining the design requirements and constraints for the network. Best practices include careful selection of the search space and algorithm, proper evaluation and validation techniques, regularization or ensuring that the results are reproducible. Finally, the limitations of NAS are reviewed, such as the high computational cost and the difficulty in interpreting the resulting architectures.

2 Practical Considerations

In defining or using a NAS method, several inherent characteristics must be considered, including the search space, evaluation strategy, and optimization or search algorithm. Additionally, practical considerations such as computational resources, baselines, the requirement for human expertise, and the choice of tools and frameworks are vital aspects to be addressed.

2.1 Search Space

A search space is the set of possible network architectures a NAS can generate and evaluate. Defining it involves setting a set of operations (convolutional layers, fully connected, recurrent, etc.) and the connections between them to create valid architectures. The search space directly affects NAS performance: a search space that is too complex would prohibitively increase compute time and resource usage; one that is too small could miss optimal architectures. The design depends on the use case or the network topology sought. The simplest form of search space is the sequential search space, which comprises a linear sequence of layer-wise operations. In contrast, hierarchical search spaces [25] are more intricate, involving multiple levels of operations. Each level represents a set of operations that can be interconnected to create a subnetwork. Within hierarchical search spaces, cell-based search spaces are a specific type characterized by a repeating cell structure. Each cell serves as a self-contained network unit that can be stacked together to construct a larger network. Other search spaces worth mentioning are the continuous search space of DARTS [26]. There are two main types of search spaces: rigid (fixed length) and dynamic (variable length). Rigid search spaces have predetermined architectures with a fixed number of layers and nodes, offering faster and more efficient search. In contrast, dynamic search spaces allow the addition or removal of layers and nodes, providing greater flexibility and potential for more powerful architectures. However, dynamic search spaces require more computational resources compared to rigid search spaces. Note that the characterisation of the search space is closely relate to the choice of optimization or search algorithm.

2.2 Evaluation Strategy

A evaluation strategy is a method used to evaluate candidate architectures generated by a NAS algorithm during the search process. Traditional Evaluation Methods (TEM) comprise performance evaluation methods that train the searched Deep Neural Networks (DNNs) per iteration to evaluate them. This can be computationally expensive since the evaluation time for TEM is the sum of the time to train all searched architectures. In contrast, Efficient Evaluation Methods (EEMs) aim to reduce the computational cost of NAS by reducing the number of DNNs that need to be trained during the performance evaluation process. A comprehensive analysis of various EEMs is presented in [55]. These strategies include N-shot evaluation methods, few-shot evaluation methods, one-shot evaluation methods, and zero-shot evaluation methods [32]. The paper provides a detailed analysis of the design principles and strengths and weaknesses of each method, the evaluation metrics, benchmark datasets, and comparison results of various EEMs on them.

2.3 Optimization Algorithm

NAS techniques comprise a variety of optimization algorithms. The most widespread are Evolutionary Algorithms, Reinforcement Learning, Bayesian optimization and Gradient-Based Optimization [15].

Evolutionary algorithms (EA) initialize a population of candidate architectures, evolve them subject to certain constraints and evaluation criteria through a process of mutation, recombination, and selection, and evaluate their fitness on a validation set, selecting the best-performing architectures to be parents for the next generation, repeating this process until a satisfactory architecture is found or a stopping criteria is met. Some classic examples are NEAT [42], CoDeepNEAT [30], Genetic CNN [53], and recent one AESENAS [9] .

In the general procedure of Reinforcement Learning (RL) in NAS, a neural network architecture is initially initialized with random or fixed structures, and then it is trained on a particular task to produce a performance metric that acts as a reward signal. The RL agent then updates its policy based on the obtained reward and chooses the next architecture to evaluate. This is a iterative process repeated until a suitable architecture is found or the search budget is exhausted. Some RL-based methods are NASNet [61], ENAS [35,60].

Bayesian Optimization (BO) maps an architecture to its validation error after training for several epochs. BO utilizes a surrogate to model the objective function based on previous architectures and their validation errors. BO selects the next architecture to evaluate by maximizing an acquisition function, which balances exploration and exploitation. BO has been shown to be effective in neural architecture search, particularly in small search spaces. Examples includes Interpretable NAS [40], BANANAS [48] or BayesNAS [58].

Gradient-based optimization iteratively updates architecture parameters using the gradient of a performance metric. The process continues until convergence or a stopping criterion is reached. Examples includes DARTS [26], SNAS [54], SMASH [3], ProxylessNAS [4], AlphaNet [45], etc. [21,27].

Optimization algorithm choice depends on the search space and evaluation strategy. Non-convex or discrete variable-rich spaces may require alternative methods like evolutionary algorithms or reinforcement learning instead of gradient-based techniques.

2.4 Computational Resources

NAS techniques may require high-performance computing clusters or cloud computing services due to their computational demands. Factors such as the size of the search space, assessment approach, and available budget should be considered when selecting computational resources. Acceleration mechanisms such as parallelization or sampling methods are recommended [19]. Efforts have been made to reduce resource consumption, with DARTS achieving a search time of 2–3 GPU days [26]. A performance comparison between state-of-the-art NAS including GPU days is presented in [38].

2.5 NAS Baselines

Baselines establish a simple point of reference for evaluating the performance of a method. The searched architectures are expected to perform better than the baseline model. Random search is often a proper baseline for NAS methods [10,22]; this involves randomly sampling architectures from the search space and evaluating their performance. Local search is another strong baseline that outperformed many state-of-the-art NAS algorithms across three popular benchmark datasets [49]. Common benchmark datasets for NAS are NAS-Bench-101, NAS-Bench-201, DAS-training-bench or IMB-NAS.

2.6 Human Expertise

Despite the automation of NAS, human expertise remains vital in the process. Formulating the design choices, such as the number of layers, channels, and resource constraints, into a search space allows humans to focus on creative work while leaving the search to NAS algorithms. However, it is important to consider that training techniques, such as learning rate schedules, regularization strategies, and data augmentation strategies, also significantly impact performance and may have a greater influence than architecture design. Therefore, a robust training pipeline is essential for developing and evaluating different architectures.

2.7 Tools and Frameworks

Several frameworks and tools have been developed to ease NAS process, implementation and research. AutoKeras simplifies the process of designing neural network architectures by automating the choice of the structure of layers, number of neurons, and other hyperparameters. Its advantages includes user-friendliness, reduced dependency on specialized domain knowledge, and the potential to achieve cutting-edge performance effortlessly [18]. Google Cloud's Vertex AI Neural Architecture Search finds the most accurate neural architectures while accommodating constraints like latency and memory. The key advantages of Vertex AI NAS include improved accuracy for complex tasks, reduced latency, optimized power consumption for mobile devices, automated neural network design, and customizable search constraints [1]. Neural Network Intelligence (NNI), developed by Microsoft, automates feature engineering, neural architecture search, hyperparameter tuning, and model compression for deep learning. It provides a unified architecture to facilitate NAS innovations and apply state-of-the-art algorithms to real-world problems. The benefits of NNI include its lightweight and flexible nature, making it suitable for various tasks and hardware platforms [29]. Auto-PyTorch optimizes both the neural architecture and hyperparameters through a combination of meta-learning, ensembling, and multi-fidelity optimization. The benefits of Auto-PyTorch include its ability to achieve state-of-the-art performance on several tabular benchmarks and its ease of use [59]. NASLib provides a modular and flexible codebase for NAS research, enabling fair comparisons of diverse NAS methods. The benefits of NASLib include its modularity and extensibility [50].

3 Case Studies

NAS methods are application domain dependent and specific techniques have been developed for this purpose: image classification [43,46], natural language processing [17,20,47], speech recognition [16,28,52], graph structures [5,12,13], etc. [7,23,37]. NAS algorithms can optimize CNN designs for mobile devices: MnasNet achieves 75.2% top-1 accuracy with 78ms latency, outperforming MobileNetV2 and NASNet. MobileNetV3-Small has 4.6% higher accuracy and 5% reduced latency compared to MobileNetV2. EfficientNet-B7 achieves 84.4% top-1 accuracy and 97.1% top-5 accuracy, while being smaller and more efficient than other CNNs. NAS algorithms enable efficient architectures like MobileNetV3 and EfficientNet for improved mobile performance.

Despite NAS research is a field that is gaining momentum [19], real-world applications are not being reporting at the same rate. Most recent applications include classification, segmentation and reconstruction of medical images [44,57]. Some notable NAS applications are breast cancer detection from histopathology images [34], brain tumor classification [8] or pneumonia diagnosis from chest X-rays [14]. A industry successful application is the detection of bearing fault diagnosis and remaining useful life prediction [41].

4 Best Practices

When performing Neural Architecture Search (NAS), there are several practical considerations and best practices that researchers and practitioners should keep in mind to ensure the effectiveness and efficiency of the search process.

Basic suggestions are to clearly define objectives and constraints for the search, choose a efficient and appropriate search space (both expressive enough to capture a wide range of architectures and small enough to be computationally tractable) and search algorithm for the problem (depends on the specific problem being addressed), regularize the search to avoid overfitting and try transfer learning to leverage existing architectures and interpret the architectures.

Recommendations for improving efficiency include the use of early stopping, the use of weight sharing —which involves sharing weights between different architectures to reduce the number of parameters that need to be trained— or the prune of unimportant operations or connections during the search process [11]. Stability can be improved by using weights warm-up —which involves gradually increasing the weights of the network during training—, by using progressive shrinking —which involves gradually reducing the size of the network during training—, or by using learning rate scheduling —which involves adjusting the learning rate during training to prevent instability—. Additionally, batch normalization and layer normalization are suggested [11].

Finally, some best practices for scientific research on neural architecture search are to report all experimental details, including hyperparameters and random seeds, use appropriate statistical tests to compare results, conduct ablation studies to analyze the contribution of different components, use multiple datasets to evaluate generalization performance of the NAS, report both mean and variance of results across multiple runs, provide a clear description of the baseline methods used for comparison, conduct sensitivity analysis to evaluate the robustness of results, provide open-source code and documentation for reproducibility, use a unified benchmark suite for fair comparison across methods and state the limitations [24].

5 Limitations and Future Directions

NAS research is still evolving with inherent limitations in various methods. Evolutionary algorithms are robust but time-consuming, RL-based methods ensure performance but also require excessive search time, and gradient-based methods have faster search time but consume excessive memory [19].

The major drawback of NAS is the excessive computational cost [2]; various solutions have been proposed such as incorporating hardware-awareness into the search process, using early-exit options to improve energy efficiency, and combining NAS with pruning to reduce the search space [2,33]. Despite these efforts, there is still much research needed to improve the efficiency and effectiveness of NAS. Transferability is also one of the challenges of NAs, since the architectures found by a NAS may not generalize well to new tasks or datasets [39]. Some studies built supernet optimization that assembles all the architectures as its sub-models by using weight sharing, which may improve transferability in NAS [6]. Additionally, NAS faces interpretability challenges [38] due to complex architectures found or the difficulty of analyzing the search process, presenting difficulties for researchers to improve performance and explain models to end-users. Solutions include visualization techniques and incorporating interpretability into the search process via interpretable RL policies. However, further research is needed to enhance the interpretability of NAS algorithms [39]. Finally, there is a growing interest in exploring new types of architectures [31,36]. In addition, the proposed best practices should be validated in the future, preferably on real domains in order to obtain valuable feedback.

6 Conclusion

Neural Architecture Search (NAS) optimizes deep learning models and shows effectiveness in medical applications. This article highlights practical considerations, best practices, and limitations of NAS. Best practices involve defining the search space, using appropriate evaluation metrics, incorporating human expertise, regularizing the search, and ensuring experiment reproducibility. NAS limitations include resource requirements, interpretability challenges, and transferability. Future NAS research aims to develop efficient algorithms, improve interpretability, and explore graph structures and industry applications. NAS has transformative potential in deep learning, but challenges remain.

Acknowledgements. This research has been funded through the call of the public business entity red.es, for 2021 grants for research and development projects in artificial intelligence and other digital technologies and their integration into value chains with Code: C005/21-ED, Funded by the European Union NextGenerationEU.

References

1. About vertex AI neural architecture search. https://cloud.google.com/vertex-ai/docs/training/neural-architecture-search/overview. Accessed 21 May 2023
2. Benmeziane, H., Maghraoui, K.E., Ouarnoughi, H., Niar, S., Wistuba, M., Wang, N.: Hardware-aware neural architecture search: survey and taxonomy. In: International Joint Conference on Artificial Intelligence (2021)
3. Brock, A., Lim, T., Ritchie, J.M., Weston, N.: SMASH: one-shot model architecture search through hypernetworks (2017)
4. Cai, H., Zhu, L., Han, S.: ProxylessNAS: direct neural architecture search on target task and hardware (2019)
5. Cai, S., Li, L., Deng, J., Zhang, B., Zha, Z.J., Su, L., Huang, Q.: Rethinking graph neural architecture search from message-passing (2021)
6. Cha, S., Kim, T., Lee, H., Yun, S.Y.: SuperNet in neural architecture search: a taxonomic survey. ArXiv abs/2204.03916 (2022)
7. Chen, D., Chen, L., Shang, Z., Zhang, Y., Wen, B., Yang, C.: Scale-aware neural architecture search for multivariate time series forecasting (2021)
8. Chitnis, S., Hosseini, R., Xie, P.: Brain tumor classification based on neural architecture search. Sci. Rep. **12**(1), 19206 (2022). https://doi.org/10.1038/s41598-022-22172-6
9. Chu, J., Yu, X., Yang, S., Qiu, J., Wang, Q.: Architecture entropy sampling-based evolutionary neural architecture search and its application in osteoporosis diagnosis. Complex Intell. Syst. **9**(1), 213–231 (2023). https://doi.org/10.1007/s40747-022-00794-7
10. Elsken, T., Metzen, J.H., Hutter, F.: Neural architecture search: a survey (2019)
11. Elsken, T., Staffler, B., Zela, A., Metzen, J.H., Hutter, F.: Bag of tricks for neural architecture search (2021)
12. Gao, Y., Yang, H., Zhang, P., Zhou, C., Hu, Y.: GraphNAS: graph neural architecture search with reinforcement learning (2019)

13. Guan, C., Wang, X., Chen, H., Zhang, Z., Zhu, W.: Large-scale graph neural architecture search. In: Chaudhuri, K., Jegelka, S., Song, L., Szepesvari, C., Niu, G., Sabato, S. (eds.) Proceedings of the 39th International Conference on Machine Learning. Proceedings of Machine Learning Research, vol. 162, pp. 7968–7981. PMLR (2022). https://proceedings.mlr.press/v162/guan22d.html
14. Gupta, A., Sheth, P., Xie, P.: Neural architecture search for pneumonia diagnosis from chest X-rays. Sci. Rep. **12**(1), 11309 (2022). https://doi.org/10.1038/s41598-022-15341-0
15. He, C., Ye, H., Shen, L., Zhang, T.: MileNAS: efficient neural architecture search via mixed-level reformulation (2020)
16. Hu, S., Xie, X., Liu, S., Geng, M., Liu, X., Meng, H.: Neural architecture search for speech recognition (2020)
17. Jiang, Y., Hu, C., Xiao, T., Zhang, C., Zhu, J.: Improved differentiable architecture search for language modeling and named entity recognition. In: Proceedings of the 2019 Conference on Empirical Methods in Natural Language Processing and the 9th International Joint Conference on Natural Language Processing (EMNLP-IJCNLP), pp. 3585–3590. Association for Computational Linguistics, Hong Kong (2019). https://doi.org/10.18653/v1/D19-1367, https://aclanthology.org/D19-1367
18. Jin, H., Song, Q., Hu, X.: Auto-keras: an efficient neural architecture search system (2019)
19. Kim, Y., Yun, W.J., Lee, Y.K., Jung, S., Kim, J.: Trends in neural architecture search: Towards the acceleration of search (2021)
20. Klyuchnikov, N., Trofimov, I., Artemova, E., Salnikov, M., Fedorov, M., Burnaev, E.: NAS-bench-NLP: neural architecture search benchmark for natural language processing. IEEE Access 1 (2020)
21. Li, C., et al.: BossNAS: exploring hybrid CNN-transformers with block-wisely self-supervised neural architecture search (2021)
22. Li, L., Talwalkar, A.: Random search and reproducibility for neural architecture search (2019)
23. Li, Y., Hao, C., Li, P., Xiong, J., Chen, D.: Generic neural architecture search via regression (2021)
24. Lindauer, M., Hutter, F.: Best practices for scientific research on neural architecture search. J. Mach. Learn. Res. **21**(1), 9820–9837 (2020)
25. Liu, H., Simonyan, K., Vinyals, O., Fernando, C., Kavukcuoglu, K.: Hierarchical representations for efficient architecture search (2018)
26. Liu, H., Simonyan, K., Yang, Y.: DARTS: differentiable architecture search (2019)
27. Luo, R., Tian, F., Qin, T., Chen, E., Liu, T.Y.: Neural architecture optimization. In: Bengio, S., Wallach, H., Larochelle, H., Grauman, K., Cesa-Bianchi, N., Garnett, R. (eds.) Advances in Neural Information Processing Systems, vol. 31. Curran Associates, Inc. (2018)
28. Mehrotra, A., et al.: NAS-bench-ASR: reproducible neural architecture search for speech recognition. In: International Conference on Learning Representations (2021)
29. Microsoft: Neural network intelligence (2021). https://github.com/microsoft/nni. If you use this software, please cite it as above
30. Miikkulainen, R., et al.: Evolving deep neural networks (2017)
31. Moser, B.B., Raue, F., Hees, J., Dengel, A.: DartsReNet: exploring New RNN cells in ReNet architectures. In: Farkaš, I., Masulli, P., Wermter, S. (eds.) ICANN 2020. LNCS, vol. 12396, pp. 850–861. Springer, Cham (2020). https://doi.org/10.1007/978-3-030-61609-0_67

32. Ning, X., et al.: Evaluating efficient performance estimators of neural architectures. In: Beygelzimer, A., Dauphin, Y., Liang, P., Vaughan, J.W. (eds.) Advances in Neural Information Processing Systems (2021). https://openreview.net/forum?id=Esd7tGH3Spl
33. Odema, M., Rashid, N., Faruque, M.A.A.: EExNAS: early-exit neural architecture search solutions for low-power wearable devices. In: 2021 IEEE/ACM International Symposium on Low Power Electronics and Design (ISLPED), pp. 1–6 (2021)
34. Oyelade, O.N., Ezugwu, A.E.: A bioinspired neural architecture search based convolutional neural network for breast cancer detection using histopathology images. Sci. Rep. **11**(1), 19940 (2021). https://doi.org/10.1038/s41598-021-98978-7
35. Pham, H., Guan, M.Y., Zoph, B., Le, Q.V., Dean, J.: Efficient neural architecture search via parameter sharing (2018)
36. Qian, G., et al.: When NAS meets trees: an efficient algorithm for neural architecture search (2022)
37. Rakhshani, H., et al.: Neural architecture search for time series classification. In: 2020 International Joint Conference on Neural Networks (IJCNN), pp. 1–8 (2020). https://doi.org/10.1109/IJCNN48605.2020.9206721
38. Ren, P., et al.: A comprehensive survey of neural architecture search: Challenges and solutions (2021)
39. Robles, J.G., Vanschoren, J.: Learning to reinforcement learn for neural architecture search (2019)
40. Ru, B., Wan, X., Dong, X., Osborne, M.: Interpretable neural architecture search via Bayesian optimisation with Weisfeiler-Lehman kernels (2021)
41. Ruan, D., Han, J., Yan, J., Gühmann, C.: Light convolutional neural network by neural architecture search and model pruning for bearing fault diagnosis and remaining useful life prediction. Sci. Rep. **13**(1), 5484 (2023). https://doi.org/10.1038/s41598-023-31532-9
42. Stanley, K.O., Miikkulainen, R.: Evolving neural networks through augmenting topologies. Evol. Comput. **10**(2), 99–127 (2002). https://doi.org/10.1162/106365602320169811
43. Tao, T.M., Kim, H., Youn, C.H.: A compact neural architecture search for accelerating image classification models. In: 2021 International Conference on Information and Communication Technology Convergence (ICTC), pp. 1713–1718 (2021). https://doi.org/10.1109/ICTC52510.2021.9620797
44. Vo-Ho, V.K., Yamazaki, K., Hoang, H., Tran, M.T., Le, N.: Chapter 19 - neural architecture search for medical image applications. In: Nguyen, H.V., Summers, R., Chellappa, R. (eds.) Meta Learning With Medical Imaging and Health Informatics Applications, The MICCAI Society book Series, pp. 369–384. Academic Press (2023). https://doi.org/10.1016/B978-0-32-399851-2.00029-6, https://www.sciencedirect.com/science/article/pii/B9780323998512000296
45. Wang, D., Gong, C., Li, M., Liu, Q., Chandra, V.: AlphaNet: improved training of supernets with alpha-divergence (2021)
46. Wang, W., Zhang, X., Cui, H., Yin, H., Zhang, Y.: FP-DARTS: fast parallel differentiable neural architecture search for image classification. Pattern Recognit. **136**, 109193 (2023). https://doi.org/10.1016/j.patcog.2022.109193, https://www.sciencedirect.com/science/article/pii/S0031320322006720
47. Wang, Y., et al.: TextNAS: a neural architecture search space tailored for text representation (2019)
48. White, C., Neiswanger, W., Savani, Y.: BANANAS: Bayesian optimization with neural architectures for neural architecture search (2020)

49. White, C., Nolen, S., Savani, Y.: Exploring the loss landscape in neural architecture search (2021)
50. White, C., Zela, A., Ru, R., Liu, Y., Hutter, F.: How powerful are performance predictors in neural architecture search? Adv. Neural Inf. Process. Syst. **34** (2021)
51. Wistuba, M., Rawat, A., Pedapati, T.: A survey on neural architecture search (2019)
52. Wu, X., Hu, S., Wu, Z., Liu, X., Meng, H.: Neural architecture search for speech emotion recognition (2022)
53. Xie, L., Yuille, A.: Genetic CNN (2017)
54. Xie, S., Zheng, H., Liu, C., Lin, L.: SNAS: stochastic neural architecture search (2020)
55. Xie, X., Song, X., Lv, Z., Yen, G.G., Ding, W., Sun, Y.: Efficient evaluation methods for neural architecture search: a survey (2023)
56. Yao, X., Liu, Y.: A new evolutionary system for evolving artificial neural networks. IEEE Trans. Neural Netw. **8**(3), 694–713 (1997). https://doi.org/10.1109/72.572107
57. Yu, Q., Yang, D., Roth, H., Bai, Y., Zhang, Y., Yuille, A.L., Xu, D.: C2FNAS: coarse-to-fine neural architecture search for 3D medical image segmentation. In: Proceedings of the IEEE/CVF Conference on Computer Vision and Pattern Recognition, pp. 4126–4135 (2020)
58. Zhou, H., Yang, M., Wang, J., Pan, W.: BayesNAS: a Bayesian approach for neural architecture search (2019)
59. Zimmer, L., Lindauer, M., Hutter, F.: Auto-pytorch: multi-fidelity metalearning for efficient and robust AutoDL. IEEE Trans. Pattern Anal. Mach. Intell. **43**(9), 3079–3090 (2021). https://doi.org/10.1109/TPAMI.2021.3067763
60. Zoph, B., Le, Q.V.: Neural architecture search with reinforcement learning (2017)
61. Zoph, B., Vasudevan, V., Shlens, J., Le, Q.V.: Learning transferable architectures for scalable image recognition (2018)

Leveraging Smart City Services for Citizen Benefits Through a Unified Management Platform

Francisco Pinto-Santos[1](✉), Juan Antonio González-Ramos[2], Sergio Alonso-Rollán[1], and Ricardo S. Alonso[1]

[1] Air Institute, IoT Digital Innovation Hub (Spain), 37188 Carbajosa de la Sagrada, Salamanca, Spain
{franpintosantos,salonso,ralonso}@air-institute.com

[2] Servicios Informáticos, Universidad de Salamanca, Salamanca, Spain
juanan@usal.es

Abstract. Smart cities offer numerous benefits to citizens by utilizing advanced technologies and data-driven approaches to improve urban life quality. In this paper, we propose a novel platform for managing heterogeneous information from multiple sources, such as relational and non-relational databases, web services, and IoT devices. The platform preprocesses and presents the data dynamically and interactively in real time, providing city administrators with a comprehensive backend system for asset management, status monitoring, and results delivery. We describe the design and implementation of the platform, highlighting its features and benefits for citizens and city administrators. The proposed platform aims to streamline the integration and utilization of smart city services, ultimately enhancing the urban living experience and fostering more sustainable and efficient cities.

Keywords: Smart cities · Services for Citizen · Software platform

1 Introduction

Urbanization is a global trend that has been steadily growing over the past decades. As cities continue to expand and attract more inhabitants, there is an increasing need for efficient management of resources, services, and infrastructure to ensure the well-being of citizens and the sustainability of urban environments. In response to these challenges, the concept of smart cities has emerged, which aims to leverage the potential of information and communication technologies (ICT) and data-driven solutions to optimize the functioning of cities and enhance the quality of life for their inhabitants.

Smart cities integrate a wide range of technologies and systems, including the Internet of Things (IoT), sensors, big data analytics, and artificial intelligence (AI), to collect, process, and analyze data from diverse sources. This data can

S. Ossowski et al. (Eds.): DCAI 2023, LNNS 740, pp. 175–183, 2023.
https://doi.org/10.1007/978-3-031-38333-5_18

be used to monitor and manage various aspects of urban life, such as traffic, waste management, energy consumption, and public safety, among others. The insights gained from this data enable city administrators to make more informed decisions, improve the efficiency of city services, and foster a more sustainable and livable urban environment.

However, the implementation of smart city initiatives often faces several challenges, particularly in terms of data management and integration. The diverse nature of data sources, including relational and non-relational databases, web services, and IoT devices, can lead to difficulties in accessing, processing, and analyzing the data. Furthermore, the large volumes of data generated in urban environments require scalable and efficient solutions for data storage, preprocessing, and analysis.

To address these challenges, there is a need for comprehensive platforms that can facilitate the integration and utilization of heterogeneous data sources in smart cities. These platforms should provide city administrators with the necessary tools and capabilities to manage diverse data sources, preprocess and transform the data, and deliver meaningful insights and services to citizens. Furthermore, such platforms should be user-friendly and accessible to both technical and non-technical users, enabling them to easily access and interact with the data and services provided by the smart city.

In this paper, we present a novel platform for managing heterogeneous data sources in smart cities and delivering dynamic, interactive insights to citizens. Our platform addresses the challenges of integrating and utilizing diverse data sources, providing city administrators with a comprehensive backend system for managing city assets and services, and enabling citizens to benefit from the wide range of services offered by smart cities.

The proposed platform has been implemented using various programming languages and technologies tailored to the specific needs of each component, including TypeScript, Vue.js, and Node.js for the web application development, and Java and Python for data preprocessing and real-time data processing components. The platform has been successfully deployed and tested in a pilot smart city project, demonstrating its effectiveness in streamlining the integration and utilization of smart city services.

The remainder of this paper is structured as follows: Sect. 2 provides an overview of the current state of smart cities and the challenges associated with managing heterogeneous data sources. Section 3 presents our proposed platform and its key features, including the use of various programming languages and technologies to address the challenges of data integration and utilization. Section 4 discusses the results of our pilot implementation, highlighting the platform's effectiveness in enhancing the urban living experience and fostering more sustainable and efficient cities. Finally, Sect. 5 concludes the paper and outlines future work to further improve and expand the capabilities of our platform.

2 Background

The rapid growth of urbanization worldwide has led to a surge in the demand for innovative solutions that can address the challenges of managing resources, infrastructure, and services in urban environments. Smart cities have emerged as a promising approach to tackle these challenges, utilizing information and communication technologies (ICT), data analytics, and artificial intelligence (AI) to optimize the functioning of cities and improve the quality of life for their inhabitants.

2.1 Smart City Technologies

Smart cities integrate a wide range of technologies and systems to facilitate the collection, processing, and analysis of data from diverse sources. Some of the key technologies employed in smart cities include:

- **Internet of Things (IoT):** IoT devices, such as sensors and actuators, are widely used in smart cities [1] to collect real-time data about various aspects of urban life, including traffic [14], air quality, energy consumption [3,13], industry [15], and waste management [6]. These devices enable continuous monitoring and control of city services [4], contributing to more efficient and sustainable urban environments [8].
- **Big Data Analytics:** Smart cities generate vast amounts of data, which can be analyzed using big data analytics techniques to extract valuable insights and patterns [9], for example, from social media sources [16]. These insights can be used by city administrators to make more informed decisions and improve the efficiency of city services.
- **Artificial Intelligence (AI):** AI techniques, such as machine learning, natural language processing, and computer vision, can be employed to process and analyze the data collected by smart cities [7]. AI can help identify patterns and trends in the data, enabling more accurate predictions and the development of more effective solutions to urban challenges.
- **Cloud Computing:** Cloud computing provides scalable and flexible computing resources, allowing smart cities to store, process, and analyze large volumes of data in a cost-effective manner [5]. Cloud-based platforms and services also facilitate the integration and sharing of data between different city stakeholders and systems, contributing to more holistic and coordinated urban management.
- **Geospatial Technologies:** Geospatial technologies, such as geographic information systems (GIS) and remote sensing, play a crucial role in smart cities by providing spatial information and analysis capabilities [12]. These technologies enable city administrators to visualize and analyze urban data in a spatial context, supporting more effective planning and decision-making processes.

2.2 Challenges in Managing Heterogeneous Data

Despite the significant potential of smart city technologies, several challenges need to be addressed to fully realize their benefits [2]. One of the primary challenges lies in the management and integration of heterogeneous data sources, which can include:

- **Relational and Non-relational Databases:** Smart cities often rely on a mix of relational and non-relational databases to store and manage various types of data, such as structured, semi-structured, and unstructured data. Integrating these diverse data sources can be complex and time-consuming, requiring the development of custom solutions and data transformation processes.
- **Web Services:** Many smart city applications depend on web services to access real-time data and functionality provided by external systems and organizations. Integrating these web services into smart city platforms can involve challenges related to data format, access control, and service reliability.
- **IoT Devices:** IoT devices generate large volumes of real-time data, which need to be ingested, processed, and analyzed in an efficient and scalable manner. Ensuring the interoperability and seamless integration of IoT devices with smart city platforms is essential for the effective functioning of these systems.
- **Data Volume and Velocity:** The sheer volume and velocity of data generated by smart cities can pose significant challenges for data management and analysis. Traditional data processing and storage solutions may struggle to keep up with the demands of handling large-scale, real-time data, necessitating the adoption of more advanced technologies, such as big data analytics and cloud computing.
- **Data Quality:** Ensuring the quality and accuracy of data collected from various sources is crucial for the success of smart city initiatives. Inaccurate or incomplete data can lead to incorrect insights and decision-making, potentially undermining the effectiveness of smart city solutions. Addressing data quality issues requires robust data validation, cleaning, and enrichment processes.
- **Data Integration:** Integrating data from diverse sources and formats can be a complex and time-consuming task, requiring the development of data transformation processes and custom integration solutions. Effective data integration is essential for enabling a holistic view of urban systems and facilitating the sharing of information between different city stakeholders.
- **Data Security and Privacy:** The collection and processing of large volumes of data, including personal and sensitive information, raises significant concerns regarding data security and privacy. Smart city platforms must implement robust security measures and privacy-preserving techniques to protect the confidentiality, integrity, and availability of data, and to comply with relevant data protection regulations.

2.3 Open Data Platforms

Open data platforms play a crucial role in fostering transparency, collaboration, and innovation in smart cities. These platforms enable the publication, sharing, and reuse of data by various stakeholders, including citizens, businesses, researchers, and public authorities. A number of open data platforms have been developed and deployed in cities around the world, with some of the most prominent examples including:

- **CKAN:** The Comprehensive Knowledge Archive Network (CKAN) is an open-source data management system designed to facilitate the publication and sharing of open data. CKAN provides a rich set of features, including data cataloging, metadata management, search and discovery, data visualization, and API access, making it a popular choice for many open data initiatives.
- **Socrata:** Socrata is a cloud-based open data platform that offers a suite of tools and services for managing, sharing, and analyzing data. Socrata provides features such as data hosting, API management, data visualization, and data access controls, enabling organizations to easily publish and share their data with a wide range of users.
- **OpenDataSoft:** OpenDataSoft is a cloud-based open data platform designed to simplify the process of publishing, sharing, and visualizing data. The platform offers features such as data cataloging, metadata management, data transformation, and API access, allowing organizations to quickly and easily make their data accessible to external users.
- **ArcGIS Open Data:** ArcGIS Open Data is a component of the ArcGIS platform that enables organizations to publish and share their geospatial data as open data. The platform provides tools for data cataloging, metadata management, search and discovery, and data visualization, making it a popular choice for open data initiatives involving spatial information.

While these open data platforms have made significant progress in facilitating the publication and sharing of data, there is still room for improvement, particularly with regard to addressing the challenges associated with managing heterogeneous data sources and ensuring data security and privacy. As the demand for open data continues to grow, it is essential to develop innovative solutions and enhancements to existing platforms to meet these evolving requirements.

3 Proposal

Our proposed platform aims to streamline the integration and utilization of smart city services by providing city administrators with a comprehensive backend system for managing diverse data sources and delivering actionable insights to citizens. The platform has been designed using a combination of cutting-edge web development technologies and high-performance data preprocessing tools. The key features, components, and technology choices of the platform include:

- Web application development: The frontend of the platform is developed using TypeScript and Vue.js, which are modern, flexible, and powerful web development frameworks. These technologies enable the creation of responsive and user-friendly interfaces, making it easier for city administrators to interact with and manage city assets and services. The backend of the platform is built using Node.js, which is known for its high performance, scalability, and ease of integration with various data sources and services.
- Data preprocessing and transformation: The platform includes advanced data preprocessing and transformation capabilities, which are implemented using Java and Python. These languages were chosen due to their efficiency and the wide array of available libraries for data manipulation and transformation, allowing city administrators to clean, normalize, and enrich data from diverse sources. This ensures that the data is accurate, consistent, and suitable for analysis and visualization.
- Web application development: To create an intuitive and user-friendly web application for city administrators and citizens, we have employed TypeScript, Vue.js, and Node.js. TypeScript offers strong typing and improved maintainability, while Vue.js, a lightweight and versatile JavaScript framework, enables the development of responsive and interactive user interfaces. Node.js serves as a scalable and efficient backend solution that can handle numerous simultaneous connections, making it an ideal choice for the platform's server-side requirements.
- Real-time data processing and analysis: The platform supports real-time data processing and analysis, enabling city administrators to monitor the status of city assets and services in real-time and respond to emerging issues and opportunities more effectively. Java and Python are used for real-time data processing, leveraging their performance and extensive libraries for efficient data handling and analysis.
- Interactive data visualization: The platform provides a range of interactive data visualization tools, implemented using Vue.js and D3.js, a powerful data-driven documents library. This combination allows city administrators to create dynamic and engaging visual representations of city assets and services, making it easier for citizens to access and understand the data.
- User management and access control: The platform includes robust user management and access control features, ensuring that sensitive data is protected and only accessible to authorized users. This is achieved through the use of Node.js for server-side authentication and authorization, along with secure client-side libraries for managing user sessions and access control on the frontend.

By leveraging these technologies, our proposed platform addresses the challenges associated with managing heterogeneous data sources in smart cities and provides an effective solution for city administrators and citizens to better utilize smart city services.

4 Results

We have implemented the proposed platform using the selected technologies and programming languages, including TypeScript, Vue.js, and Node.js for the web application development, as well as Java and Python for the data preprocessing and real-time data processing components. The platform has been successfully deployed and tested in a pilot smart city project, yielding the following results:

- Seamless integration of heterogeneous data sources: The platform's data ingestion and integration capabilities have enabled city administrators to easily incorporate data from a wide variety of sources, including relational and non-relational databases, web services, and IoT devices. This has facilitated a more comprehensive understanding of city assets and services, allowing administrators to make more informed decisions and better allocate resources.
- Enhanced data quality and consistency: The platform's advanced data preprocessing and transformation features have helped to ensure that the data is accurate, complete, and consistent. This has led to more reliable and meaningful insights for both city administrators and citizens, ultimately improving the effectiveness of smart city services.
- Real-time monitoring and decision-making: The platform's real-time data processing and analysis capabilities have allowed city administrators to monitor the status of city assets and services more closely, enabling them to respond more effectively to emerging issues and opportunities. This has led to more efficient and proactive management of city resources, ultimately benefiting citizens through improved services and reduced costs.
- Increased citizen engagement and access to information: The platform's interactive data visualization tools, developed using Vue.js and D3.js, have made it easier for citizens to access and understand the data related to city assets and services. This has fostered greater citizen engagement with smart city initiatives, empowering citizens to take advantage of the benefits offered by these services and contribute to the ongoing improvement of urban life.
- Secure and efficient user management and access control: The platform's user management and access control features, implemented using Node.js and secure client-side libraries, have helped to ensure that sensitive data is protected and only accessible to authorized users. This has maintained the privacy and security of citizens' data while still enabling them to benefit from the insights and services provided by the platform.

These results demonstrate the effectiveness of our platform in streamlining the integration and utilization of smart city services, ultimately enhancing the urban living experience and fostering more sustainable and efficient cities. The use of diverse technologies and programming languages, tailored to the specific needs of each component, has contributed to the platform's robust performance and seamless integration with various data sources and systems.

5 Conclusion and Future Work

In this paper, we have presented a novel platform for managing heterogeneous data sources in smart cities and delivering dynamic, interactive insights to citizens. Our platform addresses the challenges of integrating and utilizing diverse data sources, providing city administrators with a comprehensive backend system for managing city assets and services, and enabling citizens to benefit from the wide range of services offered by smart cities.

The results from our pilot implementation have demonstrated the platform's effectiveness in streamlining the integration and utilization of smart city services, ultimately enhancing the urban living experience and fostering more sustainable and efficient cities. In the future, we plan to extend the platform's capabilities, incorporating additional data sources and technologies, and exploring new ways to leverage the platform's features to benefit both city administrators and citizens. Additionally, we will investigate the potential for integrating advanced analytics and machine learning techniques into the platform, further enhancing its ability to deliver meaningful and actionable insights for smart city initiatives.

Acknowledgments. This work has been partially supported by the Institute for Business Competitiveness of Castilla y León, and the European Regional Development Fund under grant CCTT3/20/SA/0002 (AIR-SCity project).

References

1. Bellini, P., Nesi, P., Pantaleo, G.: IoT-enabled smart cities: a review of concepts, frameworks and key technologies. Appl. Sci. **12**(3), 1607 (2022)
2. Bokhari, S.A.A., Myeong, S.: Use of artificial intelligence in smart cities for smart decision-making: a social innovation perspective. Sustainability **14**(2), 620 (2022)
3. Canizes, B., Pinto, T., Soares, J., Vale, Z., Chamoso, P., Santos, D.: Smart city: a GECAD-BISITE energy management case study. In: De la Prieta, F., et al. (eds.) PAAMS 2017. AISC, vol. 619, pp. 92–100. Springer, Cham (2018). https://doi.org/10.1007/978-3-319-61578-3_9
4. Casado-Vara, R., Martin-del Rey, A., Affes, S., Prieto, J., Corchado, J.M.: IoT network slicing on virtual layers of homogeneous data for improved algorithm operation in smart buildings. Future Gener. Comput. Syst. **102**, 965–977 (2020)
5. Casado-Vara, R., Prieto-Castrillo, F., Corchado, J.M.: A game theory approach for cooperative control to improve data quality and false data detection in WSN. Int. J. Robust Nonlinear Control **28**(16), 5087–5102 (2018)
6. Chamoso, P., González-Briones, A., Rodríguez, S., Corchado, J.M.: Tendencies of technologies and platforms in smart cities: a state-of-the-art review. Wirel. Commun. Mob. Comput. (2018)
7. Chamoso, P., González-Briones, A., Rivas, A., De La Prieta, F., Corchado, J.M.: Social computing in currency exchange. Knowl. Inf. Syst. **61**(2), 733–753 (2019). https://doi.org/10.1007/s10115-018-1289-4
8. Corchado, J.M., et al.: Deepint. net: a rapid deployment platform for smart territories. Sensors **21**(1), 236 (2021)

9. Corchado, J.M., Pinto-Santos, F., Aghmou, O., Trabelsi, S.: Intelligent development of smart cities: Deepint.net case studies. In: Corchado, J.M., Trabelsi, S. (eds.) SSCTIC 2021. LNNS, vol. 253, pp. 211–225. Springer, Cham (2022). https://doi.org/10.1007/978-3-030-78901-5_19
10. Dash, B., Sharma, P.: Role of artificial intelligence in smart cities for information gathering and dissemination (a review). Acad. J. Res. Sci. Publishing **4**(39) (2022)
11. García-García, L., Jiménez, J.M., Abdullah, M.T.A., Lloret, J.: Wireless technologies for IoT in smart cities. Netw. Protoc. Algorithms **10**(1), 23–64 (2018)
12. Garcia-Retuerta, D., Chamoso, P., Hernández, G., Guzmán, A.S.R., Yigitcanlar, T., Corchado, J.M.: An efficient management platform for developing smart cities: solution for real-time and future crowd detection. Electronics **10**(7), 765 (2021)
13. González-Briones, A., Chamoso, P., de Alba, F.L., Corchado, J.M.: Smart cities energy trading platform based on a multi-agent approach. In: Artificial Intelligence and Environmental Sustainability: Challenges and Solutions in the Era of Industry, vol. 4, pp. 131–146 (2022)
14. Martí, P., Jordán, J., Chamoso, P., Julian, V.: Taxi services and the carsharing alternative: a case study of Valencia city (2022)
15. Rivas, A., Fraile, J.M., Chamoso, P., González-Briones, A., Sittón, I., Corchado, J.M.: A predictive maintenance model using recurrent neural networks. In: Martínez Álvarez, F., Troncoso Lora, A., Sáez Muñoz, J.A., Quintián, H., Corchado, E. (eds.) SOCO 2019. AISC, vol. 950, pp. 261–270. Springer, Cham (2020). https://doi.org/10.1007/978-3-030-20055-8_25
16. Shoeibi, N., Shoeibi, N., Julian, V., Ossowski, S., Arrieta, A.G., Chamoso, P.: Smart cyber victimization discovery on twitter. In: Corchado, J.M., Trabelsi, S. (eds.) SSCTIC 2021. LNNS, vol. 253, pp. 289–299. Springer, Cham (2022). https://doi.org/10.1007/978-3-030-78901-5_25

Step-Wise Model Aggregation for Securing Federated Learning

Shahenda Magdy(✉), Mahmoud Bahaa, and Alia ElBolock

German International University, Cairo, Egypt
shahendamagdymohammed54@gmail.com

Abstract. Federated learning (FL) is a distributed machine learning technique that enables remote devices to share their local models without sharing their data. While this system benefits security, it still has many vulnerabilities. In this work, we propose a new aggregation system that mitigates some of these vulnerabilities. Our aggregation framework is based on: Connecting with each client individually, calculating clients' model changes that will affect the global model, and finally preventing aggregation of any client model until the accepted range of distances with other clients is calculated and the range of this client is within it. This approach aims to mitigate against Causative, Byzantine, and Membership Inference attacks. It has achieved an accuracy of over 90 percent for detecting malicious agents and removing them.

Keywords: Federated Learning · Security · Step - wise Model Aggregation

1 Introduction

In order to make predictions or decisions without being explicitly programmed, Machine learning (ML) algorithms use sample data, or training data, to build a model. These algorithms are used in a wide range of applications, such as medicine, email filtering, speech recognition, agriculture, and computer vision, where it is difficult or impossible to develop conventional algorithms to perform the required tasks. Since clients' data is typically included in exchanged gradients or models, data privacy and confidentiality are a concern in such approaches as these data may hold confidential information.

Federated learning (FL) is a machine learning technique that trains an algorithm across multiple decentralized edge devices or clients holding local data samples, without exchanging them. Training data is partitioned and trained by trainers locally and individually in federated learning. Local gradients are sent to the parameter server which aggregates those gradients and updates global model parameters accordingly. The primary benefit of employing FL approaches to machine learning is the guarantee of data secrecy and privacy. In fact, there is no external upload, concatenation, or exchange of local data. It is more difficult

S. Ossowski et al. (Eds.): DCAI 2023, LNNS 740, pp. 184–192, 2023.
https://doi.org/10.1007/978-3-031-38333-5_19

to break into the database because it is divided into local bits externally. To perform that successfully, all parties should ensure that they are dealing with a security system.

Although FL does contain differential-privacy (DP) and many works have been done in this area, it is still vulnerable to many attacks. Many of the works can serve as a powerful first line of defense against partially honest participants who wish to expose the private data of other participants. There is still a huge gap between what is actually needed and the existing privacy-preserving solutions. Participants can now view the intermediate model and provide arbitrary updates as part of the federated learning environment. For example, attackers who impersonate helpful participants may send manipulated updates that maliciously affect the characteristics of the trained model (Causative attack), or a lazy participant may lie about the volume of data records he or she uses for training, causing the algorithm to produce an inaccurate model.

Assuring the integrity of regional training procedures is essential for the effective mitigation of these model-tampered threats on federated learning. In other words, the local training algorithm's output needs to be integrated. There are established defenses to control the participants or explicit observation of the training, such as robust losses and anomaly detection data. Since neither of these presumptions holds true security for federated learning, maintaining integrity is a significant challenge; because each participants' local model parameters are the only ones that the server observes.

To mitigate this, we developed a new integrity system that disallows each client to integrate his local model with the server unless it is completely ensured that they are an honest client. We first let each client send their local model and calculate Euclidean Distance between the client model and the global model. Then we compare all the distances of all the client models and determine the accepted range of distances. Finally, we recheck the distance of each client and according to the result, the system will decide whether to aggregate with the client's model or not.

This is the first step in order to connect with clients. If any maliciousness is detected that client's connection is immediately closed and the client cannot add any updates to the server. The client is removed totally from the system. The approach is expected to mitigate against all Constructive, Byzantine, and Inference attacks.

This paper is structured as follows: Sect. 2) Related work: which will discuss the previous works that were done to secure the federated learning system. Section 3) Methodology section: explaining our approach. Section 4) Experiment section: A brief explanation of the experiment done, data sets used, and table accurately showing different results of the experiment. Evaluating these results before the conclusion is also in this Sect. 5) Finally, the conclusion and Future work.

2 Related Work

Federated learning (FL) enables machine learning models to obtain experience from different data sets located in different sites without sharing training data. This allows personal data to remain on local sites although FL learns from it. Local data samples are not shared. This improves data protection and privacy. The characteristics of the global model are shared with local data centers to integrate the global model into their ML local models. Although, there're still security challenges facing Fl. To improve that, some related works are introduced in the following.

Fang et al. developed a multi-party data protection machine learning framework in [2] that combines homomorphic encryption and federated learning. Homomorphic encryption, allows calculations to be performed on encrypted data without decrypting it. The result of the homomorphic operation after decryption is equivalent to the operation on the plain text data.

Another approach is proposed in [5] by Kanagavelu et al. that chooses a subset of FL members as members of the model aggregation committee from the entire member list. Elected committee members then use multi-party computation services to aggregate local models from all FL parties. This provided a mechanism to enable MPC-based model aggregation over the set of model tensors with large parameter sets. Using both Additive and Shamir Secret Sharing MPC protocols. Compared to traditional peer-to-peer frameworks, this is two phases. The MPC-enabled FL framework greatly reduces communication costs and improves system scalability. In the previous 2 works Combining Differential Data Protection with Secure Multiparty Computing/ Homomorphic Encryption allows them to reduce the growth of noise injection by enhancing the number of parties without sacrificing privacy while preserving a predefined confidence level.

Using HE, the proposed scheme in [3] realizes the assessment of data inequality in a way of protecting privacy. Guo et al. propose a secure aggregation protocol that then uses a zero-knowledge proof protocol to offload the task of detecting attacks in a local model from the server to the user. They developed the ZKP protocol that allows users to review models without revealing information about them and without back doors. Their framework thus allows a central server to identify trained model updates without violating the privacy guarantees of secure aggregation.

Brunetta et al. propose a novel approach called NIVA in [1], that enables the distributed aggregation of secret inputs from multiple users by multiple entrusted servers. The result of the aggregation can be verified by the public in a non-interactive manner. In the context of federated learning, for instance, NIVA can be utilized to securely compute the sum of a large number of user data in order to aggregate the model updates for a deep neural network.

A FastSecAgg approach is proposed in [4], which is computationally and communication-ally efficient. FastShare, a novel multi-secret sharing scheme band robust to client dropouts' secure aggregation protocol. FastShare achieves a trade-off between the number of secrets, privacy threshold, and dropout tolerance in order to be information secure. They demonstrate that FastSecAgg is (i) secure against the server colluding with any subset of some constant fraction, such as 10 percent of the clients in the honest-but-curious search, based on FastShare's capabilities; and (ii) is tolerant of a random subset of some constant fraction, such as less than ten percent of clients. While maintaining the same communication cost, FastSecAgg achieves a significant reduction in computation costs. In addition, it ensures security against adaptive adversaries, which can dynamically carry out client corruption during the protocol's execution.

3 Step-Wise Model Aggregation

All the previously presented works show efficient proposed secure aggregation schemes but none of them implemented efficient action against malicious users. Each user's private data is used to train the model, and only local updates are transmitted to the server. In this way, user data is protected while also being used inadvertently to create better models. In our approach, the models sent by each client will be classified as honest client models or malicious client models. Honest clients' models will be classified by 2 filters.

3.1 Filter 1: Consistency Filter

Once clients start connecting to the server, they will be added to a stack. The server will start to accept each of these clients' local models individually and in order.

Firstly, after each client sends their model, aggregation happens by converting the server's global model and the client's model to 2 matrices. The matrices will be created from the neural layers shown in Fig. 1.

Secondly, the system attempts to sum these matrices, divide them by 2 (Getting the average) then convert them to a new model matrix. The system will detect maliciousness if this summation of matrices failed due to different shapes. As shown in Fig. 2 The purpose of Filter 1 is to detect maliciousness by different shapes of matrices (occurs when using different data-sets to train the models).

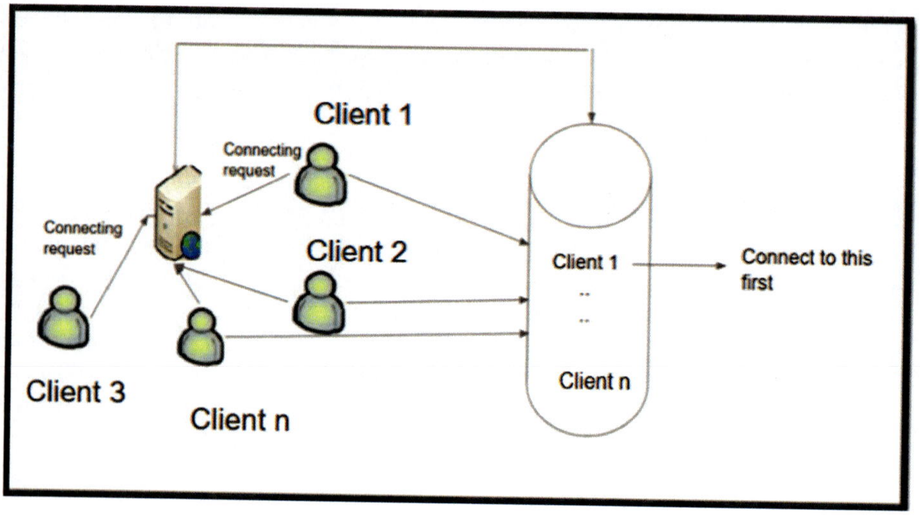

Fig. 1. Overview of Connection and Ordering of Clients.

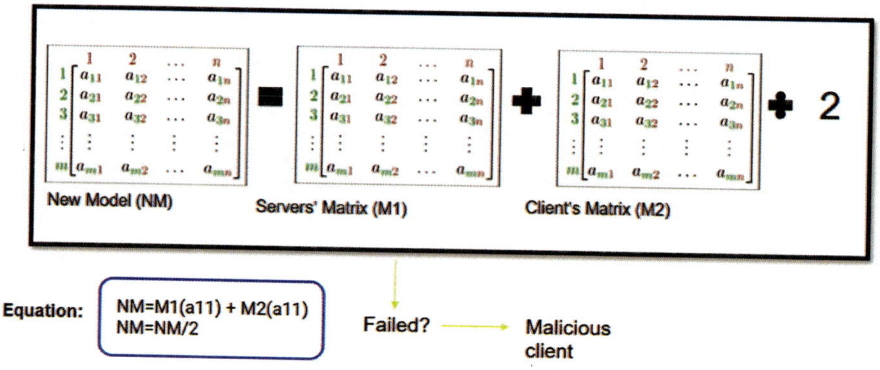

Fig. 2. Matrix Summation and Calucaltions

3.2 Filter 2: Average Distances

As shown in Fig. 3, after calculating the new model matrix, euclidean distance will be calculated between this new model matrix and the old model's matrix based on paper [6]. The process of this client will be paused, and the whole process will be repeated for all remaining clients. To calculate Euclidean distances of all other clients. Using all these distances, the mean and standard deviation (std) will be calculated. Values within range of the standard deviation added to and subtracted from the mean will be the accepted. Each client model will be checked by comparing its distance with that range. Using this approach the system will be able to detect the change each client added to the global model and claim that extreme changes resulted because of malicious updates of that model.

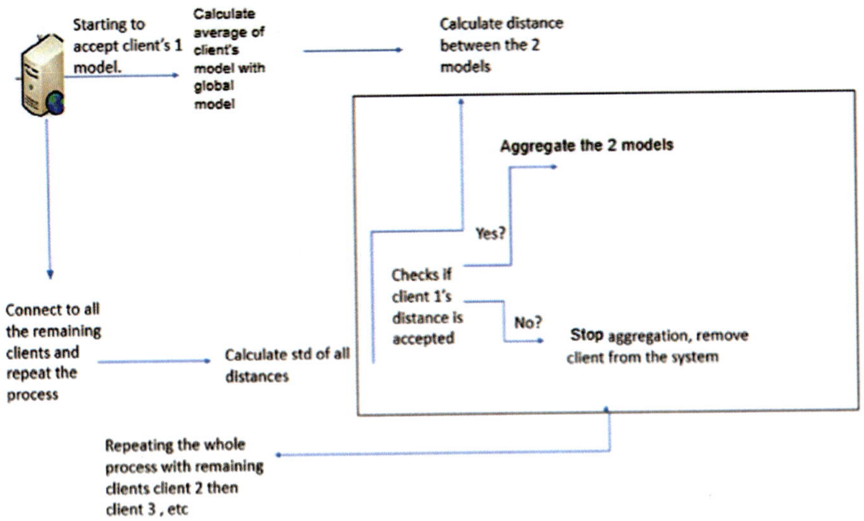

Fig. 3. Overview of the whole Filter 2

After each client's model passes both filters, the aggregation will be done by setting the global model matrix into the new model's matrix.

4 Experiment

The experiment is done using the MNIST, FMNIST, and CIFAR10 datasets.

MNIST is a dataset of handwritten digits that has 60,000 examples and 10,000 examples of handwritten digits from 0 to 9 that have been normalized and centered in a fixed-size image with dimensions (28×28). To train malicious clients that will try to corrupt the actual model, CIFAR-10 Dataset is used. CIFAR-10 is a dataset of images that are used to help computers recognize objects. This dataset will highly affect the MNIST-based model as it contains objects' images and MNIST is based on handwritten digits' images. It includes 60,000 images that are each 32×32 pixels in size.

Fashion MNIST (FMNIST) is a dataset of Zalando's article images-consisting of a training set of 60,000 examples and a test set of 10,000 examples. Each example is a 28×28 labeled fashion grey-scale image, associated with a label from 10 classes. Moreover, some random integer arrays were also used to train some of the malicious clients' models.

4.1 Experiment Scenario

Different experiments were done, one using 4 clients and the other using 8 clients. Auto - random clients algorithm was done. Different models were implemented and the machine will choose a random model for each of the clients Fig. 4.

Honest clients models were trained using Keras MNIST library with different distribution for each client. Malicious clients models were trained using Keras Cifar10, FMNIST, and others trained using random inputs arrays.

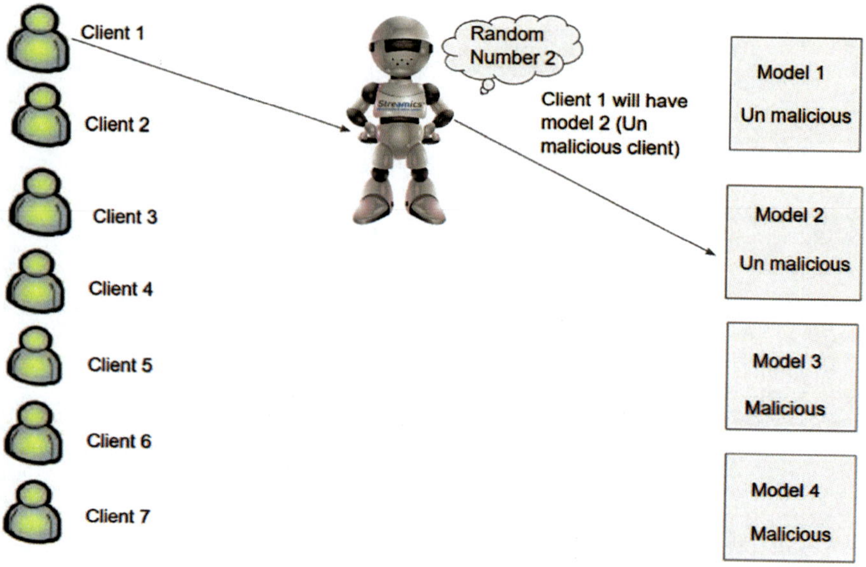

Fig. 4. The malicious random clients' machine.

4.2 Results

Results show that Filter 1 failed to detect 33% of malicious clients. After applying Filter 2, this rate effectively dropped to 4.7%, however it led to false detection of honest clients as malicious by about 11%; as malicious clients' distances affected calculations of mean and std (Table 1).

Filter 1 only took 22.5 s to aggregate all 8 clients. While using both filters caused it to take an average of 1 min and 2.7 s to complete the clients' aggregations. Figure 5a shows the effect of malicious client's enhancement in number on false - negative' rates. Figure 5b shows the effect of malicious clients' enhancement in number on false-positive rates. These show that after using filter 2 false - negative rates are limited and enhancement of malicious clients' numbers can't highly affect it. While false-positive rates are increasing with the enhancement of malicious clients' numbers.

Table 1. Overview of results

Filter Used	No. of Clients	Detected Clients	Actual Malicious Clients	Time Taken
Filter 1	4 Clients	Client 1 Client 3	Client 1 Client 2 Client 3 Client 4	18.4 s
	8 Clients	Client 1 Client 2 Client 3 Client 4	Client 1 Client 2 Client 3 Client 4 Client 5 Client 6	20.9 s
Filter 1 & 2	4 Clients	Client 1 Client 2	Client 1 Client 2	56 s
		Client 2	Client 2 Client 3	1 min
	8 Clients	Client 1 Client 2 Client 3 Client 4	Client 1 Client 2 Client 4	43.9 s

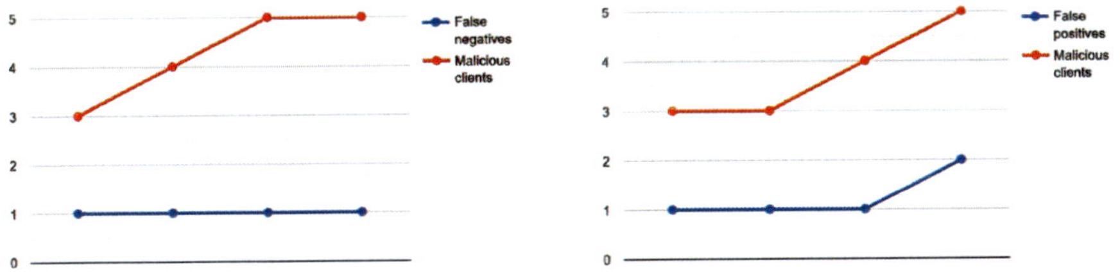

(a) (b)

Fig. 5. Affect on false negative and false positive rates

5 Conclusion

In this work, we have proposed a very efficient step-wise aggregation system that highly secures federated learning systems. Compared to existing approaches, our system guarantees highly efficient global model updates as it prevents any client from updating the global model until it is ensured that he/she sends appropriate model updates and compares his updates with other clients' updates. And ensure that these updates won't corrupt/change the global model. Although the excessive presence of malicious models/clients can affect it, still its effect is very limited and the presence of malicious models/clients in the presence of other honest clients is detected and removed easily. We guarantee that this approach effectively mitigates against Byzantine, Causative, and Membership Inference attacks.

5.1 Future Work

For future work, we plan to work on more maliciousness presence and try to decrease its effect on false positive results and totally clean its effect on false negative results. Moreover, work on increasing the efficiency of the whole approach to take less time and processes.

References

1. Brunetta, C., Tsaloli, G., Liang, B., Banegas, G., Mitrokotsa, A.: Non-interactive, secure verifiable aggregation for decentralized, privacy-preserving learning. In: Baek, J., Ruj, S. (eds.) ACISP 2021. LNCS, vol. 13083, pp. 510–528. Springer, Cham (2021). https://doi.org/10.1007/978-3-030-90567-5_26
2. Fang, H., Qian, Q.: Privacy preserving machine learning with homomorphic encryption and federated learning. Future Internet **13**(4), 94 (2021)
3. Guo, J., Liu, Z., Lam, K.-Y., Zhao, J., Chen, Y., Xing, C.: Secure weighted aggregation for federated learning. arXiv preprint arXiv:2010.08730 (2020)
4. Kadhe, S., Rajaraman, N., Koyluoglu, O.O., Ramchandran, K.: FastSecAgg: scalable secure aggregation for privacy-preserving federated learning. arXiv preprint arXiv:2009.11248 (2020)
5. Kanagavelu, R., et al.: Two-phase multi-party computation enabled privacy-preserving federated learning. In: 2020 20th IEEE/ACM International Symposium on Cluster, Cloud and Internet Computing (CCGRID), pp. 410–419. IEEE (2020)
6. Zhang, Z., Cao, X., Jia, J., Gong, N.Z.: FLDetector: defending federated learning against model poisoning attacks via detecting malicious clients. In: Proceedings of the 28th ACM SIGKDD Conference on Knowledge Discovery and Data Mining, pp. 2545–2555 (2022)

Federated Genetic Programming: A Study About the Effects of Non-IID and Federation Size

Bruno Ribeiro, Luis Gomes, Ricardo Faia, and Zita Vale(✉)

GECAD Research Group on Intelligent Engineering and Computing for Advanced Innovation and Development, LASI—Intelligent Systems Associate Laboratory, Polytechnic of Porto, R. Dr. António Bernardino de Almeida, 431, 4200-072 Porto, Portugal
{brgri,lfg,rff,zav}@isep.ipp.pt

Abstract. The privacy and security of users' data have been a concern in the last few years. New techniques like federated learning have appeared to allow the training of machine learning models without sharing the users' personal data. These systems have a lot of variables that can change the outcome of the models. Studies have been made to explore the effect of these variables and their importance. However, those studies only focus on machine learning models, namely deep learning. This paper explores the impact of the federation size in unbalanced data using a genetic programming model for image classification. The results show that the federation size can have impact on the contributions of each client, and the dataset size can influence the quality of the individuals.

Keywords: Federated Learning · Genetic Programming · Non-IID · Image Classification

1 Introduction

Federated learning (FL) has been one of the most recent artificial intelligence (AI) methodologies that have been rising interest among the research community [1]. This is especially due to the General Data Protection Regulation (GDPR) [2] that was implemented in Europe to combat the malicious use of user's personal data. The key point of FL is to train AI models among a group of clients with similar objectives without the need to share personal data. There can be a lot of different factors that can change the overall performance of the system, e.g., the number of clients, the data distribution among the federation and the AI model used [3]. Despite existing already some research towards studying the effects of these configurations in the performance of FL systems, its focused only on deep learning models.

Genetic programming (GP) is an algorithm where the evolutionary theory is the key concept [4]. The idea is that several computer programs can evolve through time to achieve a desired solution. This is an old methodology that has been more recently used in machine learning tasks like image classification (IC) [5]. GP is a good candidate to be

S. Ossowski et al. (Eds.): DCAI 2023, LNNS 740, pp. 193–202, 2023.
https://doi.org/10.1007/978-3-031-38333-5_20

used in a federation setting given that FL offers the distribution of computation power and data privacy.

The aim of this paper is to study the impacts that FL configurations when using GP as the global model. The configurations are related to the number of clients in the federation and the data distribution among the clients, which is known as non-identically and independently distributed (non-IID). To study these implications, an existing model was adopted, Flexible Genetic Programming (FGP) [6]. Despite the use of GP in this methodology it is possible to apply other models as well, such as evolutionary algorithms. Three scenarios with the same configurations except for the federation size were developed. In each scenario, the federation will cooperate to solve an IC problem based on the Federated Extended MNIST (FEMNIST) dataset. The results are analyzed and further discussed.

The present paper is structured in the following sections: Introduction, where an overview of the themes talked in this paper is presented, the respective summary of the methodology and the paper structure; Related works, which will describe the current work regarding GP and FL; Methodology, where the methods used in this paper are described; Case Study, where the description of the study conducted is given; Results, where the results are discussed; finally, Conclusion, outlines the main conclusions of this paper and talks about future work.

2 Related Works

FL is a recent AI field that was initially discovered by Google in 2016 [7]. This interest in training AI models in a distributed private environment arose after European Union regulate the malicious use of users' data [2]. With data privacy and protection in mind, FL has since been explored to provide solutions to this problem [8]. Because FL has several variables in the system that can change the performance of the model, researchers have been studying the impact of these variables in the system and how to mitigate its negative effects [3]. In [9] the authors investigate the effects of non-IID on the quality of a deep learning model in an FL setting. Another study [10] suggests that clustering the clients based on their data characteristics can impact positively the performance of deep learning models. The authors of this paper [11] compared the performance of some machine learning models in FL settings with several architectures with the performance of centralized learning. Today's research is focusing more on the FL settings based on machine learning models. However, this field has yet to study the impacts of the FL configurations in other models, like GP.

Even though GP has not been explored as being the model of an FL system, there is a research field called distributed evolutionary algorithms that studies evolutionary algorithms, including GP, in distributed settings, with the purpose to alleviate the computational effort and help accelerate the evolution process [12]. In [13] the authors use a distributed approach to solve classification, symbolic regression, and n-parity problems with a cartesian GP algorithm. To solve a large number of equations in a more efficient way, the authors of [14] developed a new distributed technique that decomposes the problem into smaller problems and is distributed in a cluster of computers. In another study [15] new algorithms are presented that solve multi-objective problems by dividing them into smaller problems and solving them in a distributed way. In this context, FL can be a good approach to GP models, because not only answers the same problem as the distributed GP does but also helps maintain the privacy and protection of the users' data, which is a field that has yet to be explored.

The use of FL in real applications has a significant positive potential, namely where distributed knowledge systems are required to address complex problems. This is the case of smart grids, where distributed entities and users need to manage energy resources e an intelligent manner [16]. Previous works addressed individual knowledge, supported by machine learning, for energy forecast [17], context identification [18], and resources optimization [19]. Although sometimes is not properly addressed, in smart grids, the compliance with GDPR is an important task [20]. This way, the use of FL in smart grids can really create a significant impact. This paper will test and assess the quality of FL overall result considering different number of clients. To correctly validate the use of FL in a controlled environment using an already tested case study regarding IC.

3 The New Federated Learning Model

In this methodology, FGP was chosen to be the global model of the FL system. FGP is a GP algorithm that is used for IC problems. In this algorithm, the program follows a tree structure and is meant to be flexible. The authors defined four different types of layers according to the type of image processing functions there are pooling layer, filtering layer, extracting layer, and concatenation layer. Each node in the program tree represents an image processing function and the leaves can be randomized values or images depending on the function being fed. At the root of the program structure, there is a linear SVM classifier that is trained to classify the images based on the features given by the FGP program. The fitness is the error classification rate. Throughout the execution of FGP, a hall of fame (HoF) with the best individuals is maintained. The HoF is an ordered list according to the individuals' error. For a more detailed explanation of FGP please read its original work in [6].

In this FL system, there are three main phases in the process: the initialization phase, the fitness phase, and the evaluation phase. A round is constituted of the fitness and evaluation phases, meaning that the initialization only occurs in the beginning. In a deep learning FL system, there is normally a global model shared among the federation, and in some cases local models in each client. In this methodology, in the context of GP, the FL system will have a global HoF on the server side, and a local HoF for each client. The diagram of the FL process can be seen in Fig. 1.

The first step in the process is the initialization phase. The initialization phase consists of several steps, starting by initializing the federation. The server waits for enough clients to join the federation before starting the process. In this phase, the federation can decide to initialize the model, in the case of GP models one population can be initialized globally. This can be done by asking one client to do it, or the server can do it as well. In the implementation of this paper, no population was initialized to incentive the variety of solutions. Each client will produce their own population. After these steps, the fitness phase can start.

The fitness phase can be divided into three subphases: (i) client selection, (ii) fitness, and (iii) fitness aggregation. The client selection is, as the name suggests, when a group of

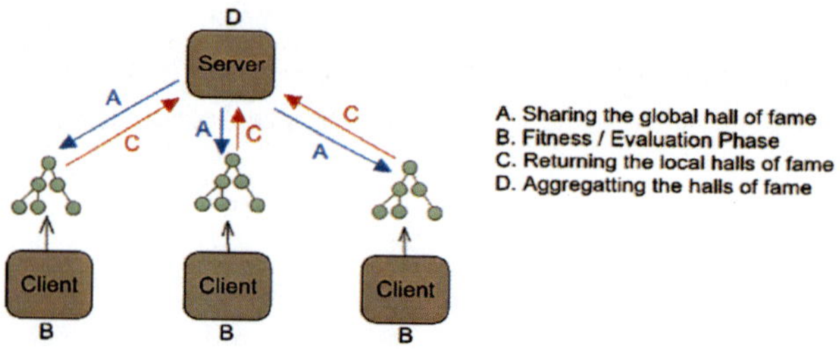

Fig. 1. Federated Learning process of the proposed methodology.

clients inside the federation is selected to participate in the fitness subphase. It starts with the server choosing which clients are going to be used. Normally this is done according to some pre-defined criteria. In this case, the criteria are based on the quantity of data that the client has. The fitness subphase will start with the server sending an instruction to the client to start training the model. In the first round, the initial population is the one created in the initialization phase, if any. In the next rounds, the initial population will be the global HoF maintained by the server, so the clients can start the evolution with the best individuals globally. The clients will then use an initial population, if any, and start to evolve their population throughout some generations. During this subphase, the clients will maintain their local HoF with the best individuals. At the end of the evolution, the clients will send only their best individual to the server. After the server receives all the results from the clients, the fitness aggregation subphase begins. In this subphase, the server will aggregate all of the clients' results, the HoF, into a single one.

The evaluation phase is very similar to the fitness phase, the difference is that instead of the clients training the model, they will evaluate the individuals of this round to update the global HoF. This phase is divided into three subphases as well: (i) client selection, (ii) evaluation, and (iii) evaluation aggregation. In the client selection, the server will start by selecting the clients that will evaluate. Here is also a predefined criterion to be used when selecting the clients. In this implementation, all the clients will be chosen even the ones that did not train the model in the fitness phase. This will help the model generalize better. In the evaluation subphase, the server will send the HoF, created previously in the fitness aggregation subphase, to the clients so everyone can evaluate them. The clients will then start to evaluate those individuals. When they finish evaluating the individuals, they send them to the server with the fitness values updated. In the evaluation aggregation subphase, the server will aggregate the evaluations and update the global HoF based on those evaluations. After this subphase, a new round starts from the fitness phase.

The fitness aggregation algorithm used in this implementation is used to aggregate the clients' local HoF into one HoF, in the fitness aggregation subphase. The server will have the best individuals for each client. Because it would very time-consuming and require a lot of computer resources to evaluate all the individuals, especially with huge federations, only a few can be selected. The criteria used in this selection is to use the local error of each individual to create a new HoF. This means that the individuals' error in this HoF will not reflect the error of the global federation. This HoF is temporary and does not interfere with the global HoF. After filtering and reducing the number on individuals, these will then be used in the evaluation phase.

In the evaluation aggregation algorithm, the idea is for the server to aggregate the HoFs the clients sent with the updated fitness values so they can be used to update the global HoF afterward. So, in the client evaluation phase, they will all evaluate the same individuals. The server will receive for each individual the fitness values regarding each client evaluation dataset. In addition to the individuals, the clients also send the size of the dataset they used to evaluate. This means that it is possible to calculate the global error classification rate of the individual as if the federation was working as one big dataset, making its evaluation precise. After these calculations, the individuals are updated with the global error and then used to update the global HoF.

4 Case Study

To test and analyze the impact of the number of clients in a non-IID setting using a GP-based FL system, a proper dataset was used. The Federated Extended MNIST (FEMNIST) was used in this case study [17]. In addition to the images of the 10 digits, from the original MNIST, it also has images of the lower-case letters and upper-case letters, making a total of 62 possible labels. The dataset is composed of 805,263 samples, produced by a total of 3,550 different writers. This dataset is divided by writers. The distribution of each writer is different in both size and label skewness. This dataset was provided by an open-source Python library called LEAF. LEAF also provides a benchmark baseline to test novel FL solutions described by its authors in the original paper [21].

Table 1 characterizes the three different scenarios of the case study. The main differences are the number of clients in the federation and the number of clients that will participate in the fitness phase. For this case study, it was only considered writers that had between 20 and 400 instances. The precondition required for the clients to be able to participate in the fitness phase is having more than 50 instances in their local dataset. So, clients between 20 and 50 can participate in everything except in the fitness phase, and those between 50 and 400 can participate in the whole process. All clients participate in the evaluation phase. The first scenario has only one client and it participates in every phase. The second scenario has 10 clients and one of them cannot participate in the fitness phase. The third scenario has 50 clients and five of the clients are not suitable to participate in the fitness phase.

Table 1. Total clients available in each scenario.

	Scenario 1	Scenario 2	Scenario 3
Total number of clients	1	10	50
Clients not able to train	0	1	5

For the GP configurations, it was used a population size of 100 individuals, 10 generations, and a HoF of size 10. The other configurations were adapted from the original FGP paper [6]. With the purpose of following the same tradition used in the GP community regarding the number of runs, the number of rounds was chosen to be 30. The difference here is that the global HoF is passed throughout the rounds.

5 Results

To analyze better the implications that the number of clients has in a federation several different aspects were considered. Firstly, the evolution of the client in scenario one is analyzed throughout the scenarios. Then the best individuals in each scenario are analyzed. Further, it is analyzed the impacts that the clients had either in other clients or in the global HoF. And finally, the correlation between the dataset sizes and the clients' contributions is discussed.

Client 0 is the client that was present in every scenario of the case study. Regarding the error, its best individual had a 19.02% error in the first scenario, 17.49% in the second scenario, and 21.17% in the third scenario, indicating an impact of individual error when increasing the number of clients/writers. When tuned correctly, the federation can have a good impact on the error. However, an excessive number of clients can negatively impact the federation model in a non-IID setting.

The lowest errors found in the global individuals in each scenario are 19.02% for scenario one, 24.66% for scenario two, and 28.33% for scenario three. In scenario one, the best global individual is the same as the best local of client 0 given that the federation has only one client. Adding more clients to the federation in this case study increases the error of the best global HoF. The reason behind that can be linked to the increase in data instances among the federation and, with that, the number of wrong classifications, making an impact on the overall global individuals. Other reason that could influence this outcome is that for each client they only train the individuals in a limited small dataset, the same for every scenario, and then they are evaluated in a larger dataset. The increase in error in the global HoF does not mean an increase in local error in the federation.

The clients that did not train had very sparse results, meaning that some benefited from the federation, while others did not benefit at all. For scenario two there was only one client that did not train, client 1. The error of the best individual of that client was approximately 8,57%. In scenario three there were five clients that did not train, including client 1, and the results are shown in Fig. 2. Two of the clients, 10 and 1, had models with reasonable results. Client 11 had an error of approximately 41%. Clients 13 and 12 had the highest errors which indicates that they did not benefit at all from the federation. Client 1 had the smallest error, indicating that the outcomes of the federation, in both scenarios, were best suited to its dataset.

Figure 3 shows the contributions that each client made in scenario two (a) and three (b). In each figure, on the left side of the graph are the clients that produced the individuals and on the right side are the ones that received individuals. The bigger the connection the more individuals were shared between the clients. It is possible to see, in scenario two (a), that every client provided at least one individual to other client. The client that had the most impact was client 3. Because Client 1 did not train, it receives the larger part of the individuals, and does not appear on the left side.

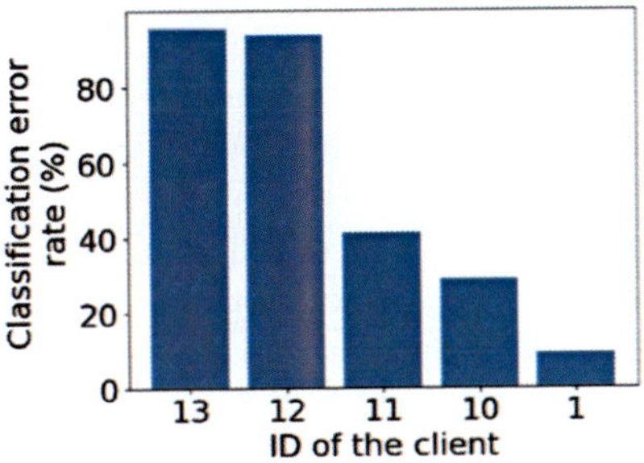

Fig. 2. Errors of the best individuals of the clients that did not train in scenario three.

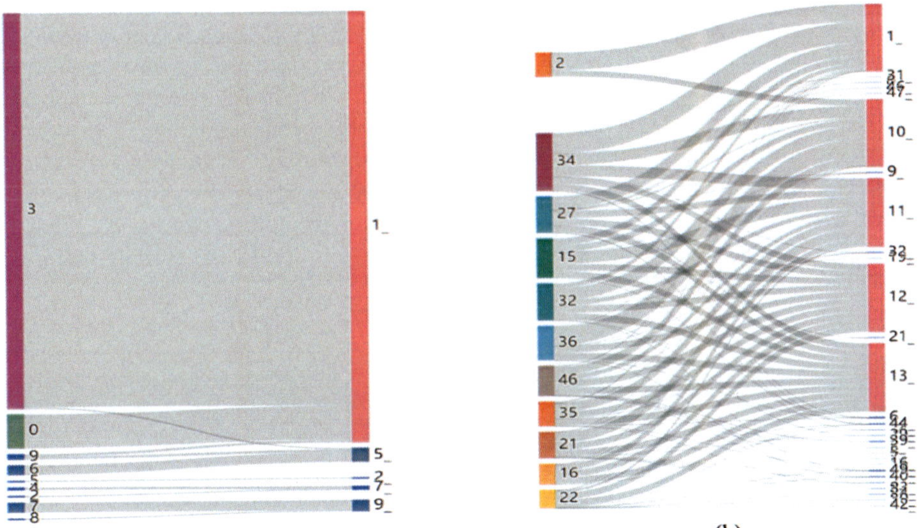

(a) (b)

Fig. 3. Impact of the contributions that the clients had between each other regarding (a) scenario two and (b) scenario three.

Regarding scenario 3 (b), not all the clients contributed individuals, however, a lot more contributions were made between the clients. In Fig. 3, the clients with the pink bar on the right side of the graph are the ones that did not train in the federation. The contributions were more balanced between the clients in this scenario. To see if the biggest contributors are correlated with the size of their dataset the following analysis was made. Using Spearman's rank correlation coefficient, it was calculated the correlation between the dataset size of each client and the number of contributions (Fig. 4). For these calculations, only the clients that trained were considered, given that those were the ones that had the possibility to contribute with individuals.

In Fig. 4, it is represented the contributions between clients in scenarios two, and three. In scenario two, regarding the local contributions, it is possible to see that the p-value (0.614) is higher than 0.05, meaning that the deviation from the null hypothesis is not statistically significant, and the null hypothesis is not rejected. As it is possible to see, the best contributor is not even in the top three of the highest datasets. Also in scenario two, regarding the global contributions, the p-value (0.143) is higher than 0.05 meaning that the deviation from the null hypothesis is not statistically significant, and the null hypothesis is not rejected. In other words, there is no indication that these two variables are correlated. Again, the highest contributor does not have the highest dataset. In scenario three, regarding the local contributions, the p-value (0.012) is lower than 0.05 meaning that the deviation from the null hypothesis is statistically significant, and the null hypothesis is rejected. In other words, there is a good indication that these two variables have a low correlation, given that the correlation (0.371) is lower than 0.5. In scenario three, regarding the global contributions, the p-value (0.001) is lower than 0.05 meaning that the deviation from the null hypothesis is statistically significant, and the null hypothesis is rejected. In other words, there is a good indication that these two variables are moderately correlated, given that the correlation (0.49) is very close to 0.5.

Although this data suggests that there is not a strong correlation between the dataset size and the number of contributions, it is possible to say that the dataset size can influence the number of contributions to some extent. Especially, in scenario three, it is possible to see that only clients with enough data have more contributions than the rest. This is probably related to the reliability of the individuals produced. In other words, the fitness value is more reliable when the individual is tested with more variety of instances. A more reliable individual is more likely to be accepted in the global HoF and by other clients.

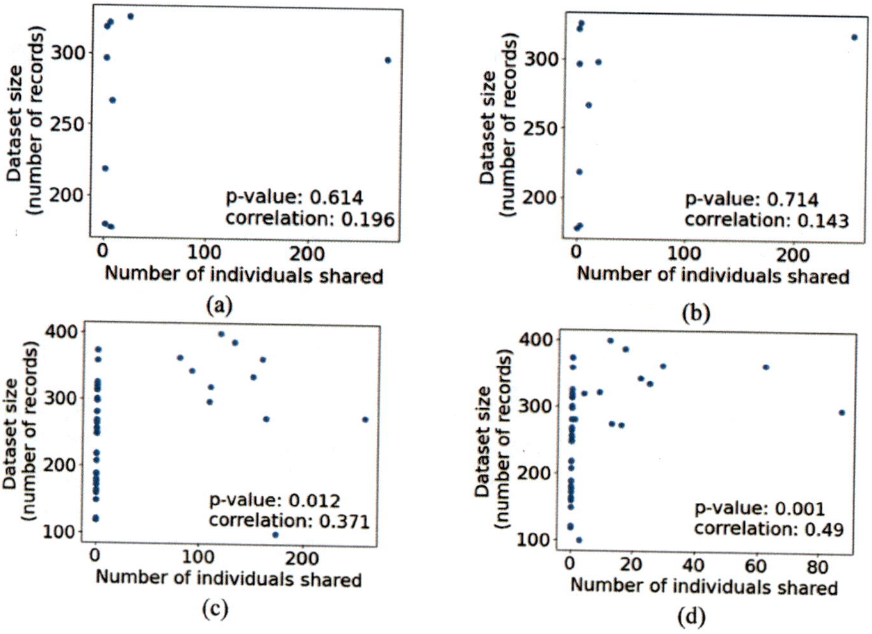

Fig. 4. Correlation between dataset sizes and number of contributions made in (a) scenario two locally (b) and globally, and in (c) scenario three locally and (d) globally.

6 Conclusions

This paper examines how different configurations of a federated learning system can impact the performance of a genetic programming model used for image classification. Three scenarios were tested, each involving non-identical and independent data, with varying numbers of clients. The findings indicate that the dataset size within each client and the federation size moderately influences the precision of the solutions. Further exploration is needed to test additional aggregation algorithms in federated learning systems with genetic programming models, in order to evaluate individuals more effectively. Furthermore, it is important to explore tasks other than image classification to determine if the results are consistent.

Acknowledgements. The present work has received funding from European Regional Development Fund through COMPETE 2020 - Operational Programme for Competitiveness and Internationalisation through the P2020 Project F4iTECH (ANI|P2020 POCI-01-0247-FEDER-181419), and has been developed under the EUREKA - CELTIC-NEXT Project F4iTECH (C2021/1-10), we also acknowledge the work facilities and equipment provided by GECAD research center (UIDB/00760/2020) to the project team.

References

1. Li, Q., et al.: A survey on federated learning systems: vision, hype and reality for data privacy and protection. IEEE Trans. Knowl. Data Eng. **35**, 3347 (2023)
2. European Union. General Data Protection Regulation. https://gdpr.eu/
3. Zhu, H., Xu, J., Liu, S., Jin, Y.: Federated learning on non-IID data: a survey. arXiv (2021)
4. Ahvanooey, M., Li, Q., Wu, M., Wang, S.: A survey of genetic programming and its applications. KSII Trans. Internet Inf. Syst. **13** (2019)
5. Khan, A., Qureshi, A.S., Wahab, N., Hussain, M., Hamza, M.Y.: A recent survey on the applications of genetic programming in image processing. arXiv (2019)
6. Bi, Y., Xue, B., Zhang, M.: Genetic programming with image-related operators and a flexible program structure for feature learning in image classification. IEEE Trans. Evol. Comput. **25**, 87 (2021)
7. McMahan, H.B., Moore, E., Ramage, D., Hampson, S., Arcas, B.A.: Communication-efficient learning of deep networks from decentralized data. In: Proceedings of the 20th International Conference on Artificial Intelligence and Statistics (2016)
8. Mahlool, D.H., Abed, M.H.: A comprehensive survey on federated learning: concept and applications. In: Shakya, S., Ntalianis, K., Kamel, K.A. (eds.) Mobile Computing and Sustainable Informatics. LNDECT, vol. 126, pp. 539–553. Springer, Singapore (2022). https://doi.org/10.1007/978-981-19-2069-1_37
9. Sheller, M.J., et al.: Federated learning in medicine: facilitating multi-institutional collaborations without sharing patient data. Sci. Rep. **10**, 12598 (2020)
10. Gong, B., Xing, T., Liu, Z., Xi, W., Chen, X.: Adaptive client clustering for efficient federated learning over non-IID and imbalanced data. IEEE Trans. Big Data **1** (2022)
11. Ge, N., Li, G., Zhang, L., Liu, Y.: Failure prediction in production line based on federated learning: an empirical study. J. Intell. Manuf. **33**, 2277 (2022)
12. Gong, Y.-J., et al.: Distributed evolutionary algorithms and their models: a survey of the state-of-the-art. Appl. Soft. Comput. **34**, 286 (2015)

13. Bremer, J., Lehnhoff, S.: Fully distributed Cartesian genetic programming. In: Advances in Practical Applications of Agents, Multi-Agent Systems, and Complex Systems Simulation. The PAAMS Collection: 20th International Conference, pp. 36–49 (2022)
14. Jahan, M., Hashem, M.M.A., Shahriar, G.A.: Distributed evolutionary computation: a new technique for solving large number of equations. Int. J. Parallel Distrib. Syst. (2013)
15. Durillo, J.J., Zhang, Q., Nebro, A.J., Alba, E.: Distribution of Computational Effort in Parallel MOEA/D. In: Coello, C.A.C. (ed.) LION 2011. LNCS, vol. 6683, pp. 488–502. Springer, Heidelberg (2011). https://doi.org/10.1007/978-3-642-25566-3_38
16. Pereira, H., Ribeiro, B., Gomes, L., Vale, Z.: CECOS: a centralized management platform supported by distributed services to represent and manage resources aggregation entities and its end-users in a smart grid context. IFAC-PapersOnLine **55**, 309 (2022)
17. Somu, N., Raman, M.R.G., Ramamritham, K.: A deep learning framework for building energy consumption forecast. Renew. Sustain. Energy Rev. **137**, 110591 (2021)
18. Tran, B., Sudusinghe, C., Nguyen, S., Alahakoon, D.: Building interpretable predictive models with context-aware evolutionary learning. Appl. Soft Comput. **132**, 109854 (2023)
19. Faia, R., Faria, P., Vale, Z., Spinola, J.: Demand response optimization using particle swarm algorithm considering optimum battery energy storage schedule in a residential house. MDPI Energies **12**, 1645 (2019)
20. Teixeira, N., Barreto, R., Gomes, L., Faria, P., Vale, Z.: A trustworthy building energy management system to enable direct IoT devices' participation in demand response programs. MDPI Electron. **11**, 897 (2022)
21. Caldas, S., et al.: LEAF: a benchmark for federated settings. arXiv (2018)

Cognitive Reinforcement for Enhanced Post Construction Aiming Fact-Check Spread

Maria Araújo Barbosa, Francisco S. Marcondes(✉), and Paulo Novais

ALGORITMI/LASI, University of Minho, Braga, Portugal
pg42844@alunos.uminho.pt, francisco.marcondes@algoritmi.uminho.pt, pjon@di.uminho.pt

Abstract. Despite the success of fact-checking agencies in presenting timely fact-checking reports on the main topics, the same success is not achieved for the dissemination of these reports. This work presents the definition of a set of heuristics applicable to messages (posts) in the microblogging environment, with the aim of increasing their engagement and, consequently, their reach. The proposed heuristics focus on two main tasks: summarisation, emotion-personality reinforcement. The results were evaluated through an experiment conducted with twenty participants, comparing the engagement of actual and generated posts. From the results of the experiment, it can be concluded that the strategy used by the generator is at least better than the one used by the fact-checking journal Snopes in its Twitter posts.

Keywords: Post Generator · Social Networks · Twitter · Fake-news · Fact-checking · Emotions · Personality · Engagement

1 Introduction

Social media platforms are very popular these days. Millions of users use *posts* to share their thoughts, opinions, news and personal information. In social media studies, the content of posts is often used as a basis for research as it provides insight into public opinion and what people are talking about [7]. Natural language processing (NLP) techniques transform the content of posts into data that can be interpreted by a computer. An example is the work of I. Singh *et al.* [4], which uses a GPT-2 and the plug-and-play language model to incorporate an emotion into the output of a generated text, ensuring grammatical correctness. L. Piccolo *et al.* [15], in turn, demonstrated a multi-agent approach to engaging Twitter users with fact-checkers; the idea is to encourage users to share verified content by educating Twitter users sharing URLs already flagged as fake.

A brief landscape of how fact-check reports are shared on Twitter is available in [8]. In short, considering 2020 data, the behaviour of `@snopes` (joined in 2008, 237.3K followers) is based on sending a fact-check link, but with a new and

S. Ossowski et al. (Eds.): DCAI 2023, LNNS 740, pp. 203–211, 2023.
https://doi.org/10.1007/978-3-031-38333-5_21

eventually witty headline for the tweet text. It uses a repackaging and retweeting strategy. @PolitiFact (joined in 2007, 673.6K followers), @Poynter (2007, 214.8K) and @factcheckdotorg (2009, 190.3K) all show similar behaviour, varying according to the style of the operator writing the post. @Channel4News (2008, 99.7K) also shows the same behaviour, but participates more in replies and some other "normal" Twitter interactions. Despite this, it has fewer followers than the others, but a higher growth rate. @APFactCheck (2011, 31.5K) is slightly more successful than the other agencies; its tweets are more instigative than journalistic, perhaps better suited to Twitter. It should be noted that it has fewer followers, although it is more successful in terms of engagement. Finally, @MBFC_News (2015, 4411) and @TheDispatchFC (2019, 1277) only tweet the title of the fact check and its link. For an illustration, refer to Fig. 1.

(a) @snopes traditional journalism

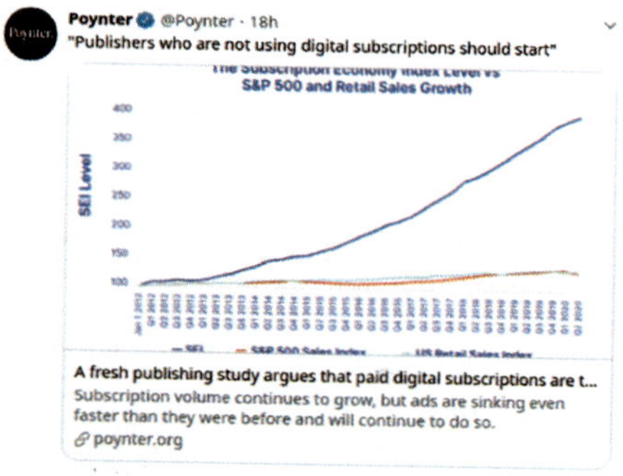

(b) @Poynter closer to "normal" tweets

(c) @APFactCheck provocative content

Fig. 1. Examples of slightly different strategies for creating tweet text

Considering the role of emotion in news [14], together with the empirical results presented in this landscape, it is possible to hypothesize that generating or tailoring microblogging posts to appeal to certain emotions and personality traits are able to improve engagement. Despite such a conception to be widespread, especially in supporting tools for digital marketing, there are few research papers testing analogous hypotheses, fewer considering mental processes [18].

In order to test this hypothesis, this paper builds a prototype post generator whose text is reinforced (or tailored) to specific emotions and mental processes. These prototype generations will be presented to a group of people in a laboratory environment along with real posts for evaluation. To reduce the scope, the prototype is focused on Twitter, and `@snopes` is chosen for comparison.

The paper is organized into two major sections. The first describing the prototype development and the second reporting the assessment and its results.

2 Tweet-Tailoring Heuristics and Prototype

For reference, a tweet is an online publication that typically consists of four different elements: *text-core*, *emoticons*, *hashtags* and *links*. The *text-core* contains the textual part of the post, comprising the relevant content/message. A *emoji* is a pictogram that can be embedded in the *text-core*. *hashtags* are search terms that can also be included in the post, and *links* are URLs to websites.

The source of the information is the Snopes fact-check reports. The information is extracted using a `beautifulsoup` scrapper, the fields searched for are: URL, date, newspaper name, title, claim, classification and the content of the report. The result is stored in a JSON file.

2.1 Summarization

The aim of this section is to obtain the post element *text-core*. To guide this procedure, the following requirements are established:

- Be consistent. Avoid contradictions and odd constructions.
- Be coherent. Have a meaningful flow of text.
- Be informative. Provide context with relevant information.
- Don't produce fake news.
- Have less than 280 characters (to meet Twitter's requirements);

In order to find the most appropriate approach, both extractive and abstractive summarization algorithms were investigated. Four extractive summarization algorithms were evaluated: 1) TextRank [11]; 2) Luhn's Heuristic Method [17]; 3) Selecting the sentence with the highest emotional value for sadness and surprise; and 4) Selecting the text in *allegation* (claim field) and *evaluation* (classification field) in the report. Five abstractive summarization models were also tested: 1) GTP-2[1] (gpt2-medium); 2) BART (facebook/bart-base); 3) BERT (bert-large-uncased); 4) XLNet (xlnet-base-cased); and 5) T5 (t5-base).

To analyse the results obtained by these models, one hundred news items were randomly selected from the dataset and each of the previous approaches was applied. The results are accessed through automated and human evaluation. The former ensures that no fake news content is generated and that the

[1] The GPT-3 is more robust version as it uses a large amount of data in the pre-training phase. However, it was not used as it is not available as open source.

maximum length of the post is respected. The human evaluation was carried out by means of a questionnaire applied to a group of eight volunteers, who were asked to rate the coherence, consistency and informative quality of each post in binary terms, in order to analyse the subjective side of the results. The verification of fake news is carried out using the tool *Fake news classifier* (https://fake-news-detection-nlp.herokuapp.com). This classifier uses the BERT model, which achieved 98% accuracy and 99% recall and precision during validation. See Table 1.

Table 1. Summary of evaluation results. The – means that the evaluation was not carried out because the approach had already been rejected.

Approach	Size (N < 280)	True context	Acceptance
T5	86%	96%	79%
BERT	60%	–	–
BART	100%	94%	25%
XLNet	60%	–	–
GTP-2	60%	–	–
Luhn's Heuristic method	23%	–	–
Text-Rank-NLTK	15%	–	–
Emotion-selection	63%	53%	–
Allegation sentence	100%	100%	83%

For extractive summarization, methods based on Luhn's heuristic method and Text-Rank-NLTK were discarded because the maximum post size is not respected in most examples. The emotion selection approach was also abandoned because it generated 37% of posts longer than 280 characters and almost half of them were classified as fake news (possibly because fake news is mostly associated with surprise and sadness). For abstractive summarization, 95% of the summaries from BERT, GPT-2 and XLNet were the same. GTP-2 was chosen to represent the results of this group. However, the GTP-2 model was then discarded as it leads to meaningless sentences (96% of the generated tweets are irrelevant, without knowledge or information related to the fact-checking news).

The results show a greater acceptance of the T5 model and the Allegation Sentence Witch as approaches chosen to obtain the *text-core* element of the post. Although the T5 model has a 14% chance of generating a post with a size greater than 280 characters, the content produced has only a 4% chance of creating a fake news. This model has a 79% acceptance rate among the study participants. The use of the allegation sentence, in addition to not producing fake content and the size always being within the expected range, has an acceptance rate of 83% in the human test carried out. On the other hand, the results of the BART model have a low acceptance among the volunteers in the study. The main reason is that in many cases the generated sentence is not complete and part of the message remains incomplete, as shown in Fig. 2a. This means that the size is respected, but the consistency and informative character is lost. Finally, Fig. 2b presents the actual tweet on this subject posted by Snopes in Twitter for comparison.

(a) BART model summarization. (b) Actual Snopes tweet.

Fig. 2. Tweet samples for reference.

2.2 Emotion Reinforcement

The strategy of emotion reinforcement is to add *emojis* and *hashtags* to the *text-core* as described in the last section, and to emphasise keywords.

Positive and negative emoticons, classified according to [13], are added to the tweet in different numbers, up to four *cf.* [16], depending on the amount of space available (respecting the 280 character limit). The post will be updated with emoticons before hashtags because they are more successful in increasing engagement rates [10]. Hashtags allow people to find tweets, especially if they are trending and relevant. By applying Latent Dirichlet Allocation (LDA) with Dirichlet-distributed topic-word distributions to the text of the fact check report, relevant topics can be extracted (skip the inter-document step) [3] and those with higher emotional levels are included in the post as a hashtag.

In addition, to increase the emotional appeal of the tweet, the most important aspects were highlighted with capital letters. This allows the user to quickly identify the content of the post, which can lead to greater engagement. To do this, the *text-core* of the post is submitted to the *KeyBERT* [5] model to identify the keywords. This doesn't add any new information to the tweet.

For a reference, Fig. 3 depicts an instance of applying the suggested reinforcements to a plain allegation extracted as described in the last section. Roughly, the emotion assessment is done by a lexical approach *cf.* [6], based on the EMOLex dataset [12].

2.3 Mental Process Reinforcement

The strategy for strengthening mental processes is based on replacing words in the *text-core* with synonyms associated with a particular mental process when-

(a) Plain allegation sentence: `anger` 0.0; `anticipation` 0.0, `disgust` 0.0, `fear` 0.0, `joy` 0.0, `negative` 0.0, `positive` 0.0, `sadness` 0.0, `surprise` 0.0, `trust` 0.0.

(b) Emotional reinforcement: `anger` 0.0; `anticipation` 0.0, `disgust` 0.0, `fear` 0.0, `joy` 0.0, `negative` 1.0, `positive` 0.0, `sadness` 0.0, `surprise` 0.0, `trust` 0.0.

Fig. 3. Emotion Reinforcement Instance

ever possible. The approach is lexical and the reference data set is the MENTALex [9]. See Fig. 4 for an example.

(a) Emotion reinforced sentence: `paranoid` 0.2, `neuroticism` 0.5, `schizoid` 0.4.

(b) Neuroticism reinforcement: `paranoid` 0.0, `neuroticism` 0.9, `schizoid` 0.1.

Fig. 4. Mental Reinforcement Instance

3 Cognitive Reinforcement Assessment and Evaluation

In order to evaluate the results delivered by the prototype and, ultimately, to test the hypothesis of this paper, an experiment was conducted with 20 participants (the recommended number for a statically significant usability study [1]), between 12 and 18 September 2022. The anonymised volunteers were accustomed Twitter users recruited at the university, from whom the authors obtained informed consent. The experiment was conducted in a laboratory environment using the microblogging simulator [2]. The number of participants was chosen according to the guidelines suggested in [1].

Participants were asked to interact with the platform in the same way they do when scrolling through their Twitter feed. The posts received voluntarily refer to ten news items. A total of 30 posts: 10 generated by the T5 model, 10 generated by the allegation sentence, and 10 posts extracted from the `@snopes` Twitter page. All posts were randomly presented to the user on the platform, without the user knowing which were generated by the prototype and which were from the `@snopes` Twitter page.

The plot in Fig. 5 shows the resulting engagement broken down by approach (T5 with cognitive reinforcement, allegation sentence with cognitive reinforcement, and actual Snopes tweets). In total, there were 434 interactions with the platform, an average of 21 interactions per participant and 14 interactions per post; in terms of interaction types, *like* was the most used by participants, followed by *follow* and *retweet*. This pattern is consistent with the one in [8].

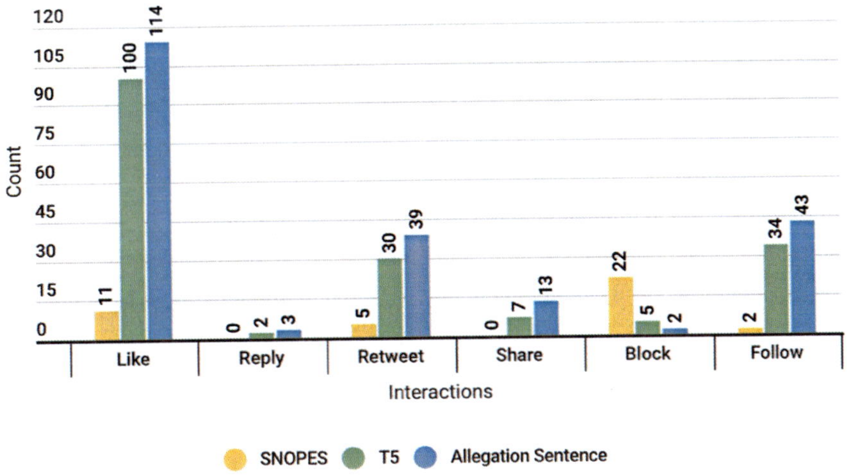

Fig. 5. Interactions count per approach

Note that in this study, unlike Twitter, the *block* interaction is associated with the post and not with the author's page. This means that when the participant *blocks* a post, it has a negative connotation towards the post. It takes an opposing viewpoint and its use shows a lack of interest in the post.

Compared to the other approaches, engagement with actual Snopes tweets is very low and includes most *block* interactions. See Fig. 2b for a comparison with Figs. 3 and 4. Broadly speaking, Snopes tweets use more formal text and fewer emoticons and hashtags compared to the proposed heuristic.

As this study was conducted in a controlled environment, it is possible to use an appropriate measure of engagement through $\frac{interactions}{visualisations}$. The idea behind this metric is that the number of interactions with a post divided by the number of times that post is presented would reveal how engaging it is. This is perhaps the simplest metric of engagement, but it is not often calculated as most online social media do not provide the number of times a post has appeared. Assuming that each volunteer was exposed to the same number of tweets, the tweet produced by the prototype has an average engagement rate of 28% (for the T5 model) and 35% (for the allegation sentence). Therefore, allegation sentence with cognitive reinforcement is the most appropriate heuristic considering the issues addressed in this paper.

4 Conclusion

This paper started from the hypothesis that enhancing tweets with emotions and personality traits would increase engagement with fact-checking tweets. In order to test this hypothesis, a set of heuristics was proposed, resulting in a prototype. The results of this prototype were then submitted for human evaluation in a laboratory environment.

Among the models studied, the T5 model and the allegation sentence produced the best results, which were then used to build the prototype. Following the literature guidelines, the use of emoticons and hashtags succeeded in reinforcing the emotional dimension. Furthermore, based on the Adaptive Personality Theory, the reinforcement of the neuroticism process helped to improve the overall result. Therefore, according to the data presented, the hypothesis is confirmed with an increase of 35% of the engagement score.

For future work, the presented heuristic needs to be divided in order to test each trait as an independent variable, with each one optimised to obtain the maximum effect. In addition, a psychological validation of the generated content will certainly contribute to the improvement of the study.

Acknowledgements. This work has been supported by FCT - Fundação para a Ciência e Tecnologia within the R&D Units Project Scope: UIDB/00319/2020;

References

1. Alroobaea, R., Mayhew, P.J.: How many participants are really enough for usability studies? In: 2014 Science and Information Conference, pp. 48–56. IEEE (2014)
2. Barbosa, M.A., Marcondes, F.S., Durães, D.A., Novais, P.: Microblogging environment simulator: an ethical approach. In: Dignum, F., Mathieu, P., Corchado, J.M., De La Prieta, F. (eds.) PAAMS 2022. LNCS, vol. 13616, pp. 461–466. Springer, Cham (2022). https://doi.org/10.1007/978-3-031-18192-4_38
3. Blei, D.M., Ng, A.Y., Jordan, M.I.: Latent Dirichlet allocation. J. Mach. Learn. Res. **3**(Jan), 993–1022 (2003)
4. Goswamy, T., Singh, I., Barkati, A., Modi, A.: Adapting a language model for controlled affective text generation. In: Proceedings of the 28th International Conference on Computational Linguistics, pp. 2787–2801 (2020)
5. Grootendorst, M.: KeyBERT: minimal keyword extraction with BERT (2020). https://doi.org/10.5281/zenodo.4461265
6. Jurafsky, D., Martin, J.H.: Speech and Language Processing. Draft, 3rd edn. (2023). https://web.stanford.edu/jurafsky/slp3/
7. Lal, S., Tiwari, L., Ranjan, R., Verma, A., Sardana, N., Mourya, R.: Analysis and classification of crime tweets. Procedia Comput. Sci. **167**, 1911–1919 (2020)

8. Marcondes, F.S.: A fact-checking profile on twitter (2020). Data can be found in ALGORITMI Centre. University of Minho, Braga, Portugal
9. Marcondes, F.S., Barbosa, M.A., Queiroz, R., Brito, L., Gala, A., Durães, D.: MentaLex: a mental processes lexicon based on the essay dataset. In: Bramer, M., Stahl, F. (eds.) SGAI-AI 2022. LNCS, vol. 13652, pp. 321–326. Springer, Cham (2022). https://doi.org/10.1007/978-3-031-21441-7_25
10. Mention: Twitter report (2018). https://mention.com/en/reports/twitter/emojis/
11. Mihalcea, R., Tarau, P.: TextRank: bringing order into text. In: Proceedings of the 2004 Conference on Empirical Methods in Natural Language Processing, pp. 404–411 (2004)
12. Mohammad, S.M., Turney, P.D.: Crowdsourcing a word-emotion association lexicon. Comput. Intell. **29**(3), 436–465 (2013)
13. Novak Kralj, P., Smailovic, J., Sluban, B., Mozetic, I.: Sentiment of emojis. PloS One **10**(12), e0144296 (2015)
14. Peters, C.: Emotion aside or emotional side? Crafting an 'experience of involvement' in the news. Journalism **12**(3), 297–316 (2011)
15. Piccolo, L., Blackwood, A.C., Farrell, T., Mensio, M.: Agents for fighting misinformation spread on twitter: design challenges. In: Proceedings of the 3rd Conference on Conversational User Interfaces, pp. 1–7 (2021)
16. Sims, S.: 7 tips for using emojis in social media marketing (2017). https://www.socialmediatoday.com/marketing/7-tips-using-emojis-social-media-marketing
17. Verma, P., Pal, S., Om, H.: A comparative analysis on Hindi and English extractive text summarization. ACM Trans. Asian Low-Resourc. Lang. Inf. Process. (TALLIP) **18**(3), 1–39 (2019)
18. Zhang, H., Song, H., Li, S., Zhou, M., Song, D.: A survey of controllable text generation using transformer-based pre-trained language models. arXiv preprint arXiv:2201.05337 (2022)

Chatbot Architecture for a Footwear E-Commerce Scenario

Vasco Rodrigues[1(✉)], Joaquim Santos[1], Pedro Carrasco[3], Isabel Jesus[1,2], Ramiro Barbosa[1,2], Paulo Matos[1], and Goreti Marreiros[1,2]

[1] GECAD - Research Group on Intelligent Engineering and Computing for Advanced Innovation and Development, Polytechnic of Porto, 4200-072 Porto, Portugal
{vvmrs,jpe,isj,rsb,psm,mgt}@isep.ipp.pt

[2] LASI - Intelligent Systems Associate Laboratory, 4800-058 Guimarães, Portugal

[3] UC - University of Coimbra, Coimbra, Portugal

Abstract. The amount of information accessible to businesses grows as more individuals get access to the internet. This information can be used by businesses to make corporate decisions that can highly affect them. E-commerce is the result of businesses starting to export their operations to the internet as a result of taking advantage of this reality. To help a footwear e-commerce platform keep up with the current demands of the market in terms of accessibility and customer service, a chatbot was created to help on this request. To build this chatbot, datasets were created leading to Intent Classification and Named-Entity Recognition pre-trained models, utilizing Bert-Large, getting fine-tuned with those datasets. These models achieved a F1-score of 0.90 and 0.88 respectively in their tasks. The Sentiment Analysis functionality of TextBlob was also utilized to help the chatbot comprehend user's text polarity to reply appropriately.

Keywords: Chatbot · Intent Classification · Named-Entity Recognition

1 Introduction

As time goes by, humans have increasingly chosen to make their shopping in online markets [20]. When the COVID-19 waves hit every country, this phenomenon of buying in online markets increased astronomically. E-commerce (electronic commerce) is the name given to markets on the internet, like a normal store to buy tools. However, it is on the internet, for example, Amazon and Facebook Marketplace.

This type of market has become an essential tool for many companies, with the fact that almost every person in a developed country has access to the internet and the easy process to buy something there. As companies and the information on the internet increases, the need for a tool to help the customer get the desired product and consequently have a better service to attract more

S. Ossowski et al. (Eds.): DCAI 2023, LNNS 740, pp. 212–222, 2023.
https://doi.org/10.1007/978-3-031-38333-5_22

customers increases, making chatbots the go-to. Chatbots are a technology that allows a conversation between a person and a computer. A basic example is a chatbot that almost everyone has heard called Siri [13] and many others utilized on home assistants and smartphones. A vital aspect for customers has always been finding the ideal product online and having the option to personalize it. Customers have certain practical or aesthetic demands, and when such alternatives are accessible, they might make the difference between purchasing the products. However, the sheer number of items on the market and the plethora of variations for each can transform options' availability into a difficult and drawn-out search. The drawbacks of a broad configuration can be lessened by selecting a suitable product with a starting point predetermined based on a streamlined user search. Taking that into consideration, the objective of this work is to develop a chatbot to make E-commerce easier to access and to give the ability for the customer to ask questions and receive recommendations on some products. NLP (Natural Language Processing) and NLU (Natural Language Understanding) techniques were utilized following a Dialogue Rollout method to achieve this. The base of any chatbot is an Intent Classification model, an NLU model that can perceive the user's intentions with the sentences they type. For the NLP part, a NER (Name Entity Recommendation System) model was developed to extract entities that represent the characteristics of the products of the Footwear e-commerce platform. SA (Sentiment Analyses) is also utilized to calculate the polarity of the users' sentences to help the chatbot reply accordingly.

The rest of the paper is organized in the following order: Sect. 2 provides some related works on chatbot architectures, Sect. 3 presents the proposed chatbot architecture to solve the problem addressed in this article, and Sect. 4 shows the obtained results. In the last section, some conclusions are drawn, alongside suggestions for work to be done afterward.

2 Related Work

Chatbots are not new, so to familiarize with the technologies, methods, and strategies used to build a chatbot, some research on related work was performed. Some domains where chatbots were used are inventory management, pricing negotiations, search aid in Covid-related issues, and bureaucracy for students and administration.

A chatbot named Hebron was created to help the Covenant University Community members save time when visiting Covenant University Shopping Mall (CUSM). The time issue arises from the fact that CUSM lacks an online inventory. Thus when a member visits the store to make a purchase, and the store is out of stock, the member wastes time traveling there. The process could be streamlined by asking the built chatbot if the desired product is available and handling the payment. React.js was used to create the chatbot's UI, MySQL was used to store and transmit the list of commodities and their quantities to the user, and spaCy was utilized for the machine learning component [15].

A business should consider the importance of product pricing negotiations because they increase customer satisfaction. A chatbot was developed to facilitate pricing negotiations in an e-commerce setting. The user's query, dialogue roll-out, sentiment analysis, and natural language processing were the recommended methodologies used to construct this chatbot [22].

Reading reviews to learn more about certain product features may require much time since some reviews are so long and dense with information. This type of review is unpleasant and undesirable for e-commerce businesses. A system is being developed that can detect and identify this particular information for a user to solve this issue. This method uses WordNet to produce synonyms already discovered in other reviews and BERT (Bidirectional Encoder Representations from Transformers) to extract the entities [1].

When consumers choose an e-commerce site to make their purchases, they could encounter problems. Since billions of people use this type of platform and often report complaints, it takes time for customer service to handle these consumer complaints. Due to the congestion caused by this, customer service representatives could not respond promptly. Another problem is finding items a consumer wants by using several platforms to compare prices, warranties, and other aspects, making it a waste of time surfing and looking for the best e-commerce site to make that purchase. A chatbot was created to address this time-consuming search, with the ability to filter products in multiple e-commerce platforms and take care of issues before they must be sent to customer support. Because there were so many modules available, RASA was the framework chosen to create this chatbot [11].

It is challenging for someone not studying or working at the college to access the website and find the needed information. A chatbot is suggested to assist a person in obtaining the needed information. This information may concern admission, exams, and other things. To construct such a chatbot, WordNet was utilized to extract information from the provided text, Semantic Sentence Similarity to compare a given question to existing ones, and AIML files to store the question-and-answer pairs. Last but not least, queries that the chatbot could not respond to were recorded in a log file [7].

Due to the distinct surroundings from the valuable environment, their lives change once they begin their university studies. Due to the absence of knowledge of university administration, general queries, and guidance regarding bureaucracy, this new environment causes new students a great deal of stress. For the Brazilian public university, S.A.N.D.R.A., a chatbot, was developed to reduce this stress [18]. A database covering all 1.453 subjects and a dataset of all the FAQs on the university website were used to build the chatbot. Utilizing the Levenshtein distance query helped users avoid making mistakes. Scikit-learn and BERT were utilized to handle the user's queries.

A chatbot was created for QA (Questions and Answers) for COVID-19 material in Vietnamese due to the recent COVID-19 waves and the fact that there is data for the Vietnamese language in the NLP field [17]. BERT and ALBERT models were employed to construct QA tasks for this chatbot, and RASA was selected as the framework.

The need to sell a company's items online has increased as more and more users have chosen to shop online, but not all businesses do it. A chatbot was created to assist businesses in marketing their goods online and to help them save money by automating some human processes [4]. This chatbot was created using ManyChat, which was incorporated into a Facebook fan page.

To better understand the user and his needs to deliver the required information, the authors in [3] proposed a personalized chatbot enhanced with sentiment analysis features to provide useful and tailored information for each user. To test their chatbot architecture and viability, they created a proof-of-concept of their chatbot architecture in Customer Service for E-commerce applications.

Table 1 shows a brief summary of the related works presented previously.

Table 1. Brief Summary of Related Works

Chatbot	NLP Models/Libraries	Domain
[15]	Spacy	Inventory Management
[22]	Not-disclosed	Pricing Negotiation
[1]	WordNet and Bert	Product Recommendation
[11]	Rasa	E-Commerce platform Recommendation
[7]	WordNet	Bureaucracy aid for students
[18]	BERT and Scikit-learn	Bureaucracy aid for administration
[17]	BERT, ALBERT and Rasa	Covid-19 related issues
[4]	ManyChat	Marketing and Process Automation
[3]	Vader, Bert	Customer Service in E-commerce

3 Proposed Chatbot Architecture

Based on the evaluation of similar works given previously, we first decided to use the RASA platform to construct our chatbot. However, this tool only offers a free trial, and while you may use it for moderate traffic quantities, if those numbers rise, you must switch to a paid plan. A further aspect to consider is that spaCy, one of Rasa's default pipelines for processing user inputs, consumes a significant amount of resources. We decided to build our architecture after acknowledging that there are more powerful state-of-the-art models than spaCy for using the same resources as Bert, which we verified with some preliminary testing in addition to the related works. An architecture was planned to develop this chatbot considering the needs of the footwear e-commerce platform which uses this chatbot. A user who is a final consumer (B2C) or another business (B2B) interacts with the platform where a chatbot is available to help with their needs. This chatbot can search, list, and recommend products based on user inputs. Figure 1 shows the proposed chatbot architecture and workflow. The user's query, in addition to available information about previous purchases, available products, and user information in the footwear e-commerce database, is retrieved and provided to the chatbot. With this information, the chatbot uses intent classification and sentiment analysis to process the user query after this query is preprocessed. After the preprocessing is complete, based on the intent of the user, whether or not it is a greeting or farewell, a reply is sent back to the user through the footwear e-commerce platform. In the case of the intent being Change, See, or Remove. The query is sent to the Name Entity Recognition model, where entities are extracted. With these extracted entities, they are sent to the Hybrid Recommendation System to recommend products based on them. The chatbot then receives these recommendations and sends them to the user through the platform. This process is done utilizing Dialogue Roll-out, a method used to plan the dialogue between the chatbot and the user [22].

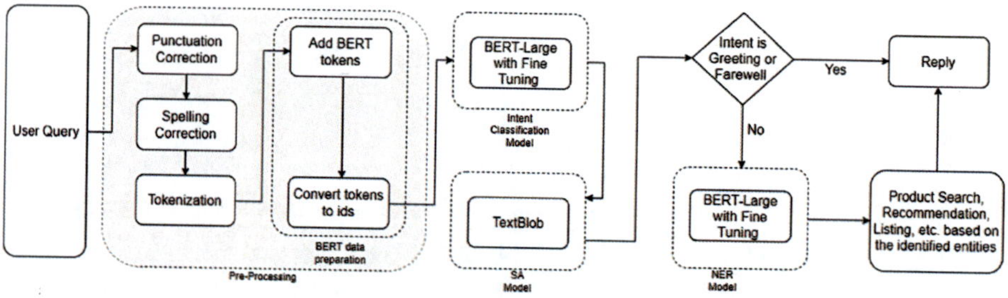

Fig. 1. System-wide Architecture

The rest of this section addresses the utilized datasets, the processing pipeline, and the proposed workflow that built the architecture of the chatbot.

3.1 Datasets Analysis

In this section, an analysis of the created and utilized datasets is performed. Due to the lack of datasets on the footwear e-commerce domain, a dataset was created to train each model used in this work.

Intent Classification. The Intent Classification dataset created for this work was based on the keywords the footwear e-commerce company gave based on their products and search query history. With this information and considering the domain of the problem, in addition to the common Greeting and Sayonara intents, we created the Change, Remove, and See intents. Change intent consists of classifying user queries where the user wants to change the criteria of the products. At the same time, Remove intent consists of classifying correctly user queries where the user wants to remove criteria of the products, and See intent consists of classifying correctly user queries where the user wants to see products. Figure 2 shows the distribution of the intents of the dataset. It is fairly balanced, but it does not have a considerable size, only 188 lines. Henceforth, this dataset should only be used for fine-tuning since it does not have enough size to train a model from scratch.

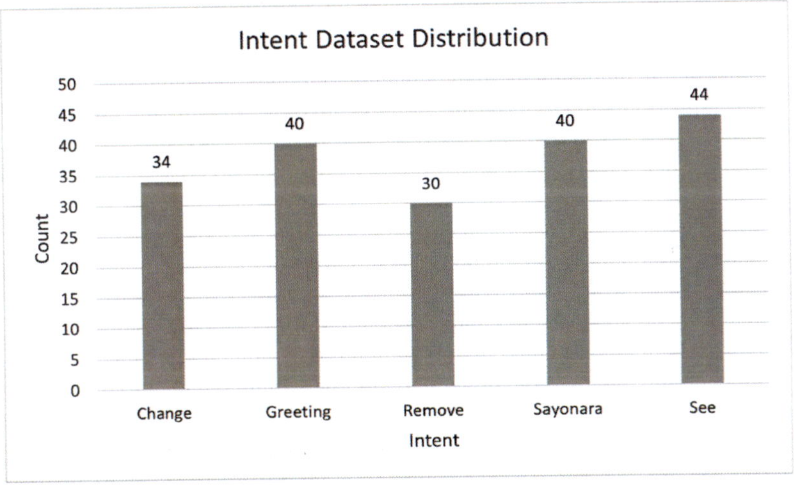

Fig. 2. Intent Dataset Distribution

Named-Entity Recognition. Regarding the dataset for the Named-Entity Recognition task, it started initially on the UBIAI platform[1], with the IOB POS format. In addition to the sentence and word, this format annotates the Part-of-Speech (POS) tag and the inside, outside, and beginning tags. POS tagging comprises assigning grammatical categories to words in a text depending on their definition and context [6]. The most widely used tagset is the Penn tag set from the Penn Treebank project, which is the most extensive set with 36 categories constructed from the Wall Street Journal dataset [21]. From the 4081 lines, which constitute 236 sentences, 3465 lines have the Outside, and the remaining lines can be seen in the distribution of the NER tags in Fig. 3. All these NER tags that the new dataset has been derived from the characteristics of the products in the footwear e-commerce platform, hence why tags like resistance level and comfort level. We also can see that most of the tags are beginning tags, meaning they are the sole words in a sentence getting tagged.

3.2 Methods

This section addresses the processing pipeline: preprocessing tools, intent classification, sentiment analysis and named-entity recognition.

[1] ubiai.tools (Retrieved October 6, 2022)

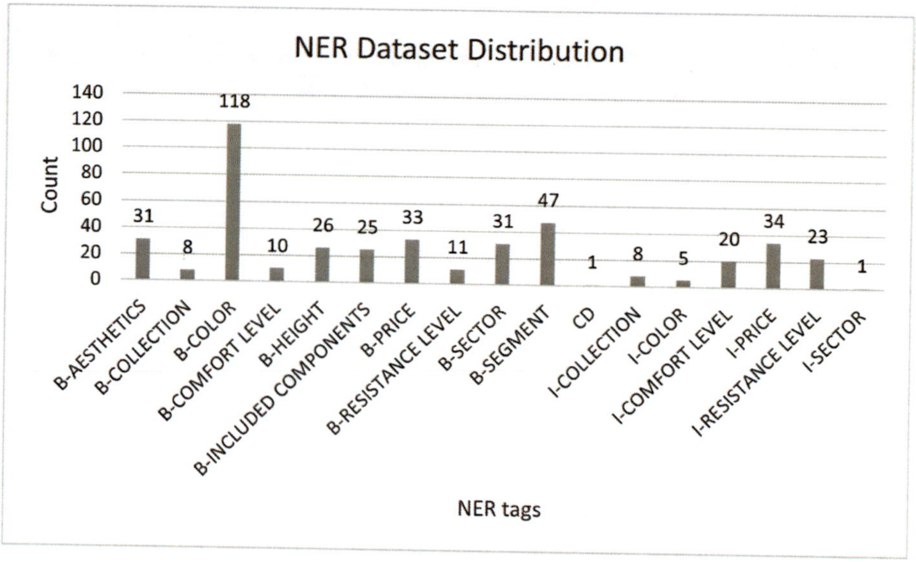

Fig. 3. NER Dataset Distribution

Pre-processing. Since textual data can be unstructured by nature, suitable preprocessing is needed to properly use them in NLP tasks to transform and represent that data in a more structured way so they can be used later on [2]. Another issue with this type of content is lots of noise, such as abbreviations, non-standard spelling, and no punctuation.

Normalization and Tokenization are used to break a text document into tokens, commonly words, for easier text manipulation [21]. It includes all sorts of lexical analysis steps like removing or correcting punctuation, number, accents, extra spacing, removing or converting emojis and emoticons, and spelling correction [2].

Regarding Spelling Correction, Strings Similarity, and Synonymous words were the methods chosen to handle misspellings of words. TextBlob library provides that functionality, but it would not be enough due to the way it corrected the errors. In some occasions, the correction would substitute the word that would be relevant for the models to classify intents or detect named entities. Taking into consideration the work in article [19] where the Levenshtein Distance algorithm is explored and further search in [12], which compares these types of algorithms, a decision was made that the most suitable algorithm for the desired objective of this work would be a Sequence-based algorithm called **Ratcliff-Obershelp similarity** [14].

Another thing we have to take into consideration is the possibility that the query may contain words that are not the same as the words that characterize the products in the footwear e-commerce platform but synonyms of these. The solution found for this aspect was the use of the WordNet library[2], which allows generating a list of synonyms for a word and that way we can compare this list to the list we have of words that characterize the products, improving the results of the models.

Intent Classification. Intent Classification [16] provides the ability to extract a sentence's intent cite. A text's intent can be classified, and the chatbot uses this information to determine the best response. The chatbot can compare the user's query's intent to intents already existent in the code once it has access to the user's intent. The Dialogue roll-out is where the already-created intents came from. Dialogue roll-out is the method used to plan the dialogue between the chatbot and the user [8]. After some preliminary tests and related works, BERT was the model used for this task. Since the dataset created was not very large, training a new model from scratch would not yield the best results, so we choose BERT-Large model and fine-tuned it for this task with the dataset we created.

Sentiment Analysis. Given a text, the Sentiment Analysis (SA) can determine the author's feelings about the written sentence [5]. The TextBlob [10] tool is used to assist in processing the query submitted to calculate the polarity of a sentence. Since the score of this tool can range from -1 to 1, we utilized the following thresholds for the sentiment of a sentence. Inferior to -0.4 is considered negative. At the same time, between -0.4 and 0.4 is neutral and superior to 0.4 for positive. Our chatbot's response considers these thresholds to reply to the user appropriately.

Named-Entity Recognition. NER (Name Entities Recognition) is a NLP technique that enables the extraction of entities from a text [9]. After initial experiments and related work, BERT was the model chosen for this objective. We chose the BERT-Large model and fine-tuned it for this task using the dataset we produced because training a new model from scratch would not provide optimal results due to the small amount of data available.

4 Results

The Bert-Large fine tuned model was used and fine tuned 2 instances of it, one for Intent Classification and the other for NER. In terms of classifying the intents of the user, it yielded the following results in Table 2. This leads to an weighted average F1-score of 0.90.

In terms of identifying the correct named-entities in sentences, the NER model yielded the following results in Table 3, leading to a weighted average F1-score of 0.88.

[2] wordnet.princeton.edu (Retrieved October 6, 2022)

Table 2. Intent Classes Metric Distribution

	Precision	Recall	F1-score	Support
Change	1.00	0.67	0.80	3
Greeting	0.80	1.00	0.89	4
Sayonara	1.00	0.86	0.92	7
See	1.00	1.00	1.00	4
Remove	0.50	1.00	0.67	1

Table 3. NER Classes Metric Distribution

	Precision	Recall	F1-score	Support
AESTHETICS	0.86	0.86	0.86	7
COLOR	0.78	0.94	0.85	34
COMFORT LEVEL	1.00	1.00	1.00	2
HEIGHT	1.00	1.00	1.00	4
INCLUDED COMPONENTS	0.86	0.90	0.88	21
PRICE	1.00	1.00	1.01	6
RESISTANCE LEVEL	1.00	1.00	1.00	1
SECTOR	0.80	1.00	0.89	8
SEGMENT	0.69	1.00	0.82	18

5 Conclusion

This work aimed to propose an architecture for a chatbot to be utilized by a footwear e-commerce platform. To achieve that, a careful analysis of the chatbot's requirements led us to create one from scratch instead of using the RASA platform. We created an Intent Classification and a Named-Entity Recognition models based on the Bert-Large pre-trained model. We fine-tuned them with the datasets we created for this specific use case based on the footwear e-commerce platform specifications. The fine-tuned Intent Classification model achieved a F1-score of 0.90, while the fine-tuned Named-Entity Recognition model achieved a F1-Score of 0.88 with our datasets. Comparing the results to [18], the F1-score is **marginally different between both works for the same amount of labels, having** 0.91 on their work.

After testing the chatbot, the models seem sufficient and capable of achieving good accuracy in their tasks. The Intent Classification model can be applied in any e-commerce branch since the labels are not domain-dependent. On the other hand, the Named-Entity Recognition model is more dependent on the business domain. However, this model can be replicated in another domain where the tags make sense, like a clothing e-commerce platform.

For future work, we plan to increase the size of the intent classification dataset and improve the user query treatment in terms of size and feature extraction.

Another consideration is testing these models in other areas of e-commerce to understand the necessary adjustments. A proof-of-concept with this chatbot is also in the works to validate our results and understand what improvements it brings to the company regarding sales, customer retention, and satisfaction.

Acknowledgements. This research was funded by National Funds through the Portuguese FCT–Fundação para a Ciência e a Tecnologia under the R&D Units Project Scope UIDB/00760/2020, UIDP/00760/2020, and OMD project (NORTE-01-0247-FEDER-179949).

References

1. Anda, S., Kikuchi, M., Ozono, T.: Developing a component comment extractor from product reviews on e-commerce sites. In: 2022 12th International Congress on Advanced Applied Informatics (IIAI-AAI), pp. 83–88 (2022). https://doi.org/10.1109/IIAIAAI55812.2022.00026
2. Denny, M., Spirling, A.: Text preprocessing for unsupervised learning: why it matters, when it misleads, and what to do about it. Polit. Anal. **26**(2), 168–189 (2018). https://doi.org/10.1017/pan.2017.44
3. El-Ansari, A., Beni-Hssane, A.: Sentiment analysis for personalized chatbots in e-commerce applications. Wirel. Pers. Commun. **129**(3), 1623–1644 (2023). https://doi.org/10.1007/s11277-023-10199-5
4. Illescas-Manzano, M., López, N., González, N., Rodríguez, C.: Implementation of chatbot in online commerce, and open innovation. J. Open Innov.: Technol. Market Complex. **7**(2), 125 (2021). https://doi.org/10.3390/joitmc7020125. https://www.mdpi.com/2199-8531/7/2/125
5. Khanvilkar, G., Vora, D.: Sentiment analysis for product recommendation using random forest. Int. J. Eng. Technol. (UAE) **7**(3), 87–89 (2018). https://doi.org/10.14419/ijet.v7i3.3.14492. http://arxiv.org/abs/1509.02437. https://www.sciencepubco.com/index.php/ijet/article/view/14492
6. Kumawat, D., Jain, V.: POS tagging approaches: a comparison. Int. J. Comput. Appl. **118**(6), 32–38 (2015). https://doi.org/10.5120/20752-3148
7. Lalwani, T., Bhalotia, S., Pal, A., Bisen, S., Rathod, V.: Implementation of a chat bot system using AI and NLP. Int. J. Innov. Res. Comput. Sci. Technol. **6**(3), 26–30 (2018). https://doi.org/10.21276/ijircst.2018.6.3.2
8. Landim, A., et al.: Chatbot design approaches for fashion E-commerce: an interdisciplinary review. Int. J. Fashion Des. Technol. Educ. **15**(2), 200–210 (2022). https://doi.org/10.1080/17543266.2021.1990417. https://www.tandfonline.com/action/journalInformation?journalCode=tfdt20
9. Li, J., Sun, A., Han, J., Li, C.: A survey on deep learning for named entity recognition. IEEE Trans. Knowl. Data Eng. **34**(1), 50–70 (2022). https://doi.org/10.1109/TKDE.2020.2981314
10. Loria, S.: TextBlob: Simplified Text Processing (2018). https://textblob.readthedocs.io/en/dev/
11. Mamatha, M., Sudha, C.: Chatbot for e-commerce assistance: based on RASA. Turk. J. Comput. Math. Educ. **12**(11), 6173–6179 (2021)
12. Mayank, M.: String similarity - the basic know your algorithms guide! (2019). https://itnext.io/string-similarity-the-basic-know-your-algorithms-guide-3de3d7346227

13. Mohamad Suhaili, S., Salim, N., Jambli, M.: Service chatbots: a systematic review. Expert Syst. Appl. **184**, 115461 (2021). https://doi.org/10.1016/j.eswa.2021.115461
14. Nalawati, R., Yuntari, A.: Ratcliff/Obershelp algorithm as an automatic assessment on e-learning. In: Proceedings of the 2021 4th International Conference on Computer and Informatics Engineering: IT-Based Digital Industrial Innovation for the Welfare of Society, IC2IE 2021, pp. 244–248 (2021). https://doi.org/10.1109/IC2IE53219.2021.9649217
15. Oguntosin, V., Olomo, A.: Development of an e-commerce chatbot for a university shopping mall. Appl. Comput. Intell. Soft Comput. **2021** (2021). https://doi.org/10.1155/2021/6630326
16. Pascual, F.: Intent Classification: How to Identify What Customers Want (2019). https://monkeylearn.com/blog/intent-classification/
17. Quy Tran, B., Van Nguyen, T., Duc Phung, T., Tan Nguyen, V., Duy Tran, D., Tung Ngo, S.: Fu Covid-19 AI agent built on attention algorithm using a combination of transformer, albert model, and rasa framework. In: 2021 10th International Conference on Software and Computer Applications, ICSCA 2021, pp. 22–31. Association for Computing Machinery, New York (2021). https://doi.org/10.1145/3457784.3457788
18. Santana, R., Ferreira, S., Rolim, V., Miranda, P., Nascimento, A., Mello, R.: A chatbot to support basic students questions. In: CEUR Workshop Proceedings, vol. 3059, pp. 58–67 (2021). http://ceur-ws.org
19. Subathra, M., Umarani, V.: AHP based feature ranking model using string similarity for resolving name ambiguity. Int. J. Nonlinear Anal. Appl. **12**(Special Issue), 1745–1751 (2021). https://doi.org/10.22075/IJNAA.2021.5862
20. Tsagkias, M., King, T., Kallumadi, S., Murdock, V., de Rijke, M.: Challenges and research opportunities in ecommerce search and recommendations. SIGIR Forum **54**(1) (2021). https://doi.org/10.1145/3451964.3451966
21. Weiss, S., Indurkhya, N., Zhang, T., Damerau, F.: Text Mining: Predictive Methods for Analyzing Unstructured Information, pp. 1–237 (2005). https://doi.org/10.1007/978-0-387-34555-0
22. Yadav, S., Pratap Singh, R., Sree, D., Vidhyavani, A.: E-commerce chatbot for price negotiation. Int. Res. J. Modernization Eng. **583**(11), 583–585 (2021). www.irjmets.com

A Geolocation Approach for Tweets Not Explicitly Georeferenced Based on Machine Learning

Thiombiano Julie[1](✉), Malo Sadouanouan[1], and Traore Yaya[2]

[1] Nazi BONI University, Bobo-Dioulasso, Burkina Faso
thiombianojulie@gmail.com

[2] Joseph KI-ZERBO University, Ouagadougou, Burkina Faso

Abstract. In this paper, we describe an inference model for deducing the location of a tweet whose geolocation information is not available in the metadata. The approach we propose is based on machine learning techniques and uses the information contained in the tweets such as the places mentioned in the tweets and the profile of the authors of the tweets. The objective of the study is to contribute to setting up an early warning system for epidemics based on the monitoring of events on social networks like twitter. For this we need to geolocate the messages even if the smartphone's GPS is not active. We trained on three models and obtained the best result with K-nearest neighbors model with an accuracy of 0.90.

Keywords: Geolocation · Machine Learning · infectious diseases · Twitter · real-time

1 Introduction

As online communication environments evolve, social media platforms are becoming increasingly popular channels for users to discuss and share information about events in their daily lives. Twitter, in particular, is a widely used platform that provides access to user tweets through APIs. Consequently, it is extensively utilized by researchers as a data source for various systems, including recommendation systems, demographic analyses, and disease surveillance, among others. Having knowledge of a user's location is therefore advantageous for these systems. However, despite the availability of tweet metadata and the option for Twitter users to voluntarily disclose their location, this information is not always accessible. Only 3% of tweets include geolocation information [1]. This limitation significantly hinders the use of a metadata-only approach for location identification. Considering that tweet content also contains valuable information, we propose a model that combines tweet metadata and textual content to geolocate the tweet.

Location prediction can be approached as either a classification or regression problem [2]. The classification method can be implemented in two ways: by

S. Ossowski et al. (Eds.): DCAI 2023, LNNS 740, pp. 223–231, 2023.
https://doi.org/10.1007/978-3-031-38333-5_23

treating known locations as distinct classes or by dividing the region into a grid and considering the center points of subregions as classes. On the other hand, the regression approach can be viewed as a straightforward dual regression task, where a model predicts the latitude and longitude of a tweet independently based on its characteristics. Previous research has shown that both tweets and their metadata can be utilized for predicting a user's location [3,4].

This paper is part of the implementation of an early alert system. At present, this system is able to identify the subject of the tweet, i.e., whether it is related to an infection or not, and the geolocation of the tweet is needed to identify possible infection hotspots. In this paper, we present an approach to predict a user's location by addressing the problem as a classification problem, predicting the location coordinates of a tweet depending on whether the tweet is related to an infection or not.

The rest of this document is organized as follows: the first section provides an overview of related work on location prediction. Section 3 describes the details of the prediction model architecture. Results on the validation set are presented in Sect. 4. Finally, Sect. 5 concludes the document with possible future work perspectives.

2 Related Work

In the literature on user location inference, different approaches have been used to predict location. There are those who focus on the processing document or user profile tweet summaries to predict the author's geolocation [5–7] are multiple projects. These works utilize geographical grids to divide the Earth's surface into regions of different sizes, whereby densely populated areas are subdivided into smaller parts, while sparsely populated areas end up as larger grid cells. Early initiatives [8,9] primarily focused on extracting indicative information from users' post content by relying on location indicative words (LIWs) that can link users to their places of residence, using various natural language processing techniques (e.g., topic models and statistical models) [10]. For example, [11] extracts bag-of-words features from user posts and [6] estimate word distributions for different regions.

Some focus solely on the textual content of tweets. [12] used a combination of geographical and text-based features to create their predictions, which improved the accuracy of their results. Incorporating both geographical and text-based features was found to enhance location prediction accuracy compared to using only text-based features. In the same vein, there are the works of [13] who used a deep learning approach to create their predictions. They found that their approach outperformed other existing methods, including those based on traditional machine learning algorithms. The authors concluded that deep learning approaches have the potential to significantly improve the accuracy of tweet location predictions.

Others also consider metadata in addition to text. In the study conducted by [14], the prediction of Twitter users' home location was explored using Long-Short Term Memory (LSTM) and BERT models. The dataset used included the textual content of tweets as well as user metadata, all in the Indonesian language. However, the input limit of 512 tokens imposed by BERT models posed a challenge in accurately classifying a user's home location. Consequently, user location prediction was performed on a per-tweet basis, and it was observed that the majority vote approach frequently led to misclassification of the user's home location. Another approach proposed by [15] takes into account both the geographical context (home location) of a tweet and its content to create predictions. They applied four machine learning techniques, namely Logistic Regression, Random Forest, Multinomial Naïve Bayes, and Support Vector Machine, with and without the utilization of the geo-distance matrix, to predict the location of a tweet using its textual content. Their extensive experiments on a large collection of Arabic tweets from Saudi Arabia with different feature sets yielded promising results with 67% accuracy.

Compared to these recent research works, our current approach incorporates a broader geographical context by leveraging geotags, predefined place names on the Twitter platform, to facilitate the association of specific locations with a tweet, along with the tweet's content and the user's profile. This multidimensional approach contributes to improving prediction accuracy and better understanding the underlying factors that contribute to the geographical information of a tweet. Furthermore, we aim to leverage the use of an ontology to enhance location detection in tweets, providing a precise semantic structure, resolving location-related ambiguities, and enriching machine learning models with additional information.

3 Approach for Geolocation

This approach is part of a global approach for an early warning system for epidemics related to infectious diseases as shown in Fig. 1. The figure illustrates the global approach. In this approach, we use a keyword dictionary for collection, which consists of concepts extracted from a domain ontology and enriched with some commonly used terms. We also use Kafka technology to streamline the real-time data collection from the Twitter stream. The data then undergo a process which includes preprocessing, annotated, and classified as either related to an infection or not. After this step, the task of predicting the location of tweets related to an infection comes into play, which will be described in the following lines.

3.1 Architecture

Basically, there are four sources of data to geolocate tweets: geotags, predefined location names on the Twitter platform (tweet place), tweet content, and user profiles. The geocoding process extracts location references from each of these possible sources. The model of the proposed geographic location estimator in the system is represented in Fig. 2. When it comes to choosing a location for mapping, the system prioritizes the following in order: the exact location where

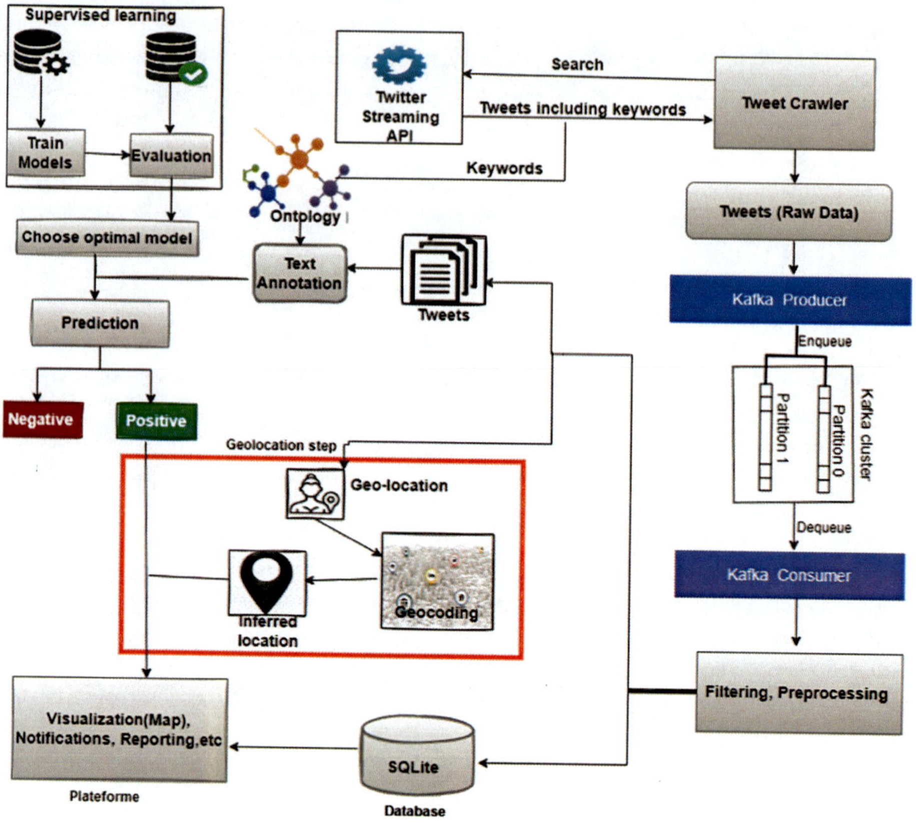

Fig. 1. The framework for implementing the proposed system

the tweet was posted (geographic coordinates), the location of the tweet, the location mentioned in the text of the tweet, and finally the location of the user's profile. We tested the architecture on three algorithms, namely Support Vector Machine, Random Forest, and K-nearest neighbors, in order to choose the most suitable model for our task. The results are presented in the next section.

3.2 Dataset

For this model, we collected a set of tweets using keywords that are actually concepts extracted from an ontology in the field of diseases. The data is then segmented, filtered, and stored in a database. We then perform data annotation followed by a learning phase where we categorize the tweets as related to an infection or not. The goal is to then locate the tweets related to an infection, which is essential for detecting infectious outbreaks.

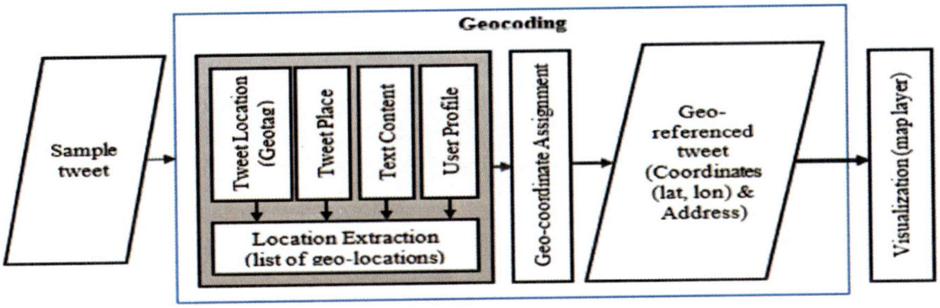

Fig. 2. Model for geographic location inference

We collected 411,157 tweets, of which only 937 have GPS coordinates. This means that the model will predict locations based on the three remaining sources considered in our approach, in addition to GPS coordinates. However, we do not exclude the 937 tweets from the dataset before experimentation. The dataset used indeed consists of the entire 411,157 tweets. Below is the distribution of the dataset according to latitude (Fig. 3) and longitude (Fig. 4).

Figure 3 and Fig. 4 represent the distribution of infectious tweets with GPS coordinates on a map, showing that the majority of the tweets are from America and Europe, with a smaller number from Africa.

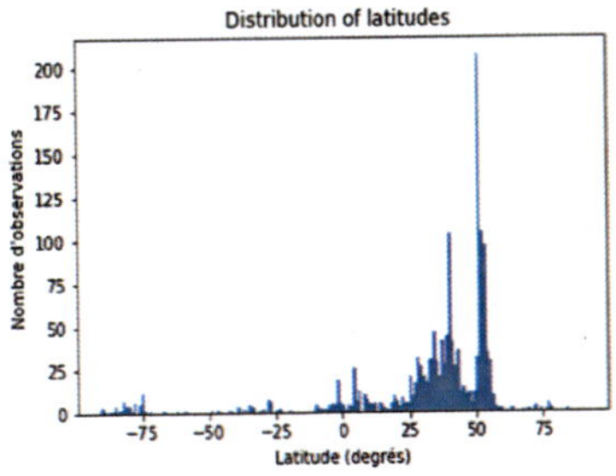

Fig. 3. Dataset distribution by latitude

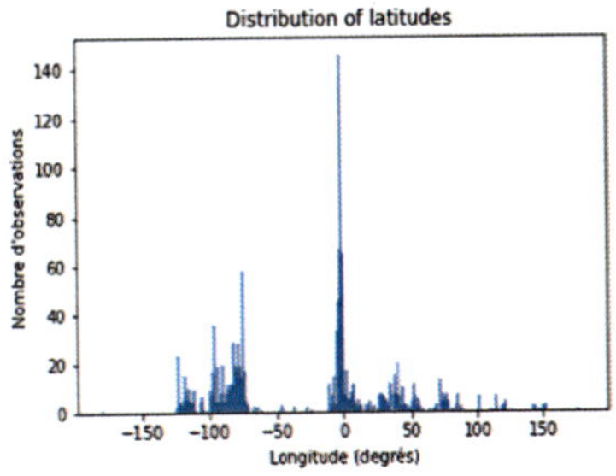

Fig. 4. Dataset distribution by longitude

4 Result and Discussion

The experiment was done on an Intel(R) Core(TM) i7-865 CPU and 1.90 Ghz CPU with 16 GB of RAM running under a Microsoft Windows 10 64-bit operating system. The models are designed using python 3.6.

We used Word2Vec vectors for embedding the content of the tweet, and searched for entities representing places that are provided by a pre-defined dictionary of locations. When a place is found, it undergoes geocoding to extract its GPS coordinates, which will be used to place the location on the map. The data was split into training and testing data using the train_test_split function from the Scikit-learn library. The split was done by setting the test set size to 20% of the initial dataset and using a pseudo-random number generator with a fixed seed of 42.

Performance of models on validation and test on data are shown in the Table 1.

Table 1. Algorithms performance evaluation

Algorithms	Precision	Recall	F_1
Support Vector Machine	0.89	0.90	0.89
K-Nearest-Neighbour	0.90	0.88	0.89
Random Forest	0.80	0.86	0.83

Based on the results obtained above between KNN and SVM, the results are roughly the same, but given our objectives, we need a model with good precision. Therefore, we consider KNN as the best-suited model. We have the best precision with KNN for $K = 5$, and for $K \leq 5$ or $K \geq 5$, we have lower precision. The infectious tweets are then placed on the map to better visualize the high-risk areas and take action on the first cases to better contain the epidemics. The figures (Fig. 5, Fig. 6) show the location of users whose message has been predicted as infectious. According to the figures, the green dots represent the locations that have been correctly predicted, while the red dots represent prediction errors.

Depending on each source we considered, there are issues related to the source. Firstly, there are few tweets with location coordinates, and secondly, with regards to the user profile, most are sarcastic and incorrect or invalid coordinates. Finally, in the textual content of the tweet, we search for place names from a dictionary of place names, which can be limiting for the extraction of location entities. However, the strength of this approach is that it combines various sources and gives us a better chance of finding a place to associate with the tweet. Our model is able to locate 90% of the tweets that have been classified as related to an infection by searching for location elements in various sources. This is already satisfactory and will allow for quick handling of these cases.

Fig. 5. Visualizing the predicted event locations.

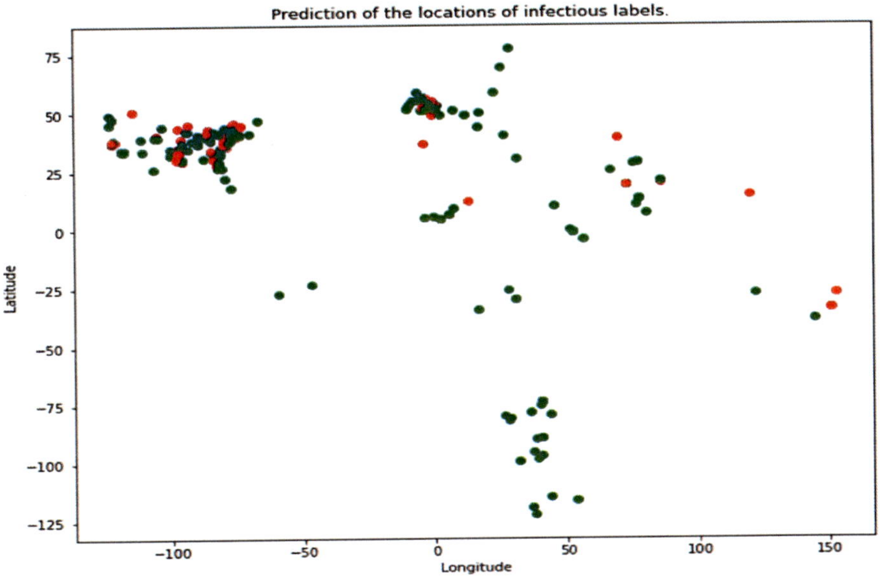

Fig. 6. Event distribution visualization

5 Conclusion and Future Work

These works are a continuation of a previous study focusing on an intelligent system for early detection of epidemic events using data collected from Twitter. This article introduces a geolocation prediction method aimed at identifying potential infectious hotspots and visualizing them on a map. To achieve this, we utilize various data sources, including geotags, predefined location names on the Twitter platform, tweet content, and user profiles. We conducted experiments testing three algorithms, with the k-nearest neighbors (K-NN) method yielding the best results. Our findings demonstrate the overall effectiveness of the model. The dataset used in this study was developed by us specifically for this purpose and has not been utilized in any other works, limiting the possibility of comparisons at present. Nevertheless, we intend to make the dataset publicly available for future research. Our upcoming work will involve employing additional datasets to determine the optimal conditions under which our approach functions effectively. And we will also use an ontology to guide the prediction process.

References

1. Jurgens, D., Finethy, T., McCorriston, J., Xu, Y.T., Ruths, D.: Geolocation prediction in twitter using social networks: a critical analysis and review of current practice. In: ICWSM (2015)
2. Mishra, P.: Geolocation of tweets with a BiLSTM regression model. In: Proceedings of the 7th Workshop on NLP for Similar Languages, Varieties and Dialects, pp. 283–289 (2020)
3. Han, B., Cook, P., Baldwin, T.: Text-based twitter user geolocation prediction. J. Artif. Intell. Res. **49**, 451–500 (2014)
4. Huang B., Carley K.M.: A hierarchical location prediction neural network for twitter user geolocation (2019). arXiv preprint arXiv:1910.12941
5. Hulden, M., Silfverberg, M., Francom, J.: Kernel density estimation for text-based geolocation. In: Proceedings of the AAAI Conference on Artificial Intelligence, vol. 29 (2015)
6. Wing, B., Baldridge, J.: Simple supervised document geolocation with geodesic grids. In: Proceedings of the 49th Annual Meeting of the Association for Computational Linguistics: Human Language Technologies, pp. 955–964 (2011)
7. Wing, B., Baldridge, J.: Hierarchical discriminative classification for text based geolocation. In: Proceedings of the 2014 Conference on Empirical Methods in Natural Language Processing (EMNLP), pp. 336–348 (2014)
8. Han, B., Cook, P., Baldwin, T.: Geolocation prediction in social media data by finding location indicative words. In: Proceedings of COLING, vol. 2012, pp. 1045–1062 (2012)
9. Roller, S., Speriosu, M., Rallapalli, S., Wing, B., Baldridge, J.: Supervised text-based geolocation using language models on an adaptive grid. In: Proceedings of the 2012 Joint Conference on Empirical Methods in Natural Language Processing and Computational Natural Language Learning, pp. 1500–1510 (2012)
10. Zhou, F., Wang, T., Zhong, T., Trajcevski, G.: Identifying user geolocation with hierarchical graph neural networks and explainable fusion. Inf. Fusion **81**, 1–13 (2022). https://doi.org/10.1016/j.inffus.2021.11.004

11. Rahimi, A., Cohn, T., Baldwin, T.: A neural model for user geolocation and lexical dialectology. arXiv preprint arXiv:1704.04008 (2017)
12. Mostafa, A., Gad, W., Abdelkader, T., Badr, N.: Pre-HLSA: predicting home location for twitter users based on sentimental analysis. Ain Shams Eng. J. **13**, 101501 (2022)
13. Mahajan, R., Mansotra, V.: Predicting geolocation of tweets: using combination of CNN and BiLSTM. Data Sci. Eng. **6**, 402–410 (2021)
14. Simanjuntak, L.F., Mahendra, R., Yulianti, E.: We know you are living in Bali: location prediction of twitter users using BERT language model. Big Data Cogn. Comput. **6**(3), 77 (2022)
15. Alsaqer, M., Alelyani, S., Mohana, M., Alreemy, K., Alqahtani, A.: Predicting location of tweets using machine learning approaches. Appl. Sci. **13**, 3025 (2023). https://doi.org/10.3390/app13053025

Classification of Scenes Using Specially Designed Convolutional Neural Networks for Detecting Robotic Environments

Luis Hernando Ríos González[1,3], Sebastián López Flórez[1,2](✉), Alfonso González-Briones[1,2], and Fernando de la Prieta[1,2]

[1] BISITE Digital Innovation Hub, University of Salamanca, Edificio Multiusos I+D+i, Calle Espejo 2, 37007 Salamanca, Spain
{lhgonza,sebastianlopezflorez,alfonsogb,fer}@usal.es

[2] Air Institute, IoT Digital Innovation Hub, 37188 Salamanca, Spain

[3] Universidad Tecnológica de Pereira, Cra. 27 N 10-02, Pereira, Risaralda, Colombia

Abstract. This article addresses the task of classifying scenes in typical indoor environments navigated by robots, using pre-trained convolutional neural network - CNN models with ImageNET and PLACES 365 datasets. The implemented models are the CNN VGG16 and RESNET50 architectures, which underwent various manipulations such as freezing and modification of intermediate layers. The performance of these networks in scene recognition tasks was analyzed, where a "scene" refers to an image that captures a partial or complete view of an indoor environment, displaying objects and their spatial relationships, and is similar to the scenes that mobile robots must perceive in their navigation task. To achieve this objective, a set of custom images JUNIO20V1, FEBR20V3, JUNIO20V3 and others, as well as 2 state-of-the-art image sets 15-Scenes and Sports, were subjected to various manipulations such as rotations, perspectives, sectioning, and lighting changes, creating a diverse image bank for both training and testing, allowing the analysis of the performance of the different modified CNN structures. The obtained results demonstrate the strengths of the different CNN structures in scene recognition tasks for application in indoor environments where mobile robots move.

Keywords: Environment classification · Scene categorization · robotics · Deep learning

1 Introduction

Among the main tasks faced by a mobile robot in interacting with the environment is the perception and recognition of its work site and in general it must focus its attention on solving recognition problems in indoor or outdoor environments. Most of the techniques used by robots for scene recognition present low performance for indoor scenes [11]. Although in recent years, amazing results have been

S. Ossowski et al. (Eds.): DCAI 2023, LNNS 740, pp. 232–241, 2023.
https://doi.org/10.1007/978-3-031-38333-5_24

achieved in robot localization and navigation based on probabilistic estimation techniques, these algorithms cannot solve problems of self-localization and calculation of its current pose (relocation), when the robot is in similar indoor environments [13]. Many approaches have been proposed to solve this problem, and scene identification is an indispensable perception capability for a mobile robot, which allows for increased levels of interaction with and understanding of its environment. As a mobile robot navigates its working environment, it can constantly provide new perspectives of the objects it perceives, allowing it to actively scan the visual world for relevant features or situations that can facilitate object, image or scene recognition [1]. Scene recognition can provide semantic information useful for tasks such as object localization, robot navigation, and manipulation. However, scene images present many difficulties for recognition, which makes identifying a specific scene in an environment not such a trivial task. Convolutional neural networks (CNNs) have been shown to perform better than hand-crafted functions. Convolutional operations and nonlinear transformations in CNN architectures can include more abstract information and more complete visual attributes than conventional hand-crafted functions.

In this work we focus on the classification of scenes in the context of robotics, selecting the images according to the type of place, for example interior environments of kitchens, living rooms, corridors and free access places. In the conformed own image base, which consists of 5 bases, JUNE20V3, FEBR19V3, FEBR19V2, FEBR19V3, IMAG18, five classes are differentiated and in turn each class has a set of characteristic objects that define it, which allows to have approximate knowledge of the place where the robot is located. This would reduce the search space to predefined locations. To classify the different scenarios, we propose to use different CNN models pre-trained with the Imagenet and Places365 datasets [17]. A comparison of the different CNN structures used is made to determine the best performing structure for each case.

The main contributions of this work are (i) Performance comparison of different modified CNN structures for the scene classification task.(ii) Use of in-house and state-of-the-art scenario image datasets for the scene classification task of mobile robot working environments.

The paper is structured as follows: Sect. 2 description of related work. Section 2.1 description of the proposed approach. Section 2.2 Description of our test data set. Section 4 Experimental results. Section 4.1 Discussions Sect. 5 Conclusions.

2 Related Work

In the solution of improving the perception of mobile robots to interact with their environment, there are many solutions described in the state of the art. There are different approaches for scene recognition in mobile robotics using computer vision and artificial intelligence. Some approaches use pattern recognition techniques to identify specific objects in the scene, while others, use image processing to build a representation of the robot's environment, allowing it to navigate autonomously and avoid obstacles. In [15] Yang et al. generated feature vectors from key point detection for image classification. Different types of CNN architectures [7] have been employed in scene classification. These CNNs have achieved acceptable performances with the Places365 dataset [17]. CNN architectures generally learn a characteristic representation of an image, but do not define with certainty the relationships of objects in the image. SegNet [2] uses a special CNN architecture to perform pixel-level segmentation (Scene Parsing). A pre-trained CNN model based on segmentation modules that can relate various forms of content in an image, which includes objects in different planes, is described in [18]. Scene classification can also be useful for location recognition. Some of the existing place recognition works include scene classification datasets [8]. MIT Place365 [17], which has 365 scene categories and is one of the largest scene classification datasets with large number of categories. In [10] presented a method based on adapting 2D Speeded Up Robust Features (SURF) image features to 3D landmarks of the environment, which can generate a prior map containing 3D Surf landmarks, which can reduce the similarity of indoor environments. In [4], the self-pose problem is addressed using Monte Carlo localization. Some work uses depth cameras to obtain the image, but these techniques are computationally very expensive [9]. In [5] a probabilistic hierarchical model is presented that allows associating objects to scenes through visual features and contextual relationships. Based on advances in convolutional neural networks, several works have been adopting visual navigation, using cameras as the main or only sensor, and performing object detection, scene understanding and feature extraction [6]. In [16], the use of semantic knowledge in context reasoning is proposed to support navigation tasks through information about relations between objects, such as shape and dimensions, to improve a priori knowledge. A visual semantic navigation model, with recurrent neural networks for memory mechanisms with memory architecture, is proposed in [12]. As an illustrative example, it is reasonable to rely on the belief that next to a refrigerator, it is usually possible to find a stove, since these objects share the same environment and are often seen together in nearby places given their related functionality. In [7]. a CNN model for object recognition, using the ImageNet image base, is presented with good results. In [14]. the advantages of sparse decomposition of convolutional features and increasing the depth of CNNs, to improve the discriminative power in scene recognition.

2.1 Proposed Approach

Scene classification and recognition has progressed using neural networks. While convolutional neural networks extract features well, they are sensitive to image changes. Accurately representing scenes while utilizing convolutional layers helps networks classify. Identifying distinguishing features reduces error and adapts to changes. Adapting to complex images requires an algorithm that deeply exploits defining objects. Preparing training images, validating the network and evaluating architectures are key parts of proposed classification methods. The paper analyzes RESNET 50 and VGG16 architectures with Places 265 and ImageNet in scene classification. Results show that some architectures are more efficient under specific conditions like rotations, perspectives, and illumination. Tests used standard and original image databases with modifications to test performance.

2.2 Description of Our Databases

For this specific instance, the challenge is to test the performance of the CNN scene classification and recognition method for images of indoor environments typical of mobile robot action environments. The images used in the implementation of the method are composed of images that vary in composition, or a variety of different scenarios. Due to the composition of this type of scene, where one can see a multiple number of objects that vary significantly in position from one image to the next, as well as objects that are repeated scene after scene, the algorithm is prevented from learning to recognize only very specific features of the images and is therefore prevented from being effective. Because some classes in one image base may be the result of the modification of images belonging to another image base, the number of objects, parts or collections of these, as well as their variation in size and perspective, are presented in the different image bases [3]. For this specific instance, the PLACES 365 and IMAGENET [8] image bases are used together with basic scene recognition features learned from architectures such as Resnet-50 or VGG-16.

An explanatory table of the different features present in the image bases is presented below (Table 1):

Table 1. Characteristics of the different image bases

Nombre	Training	Validation	Test
JUNIO20V1	Distorted Objects and Parts	Scenes	Scenes
FEBR20V3	Scenes	Objects	Objects
JUNIO20V3	Scenes	Objects	Objects
FEBR20V2	Objects	Scenes	Scenes
FEBR19V3	Scenes and Objects	Scenes and Objects	Scenes and Objects
FEBR19V2	Objects	Objects	Objects
FEBRI9V3IIMAG18	Objects	Objects	Objects
15-Scene	Scenes in Color and Monochrome	Scenes in Color and Monochrome	Scenes in Color and Monochrome
Sports	Scenes	Scenes	Scenes

2.3 Enhanced Data

When there is a limited amount of training data, the augmented data technique produces the same image through rotations, cropping and focusing. Various results and strategies can be produced randomly from the practice images.

In addition, the method advises against changing the validation images; this will teach the algorithm to recognize objects regardless of their size or orientation throughout the image.

3 Training and Evaluation of the Classification Systems Used

Classification systems were implemented using CNN VGG16 and RESNET 50 architectures for training and evaluation of the various image bases.

3.1 Image Classification Systems Using Neural Networks

CNN VGG16 and RESNET 50 architectures are imported to evaluate the effectiveness of the previously trained classification and recognition models. The imported network has 13 convolutional layers and 3 fully connected layers, and is pre-trained at this point. The only change made at this point will be to modify the number of classification categories.

Application of Pre-trained Weights in IMAGENET to the Base Case Model. The first imported architecture is CNN VGG16 pre-trained on IMAGENET ILSVRC with default parameters. VGG16 has 13 convolutional layers and 3 fully connected layers, varying the number of classification categories. IMAGENET contains 1.2 million images grouped into 1000 features. Since the own image bases contain between 5 and 15 categories that do not coincide with IMAGENET subsets, the classification layer must be modified. The CNN is trained on FEBR19V3IMAG18 with 5 categories - living rooms, kitchens and corridors grouped under interior scenes. Training uses 224×224 resolution images, a 0.001 learning rate and RMSProp optimizer. The network parameters are frozen, convolutional filters act as feature extractors using IMAGENET parameters and 5, 8 or 15 output categories, fitting the model to FEBR19V3IMAG18 images. This base model is the starting point for other bases.

Application of Pre-trained Weights in PLACES365 to the Base Case Model. Previously trained PLACES365 parameters are imported after determining model accuracy for each class and overall accuracy for each base image. PLACES365 dataset has Places365-Standard subset with 1.8 million training images and Places365-Challenge 2016 subset with 8 million training images including Places365-Standard. PLACES365 with 365 scene categories was chosen due to similarities to base images. By using PLACES365 instead of Imagenet weights, parameter tuning continued. This is the only variation compared to case 4.3.

CASE 4.3.21 Convolutional Layer Thawing. When an unfrozen layer is initialized, its parameters and statistics are randomly initialized, so its value will not be one of the imported weights. After results using IMAGENET and PLACES365 weights, some convolutional layers are unfrozen along with fully connected layers forming the new classifier. The learning rate is changed from 0.001 to 0.0001 to update final layers based on extracted features without affecting performance. This variation saw the learning rate reduced from 1E−3 to 1E−4. The last convolutional block extracts specific features, justifying selecting layers to reinitialize to improve performance. The feature map represents meaningful image details. Altering layers connected to the last block can learn patterns highly representative of images. Combined with features from previous layers, this could improve performance. conv5.3, conv5.2 or both were retrained. The final classification block parameters and output layer with softmax activation alongside the last two layers would change based on learned features. After epochs, accuracy is measured as total hits over total samples, visualizing results and model performance in terms of errors and hits.

Ranking of Images Using the Base.delo Architectures Based on the Number of Errors and Hits. The Resnet50 CNN architecture is selected, offering more layers. Creating a network from scratch requires adjusting millions of parameters, requiring many images. The pre-trained Resnet50 model accurately classifies 1000 ImageNet classes or over 300 scenes with high accuracy. This time, the network is trained directly using 365 MIT scenes. This choice may work better for indoor images because they are similar. The approach is essentially the same as previous cases, using same optimizers, learning rates and image bases to achieve overall and class accuracy, following guidelines in case 4.3.2. Identifying network components by reading architecture shows the final convolutional block named conv4.1, conv4.2 and conv4.3. The base case runs unchanged and conv4.2 and conv4.3 are reinitialized in all configurations. Results are recorded to compare performance of each test.

4 Results

The paragraph presents a comparison of the performance of VGG16 and RESNET50 architectures in the task of image classification of indoor scenes. The variations in accuracy are shown in tables, which indicate the impact of different modifications to their intermediate layers and pretraining with state-of-the-art image bases such as imageNET and PLACES365.

Upon conducting the tests, it is observed that the RESNET50 network, pre-trained with PLACES 365, performs better for the tested databases when the conv4.3 + conv4.2 layers are modified. In fact, the results indicate that its performance surpasses the base RESNET50 architecture. The underlying reason lies in the influence of the modified layers on the network's prior knowledge acquired during pre-training.

Table 2. Accuracy results obtained for the classifier using VGG16-Imagenet

Image data base	VGG16	Conv5.3	Conv5.3+Conv5.2	Conv5.2	Accuracy
JUN1020V1	77	67	76	77	[67 : 77]
FEBR20V3	65	60	60	58	[58 : 65]
JUN1020V3	62	63	55	58	[55 : 63]
FEBR20V2	96	86	87	88	[86 : 96]
FEBR19V3	93	97	96	98	[93 : 98]
FEBR19V2	90	97	99	96	[90 : 99]
FEBR19V31MAG18	95	96	95	89	[89 : 96]
15-SceNe	87	77	84	82	[77 : 87]
Sports	85	78	86	74	[74 : 86]
JUN1020V2 (8 escenas)	78	71	75	65	[65 : 78]

Table 3. Accuracy results obtained for the classifier using VGG16-Places365

Image data base	VGG16	Conv5.3+Conv5.2	Conv5.3	Accuracy
JUN1020V1	52	44	46	[44 : 52]
FEBR20V3	49	49	53	[49 : 53]
JUN1020V3	40	49	47	[40 : 49]
FEBR20V2	68	69	71	[68 : 71]
FEBR19V3	79	87	85	[79 : 87]
FEBR19V2	82	90	88	[82 : 90]
FEBR19V31MAG18	85	82	83	[82 : 85]
15-SceNe	82	83	81	[81 : 83]
Sports	79	77	82	[77 : 82]
JUN1020V2	75	66	77	[66 : 77]

Table 2 delineates the precision metrics for the VGG16 classifier pretrained on the ImageNet dataset when subjected to various modifications in the intermediate convolutional layers. The precision ranges from 57.7% to 77.2% across the test image datasets as the architecture is fine tuned by adding convolutional layers such as Conv5.3 and Conv5.2. Table 3 shows the precision results for the VGG16 classifier pretrained on the PLACES365 dataset. The precision ranges from 44.3% to 90.3% depending on the convolutional layer configuration and fine tuning.Table 4 outlines the precision attained by the ResNet50 classifier pretrained on PLACES365 dataset. The precision spans from 90.1% to 100% across the test image datasets when various convolutional layers such as Conv4.3, Conv4.2 and their combination are fine tuned.

The Sports and 15-Scene datasets, though specialized, served as a useful benchmark by enabling performance to be measured under optimal conditions. For example, ResNet50 pre-trained with PLACES365 achieves near 100% accuracy for some test datasets, demonstrating the potential effectiveness of large-scale pre-training when fine-tuning is done properly. However, the results also

Table 4. Accuracy results obtained for image classification using Resnet50-Places365

Image data base	Resnet50	Conv4.3+Conv4.2	Conv4.3	Conv4.2	Accuracy
JUN1020V1	84	88	89	84	[84 : 89]
FEBR20V3	93	96	92	92	[92 : 96]
JUN1020V3	89	93	91	91	[89 : 93]
FEBR20V2	100	100	100	100	[100]
FEBR19V3	100	99	100	100	[99 : 100]
FEBR19V2	100	100	100	100	[100]
FEBR19V31MAG18	100	96	100	100	[96 : 100]
15-SceNe	94	94	96	96	[94 : 96]
Sports	96	96	94	94	[94 : 96]
JUN1020V2	90	91	90	90	[90 : 91]

indicate that while these specialized datasets achieve their highest accuracy for some test cases like FEBR20V2, FEBR19V3, FEBR19V2, FEBR19V3, suggesting the problem is solved for these datasets, the performance for other cases like FEBR20V3, JUN1020V2, JUN1020V1, the performance is significantly below expected values, indicating the problem persists to some degree. These experiments aim to delineate the problem by fusing object and scene behavior and determining the best architecture for each specific situation. Performance varies significantly based on the test set, pre-training set and network modifications. The results quantify the effectiveness of specialized datasets under different conditions, revealing which combinations work best and where opportunities exist to improve performance for certain test cases. Better exploitation of object-scene interactions and situation-specific network architectures may help specialized datasets achieve more consistent, state-of-the-art accuracy for indoor image classification tasks related to different use cases.

4.1 Discussion

In the recognition of indoor environments, mobile robots must overcome obstacles, interact with individuals, and efficiently perform tasks, a situation that is not trivial. Although many studies show remarkable results in navigation problems, autonomous localization in indoor environments remains an unresolved and improvable challenge. Mobile robot navigation methods are progressively incorporating computer vision as a means of perception, and this is where scene recognition plays a prominent role in their interaction with various environments. Although CNNs are currently widely used techniques for scene recognition due to their potential for end-to-end pattern recognition generalization, they do not define object relationships in the image with certainty and do not provide information about the background characteristics. The CNN models obtained can relate various forms of content in an image, including objects in different planes. Given that scene classification can be useful for location recognition, the tests conducted demonstrate that model performance varies based on two important factors. Firstly, the type of architecture and its modification, and secondly, the pre-training image sets that provide a pre-trained parameter base that contributes to the optimization of the interior scene classification task. In the process of modifying intermediate layers, improved performance of the scene classification processes demonstrates the potential of CNNs for specific applications in robot navigation problems in indoor environments. Since CNNs are strongest in end-to-end pattern recognition generalization, and there exist CNN models specialized in object recognition, it is possible to implement a hybrid system that captures both scene feature information and object feature information to solve the recognition task.

5 Conclusion

In this article, a classification method for scenes in different indoor environments is presented. The method is based on the creation of a real-world image database selected with possible work environments for mobile robots, such as hallways, rooms, kitchens, etc. The implemented method uses CNN architectures, specifically VGG16 and RESNET50, which were modified in their intermediate layers and pre-trained with the parameter sets of ImageNET and PLACES365. A comparison of the implemented architectures was performed to determine which architecture had the best performance for scene classification tasks.

The obtained results show that the RESNET50 CNN network, which was manipulated in its intermediate layers and pre-trained with PLACES365 (a state-of-the-art image database of indoor environments), had better performance with the own indoor environment image databases used. The use of pre-trained parameter sets adjusted after processing more than 1 million images reduced the computational load in tests with own image databases, thereby reducing the execution time and improving classification results.

From the work carried out, it is concluded that robot navigation in indoor environments can solve the problem of self-localization in similar environments by fusing information obtained from SLAM (Simultaneous Localization and Mapping) and information obtained from scene recognition in indoor environments.

Acknowledgement. This research has been supported by the project "COordinated intelligent Services for Adaptive Smart areaS (COSASS), Reference: PID2021-123673OB-C33, financed by MCIN /AEI /10.13039/501100011033/FEDER, UE.

References

1. Aghili, F., Su, C.Y.: Robust relative navigation by integration of ICP and adaptive Kalman filter using laser scanner and IMU. IEEE/ASME Trans. Mechatron. **21**(4), 2015–2026 (2016)
2. Badrinarayanan, V., Kendall, A., Cipolla, R.: SegNet: a deep convolutional encoder-decoder architecture for image segmentation. IEEE Trans. Pattern Anal. Mach. Intell. **39**(12), 2481–2495 (2017)

3. Bosch, A., Zisserman, A., Munoz, X.: Scene classification using a hybrid generative/discriminative approach. IEEE Trans. Pattern Anal. Mach. Intell. **30**(4), 712–727 (2008)
4. Bukhori, I., Ismail, Z., Namerikawa, T.: Detection strategy for kidnapped robot problem in landmark-based map Monte Carlo localization. In: 2015 IEEE International Symposium on Robotics and Intelligent Sensors (IRIS), pp. 75–80. IEEE (2015)
5. Espinace, P., Kollar, T., Roy, N., Soto, A.: Indoor scene recognition by a mobile robot through adaptive object detection. Robot. Auton. Syst. **61**(9), 932–947 (2013)
6. Gupta, S., Davidson, J., Levine, S., Sukthankar, R., Malik, J.: Cognitive mapping and planning for visual navigation. In: Proceedings of the IEEE Conference on Computer Vision and Pattern Recognition, pp. 2616–2625. IEEE (2017)
7. Krizhevsky, A., Sutskever, I., Hinton, G.E.: ImageNet classification with deep convolutional neural networks. Commun. ACM **60**(6), 84–90 (2017)
8. Kumar, D.: Deep learning based place recognition for challenging environments. Master's thesis, University of Waterloo (2016)
9. Kuse, M., Shen, S.: Learning whole-image descriptors for real-time loop detection and kidnap recovery under large viewpoint difference. Robot. Auton. Syst. **143**, 103813 (2021)
10. Majdik, A., Popa, M., Tamas, L., Szoke, I., Lazea, G.: New approach in solving the kidnapped robot problem. In: ISR 2010 (41st International Symposium on Robotics) and ROBOTIK 2010 (6th German Conference on Robotics), pp. 1–6. VDE (2010)
11. Quattoni, A., Torralba, A.: Recognizing indoor scenes. In: 2009 IEEE Conference on Computer Vision and Pattern Recognition, pp. 413–420. IEEE (2009)
12. Santos, I.B.D.A., Romero, R.A.: A deep reinforcement learning approach with visual semantic navigation with memory for mobile robots in indoor home context. J. Intell. Robot. Syst. **104**(3), 40 (2022)
13. Thrun, S.: Probabilistic algorithms in robotics. AI Mag. **21**(4), 93 (2000)
14. Xie, L., Lee, F., Yan, Y., Chen, Q.: Sparse decomposition of convolutional features for scene recognition. In: 2017 2nd IEEE International Conference on Computational Intelligence and Applications (ICCIA), pp. 345–348. IEEE (2017)
15. Yang, J., Jiang, Y.G., Hauptmann, A.G., Ngo, C.W.: Evaluating bag-of-visual-words representations in scene classification. In: Proceedings of the International Workshop on Workshop on Multimedia Information Retrieval, pp. 197–206. ACM (2007)
16. Yang, W., Wang, X., Farhadi, A., Gupta, A., Mottaghi, R.: Visual semantic navigation using scene priors. arXiv preprint arXiv:1810.06543 (2018)
17. Zhou, B., Lapedriza, A., Khosla, A., Oliva, A., Torralba, A.: Places: a 10 million image database for scene recognition. IEEE Trans. Pattern Anal. Mach. Intell. **40**(6), 1452–1464 (2017)
18. Zhou, B., Zhao, H., Puig, X., Fidler, S., Barriuso, A., Torralba, A.: Scene parsing through ade20k dataset. In: Proceedings of the IEEE Conference on Computer Vision and Pattern Recognition, pp. 633–641 (2017)

AMIR: A Multi-agent Approach for Influence Detection in Social Networks

Chaima Messaoudi[1,2](✉), Lotfi Ben Romdhane[2], and Zahia Guessoum[1]

[1] Crestic Lab, University of Reims Champagne-Ardenne, Reims, France
{chaima.messaoudi,zahia.guessoum}@univ-reims.fr
[2] Mars Lab, University of Sousse, Hammam Sousse, Tunisia
Lotfi.Ben.Romdhane@usherbrooke.ca

Abstract. The omnipresence of social networks makes their analysis essential. Two lines of research have attracted a lot of interest in the analysis of social networks in recent years: Community detection and influence maximization. There are mainly two approaches to studying influence in a social network context: centralized and agent-based. Despite this interest, much of this research has focused on structural links between nodes, as well as interactions. But the details of the semantics of communication that takes place between the nodes of a social network have not been studied much. The main contribution of this paper is to deal with the different interactions between members of a social network. To accomplish this, different methods are used, such as opinion mining, to abstract the social network and to allow the identification of the influencers. We propose to develop a multi-agent architecture to solve the problem of detecting influential elements. We associate with each node of the social network an agent (manager of the node). We develop coordination mechanisms between these agents to conduct the voting process in order to determine the influential elements in the network.

Keywords: Social networks · Influencers · Opinion mining · Agents · Influence maximization

1 Introduction

Social Network Sites (SNS) are commonly known for posting information, posting personal activities, product reviews, online photo sharing, advertisements, and the expression of opinions and feelings [15]. These media's pervasiveness warrants their investigation. There is a lot of interest in two categories of social network analysis research in recent years: group identification and detecting the most active users. Members of social networks who have the ability to convince or influence other members to develop more social relationships are known as influencers. Feelings, emotions, and actions all have the potential to influence others. The selection of the most influential nodes is an NP-hard optimization

S. Ossowski et al. (Eds.): DCAI 2023, LNNS 740, pp. 242–253, 2023.
https://doi.org/10.1007/978-3-031-38333-5_25

problem [10]. Most existing approaches [9,12,23] are centralized. Very few studies are distributed [6,21]. Few approaches studied the semantic node-to-node contact in a social network but none of them detect the influent seed nodes based on the opinion similarity or the user profile features.

We propose a novel multi-agent approach to identify the influential elements in a social network. Our approach relies on a voting mechanism. An abstraction of the graph is proposed based on the assumption that users with similar opinions tend to influence each other mutually. So first we have to detect the dominant opinion in the user's posts. Then, for each pair of users, we determine the topics they have in common and we determine the opinions of each of them on these topics.

This paper aims to propose an approach to choose the initial seed nodes to ensure the broadcast spread is optimal. The approach is based on a multi-agent system. An abstraction of the graph is proposed based on the assumption that users with similar opinions tend to influence each other mutually. The influence power of each node relies on the actions performed by its neighbors on its publications. The previous steps will allow us to create the social graph. Which is the abstraction of the social network where each node is presented by its influence power and the link between two nodes is identified through the opinion similarity of those nodes. Once we establish the social graph, we can move to identify the candidates. These are the nodes with the highest probability of becoming influential based on the PageRank Ranking (PR) value. The aim of this step is to consider those candidates as seed nodes and to evaluate whether they can maximize the influence propagation (IP) in the network. The paper is organized as follows: Sect. 2 summarizes the related work on influence maximization in social networks. Section 3 presents the proposed methodology for influence maximization. Section 4 presents the experimental validation of our approach. Finally, Sect. 5 concludes this paper.

2 Related Work

In this section, we present the main approaches for detecting influential nodes. Those approaches can be classified into centralized approaches and agent-based approaches. The centralized approaches can be classified based on the resolution technique into communities-based approaches, PageRank-based approaches, greedy algorithm-based approaches, and heuristic approaches.

[3,23], are communities detection-based approaches. They are based on the network's structure. Communities detection involves identifying firmly linked groups of vertices called communities. There are also the PageRank-based approaches. Those approaches extend the PageRank [4] algorithm. They combine different measures with PR such as influent nodes for communities detection [2]. Another subfamily is the greedy algorithms in instance we can cite CELF [12] and CELF++ [9] that calculate the marginal gain of a node. Other approaches are heuristic approaches based on centrality [8] (global diversity and local characteristics), [1] (degree centrality) and [20].

The second family of approaches are the Agent-based approaches. [24] proposed an agent-based approach to detect influence in a hidden-structure social network using a neural network that trains an Influence Prediction Model (IPM) to represent influence. [6] proposed the Linear Threshold Ranking (LTR) by considering the degree of influence of voters on each other's. They also propose arbitrary scoring rules. They distinguish constructive and destructive scenarios for election control. [19] propose a Shapley value-based approach to tackle the influence maximization problem.

Most of existing works, consider the topology of the networks and/or the social actions along with a bunch of profile features. However, the semantic perspective of the network is not exhaustively studied in the literature. The users' interactions with their neighborhood and deeper semantic information are not extracted from the post despite its importance in network analysis. We can notice that most of the existing approaches are parameterized. They require the number k of influential nodes in a deterministic way. Also, the majority of those methods don't take the advantage that can be driven by using distributed systems such as multi-agent systems architecture.

The use of such systems reduces the complexity of proposed approaches and facilitates the presentation of coordination between users. This can unveil the communicative behavior of users in the system. Also, real-world networks are distributed by nature. Few works consider a multi-agent system to determine the influencers in social networks. However, in [13] positive and negative opinions are considered, but the innate opinion is computed based on the Gaussian distribution. Weighted Models and Trivalency Models are used instead of computed weights for edges. [6] used a random degree of influence between voters and a random preference list for each voter. This motivated us to propose our approach AMIR to overcome those shortcomings.

3 A New Multi-agent Approach for Influence Detection in Social Networks: AMIR

In this section, we present the proposed approach for detecting influential users in a social networks. Using a social network as a starting point, we abstract the graph. We determine the candidates and the set of voters for each one of them. Each voter votes for the candidates using the computed voting criteria. As a result, the node passes its decision on to the Voting Manager Agent that will compute and broadcast the results.

3.1 Social Network Abstraction into a Graph

For each network member, we associate a node. The set of nodes is represented by a labeled oriented graph (V,E,W). N is the set of nodes of the graph, V is the set of nodes, E is the set of links and W the set of weights. $W(i,j)$ relies on the opinion similarity $OS(i,j)$ between i and j. $OS(i,j)$ is computed as follow: $OS(i,j) = \sum_{k \in T_{i,j}}^{p} O_i^k * O_j^k$ Where $T_{i,j}$ is the vector of common topics of interest between i and j containing p attributes. O_i^k and O_j^k represent the opinion of users i and j on the topic k respectively. This measure comes from behavioral homophily.

Each node is also characterized by a weight that represents its influence power $\psi_i = \frac{\sum_{\beta=1}^{NB(i)} \alpha_\beta f_\beta^i}{n_1}$. Where $NB(i)$ is the set of direct neighbors of i, f_β^i is the frequency of a social action a_β performed on the publications of i. Indeed, not all actions are of equal importance. Therefore, we associate an importance to each of the actions modeled by a weighting factor.

We utilized two studies to determine the topics and opinions expressed in the tweets. [16], which enabled us to link each tweet to its corresponding topic. The second study, also conducted by [18], was employed to calculate the opinions conveyed within each topic.

The Rules of Influence. To determine the value of the opinion influence of one user on another, we define a set of rules called the influence rules, presented in Table 1.

Table 1. Influence Rules

		O_i^k		
		Negative	**Neutral**	**Positive**
O_j^k	**Negative**	α_0 No Influence	α_1 Low Influence	α_0 No Influence
	Neutral	α_1 Low Influence	α_0 No Influence	α_3 Significant Influence
	Positive	α_0 No Influence	α_1 Low Influence	α_2 Medium Influence

These rules define according to the opinion values of the two users on a topic, the likelihood that the source user i influences the target user j.

3.2 Identifying Candidate Nodes

In order to determine the influencers in the network, we first start by identifying the nodes that have high "probability" to become influencers. A general approach to measure the proximity of nodes in a network is the Personalized PageRank algorithm or PPR. The latter is recognized as one of the most effective measures that classifies nodes according to their reachability from a certain set of nodes in a network. It gives high scores to elements that are closer to the target user regarding a wide range of graph properties such as distance or the number of paths between them.

After defining the abstraction of the social network, we now have the initialization parameters for the personalized page Rank (PPR). We have already attributed the OS to the edges and we can initialize the nodes using the ψ_i formula. The formula of PPR becomes:

$$PR(i;t) = \begin{cases} \psi(i) & if\ t = 0 \\ \frac{1-d}{|V|} + d \sum_{j \in Followees(i))} \frac{PR(j;t-1)}{|Followers(j)|} & if\ t > 0 \end{cases} \quad (1)$$

3.3 Identifying the Influent Nodes

Having identified the candidate nodes in the previous step, we now proceed to identify the influential nodes. Social networks and multi-agent systems share both the structure and the scope, since they are composed of individuals connected with some kind of relationship and they are realized for accomplishing individual and/or common goals [7]. To study MAS, agents and their relations are modeled using graphs.

Since they enable information sharing between our voter agents and the ability to aggregate the results using a Voting Manager Agent, Multi-Agent Systems prove to be an excellent choice for putting our work into practice.

In the following part, we will discuss the details of the design of our minimalist yet robust framework. Our approach relies on the abstracted graph in Sect. 3.1. It associates an agent to each node of this graph. The agent has two roles: CandidateBehavior and VoteBehavior. Another agent Voting Manager Agent is used to manage the vote. This agent supervises all the election process, it collects the voting results, aggregates them and identify the influential nodes (leaders).

3.4 The Voting Process

Now that we have presented the different components of the architecture, let's move to present the overflow of our proposal. We start by presenting the voting process.

The Voting Manager Agent computes the set of Voters $Voters_k$ for a given candidate c_k. He starts by adding the candidate c_k to his set of voters. Then, all the neighbours of a voter are added to the voters' set. This step is repeated for each voter in the voters' set until we reach the maximum depth r. Each voter $voter_i$ fills out his ballot, using the following strategy:
For a voter $voter_i$ and a candidate c_k compute:

$$CV_i^k = \frac{\sum OS_j^k}{d_i} \quad (2)$$

where OS_j^k is the opinion similarity between the candidate c_k and the user $u_j \in SP_i^k$, where SP_i^k is the shortest path between a voter $voter_i$ and a candidate $c_k \in C_i$. If $CV_i^k \geq 0.5$ the candidate c_k is added to the set of approved candidates A_i.

Finally, the Voting Manager Agent collects all the ballots and computes the set of leaders L using the Majority Vote process: for a candidate $c_k \in C$, if he is present in more than 50% of his voters' Approved Candidates sets, he will be added to the list of leaders L.

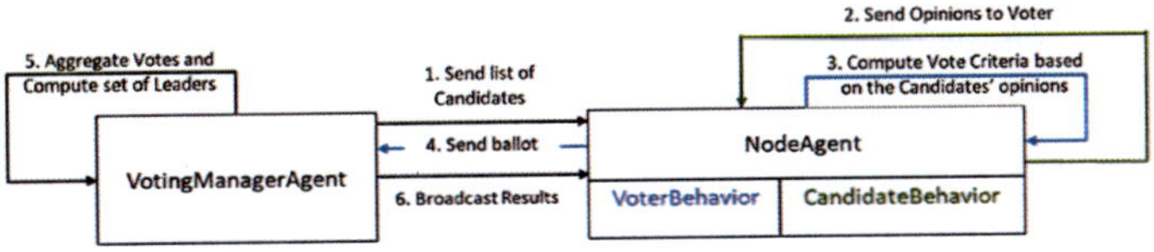

Fig. 1. AMIR: Proposed Architecture

The Fig. 1 resumes the proposed architecture described above. Our proposed multi-agent system includes cooperative agents. Before diving deeper into the works, we first have to define the types of our agents, the relation between them and their methodologies:

- The node agent: this agent manages the nodes. The agent User can have two different roles: It can be candidate that have been selected using the PPR method and/or a voter(a user in the $radius = r$). Then, it can participate in the decision making.
- The vote manager agent: it collects the voting results, aggregates the results and identify of influential nodes (leaders).

4 Influence Maximization: Experiments

In this part, we use 2 different datasets that we extracted from the original dataset using Python Digraph. We only kept the connected graphs in our dataset [17]. The reason behind this choice is the constraint of having a connected graph to study the influence maximization. This condition was highlighted in [10] as a condition for both LTM and ICM models. The obtained datasets are as follow:

- Dataset1: 1738 nodes and 7005 edges.
- Dataset2: 15654 nodes and 1069001 edges.

The experiments in this section are all conducted on a server with 8 cores 2.4 GHz (16 virtual cores) and 64 GB of RAM (32 GB RAM and 32 GB swap). To build our agent-based model, we employed the Mesa: Agent-based modeling in Python 3+ package. This package facilitates communication between agents through messages, ensuring effective interaction within the model.

4.1 Proposed Diffusion Model: OpSimTM

The influence maximization problem has been addressed primarily by LT and IC diffusion models [10] that consider the topology of the network. Neither of them considers an individual's profile, interests, or interactions. Taking into account the informational content, we propose a model called OpSimTM (Opinions similarity Threshold Model), presented in Algorithm 1, which is a standard adaptation of the LT model [10]. In OpSimTM, for a given infected node v, if the Opinion Similarity between v and u exceeds the average OS of v with it's neighbors, then v infects u.

Algorithm 1: OpSimTM

Data: L: set of leaders, $G(V, E)$: Social Graph
Result: $V_{infected}$: set of infected nodes

foreach $v_i \in L$ **do**
  $V_{infected} \leftarrow V_{infected} + \{v_i\}$;
end
foreach $v_i \in V$ **do**
  if $v_i \in V_{infected}$ **then**
    $neighbours = getNeighbours(G, v_i)$;
    foreach $neighbour \in neighbours$ **do**
      if $neighbour \notin V_{infected}$ **then**
        if $(OS_{v_i}^{neighbour} \geq avg(OS_{neighbour}^{neighbours_{neighbour}}))$ **then**
          $V_{infected} \leftarrow V_{infected} + \{neighbour\}$;
        end
      end
    end
  end
end

4.2 Results

Table 2 summarizes the obtained results for the voting strategy.

Table 2. The results of the voting strategy on dataset1 and dataset2

Dataset	number of Candidates	number of Leaders	# Infected nodes (LTM)	IP (LTM)	Infected nodes (OpSimTM)	IP (OpSimTM)
Dataset1	145	32	1111	70.36%	555	83%
Dataset2	2207	592	8143	52.01%	12523	79.99%

We can observe that the utilization of the described strategy yields significant benefits in terms of infecting the largest portion of the network. Specifically, we were able to infect 83% of Dataset1 and 79.99% of the total number of nodes in Dataset2 while working with a limited budget, represented by the seed nodes set. The seed nodes set consisted of 145 nodes for Dataset1 and 2207 nodes for Dataset2, respectively. As a starting hypothesis, we mentioned that the closer people are in the network the closer their interests and opinions are, and they can influence each other accordingly. We notice that the strategy kept consistent performance.

4.3 Comparison with Existing Approaches

Comparison Algorithms: to test the performance of our AMIR model, we will compare it with a bunch of recent models using different solving methodologies:

- CELF++ [9]: this algorithm belongs to the greedy algorithm-based approaches.
- SPIN [19]: this algorithm belongs to the agent-based approaches.
- IV greedy [22]: this algorithm is an amelioration of the greedy algorithm.
- Degree Discount [5]. The approach belongs to the heuristic approaches based on centrality measures.
- Budgeted Influence and Earned Benefit Maximization with Tags in Social Networks [3].

Comparison Results: Table 3a summarizes the results of our algorithm and its comparison with the algorithms described above in terms of influence propagation.

Table 3. Results of Comparison

(a) Number of Nodes: 1738

Approach	Diffusion Model	#seeds	#infected	IP%	Time (s)
AMIR	OpSimTM	32	1723	83	72
AMIR	LTM	32	1223	70.3	72
[5] Degree Discount	LTM	32	988	56.84	87
[22] IV Greedy	LTM	32	1008	57.99	124
[19] SPIN	ICM	32	1149	66.11	235
[9] CELF++	ICM	32	1168	67.20	287
[20]	ICM	57	999	57.47	221
[3] PRUNN	MIA	74	1261	72.55	183

Number of Nodes: 15654

Approach	Diffusion Model	#seeds	#infected	IP%	Time (s)
AMIR	OpSimTM	591	12523	79.9	3949
AMIR	LTM	591	8143	52	3949
[5] Degree Discount	LTM	591	9422	60.18	228533
[22] IV Greedy	LTM	591	6184	39.5	462480
[19] SPIN	ICM	∞	∞	∞	∞
[9] CELF++	ICM	∞	∞	∞	∞
[20]	ICM	5	1551	9.90	279896
[3] PRUNN	MIA	∞	∞	∞	∞

AMIR appears to have a higher influence propagation rate than the other approaches. Using the classical LTM as the diffusion model limited our approach since it does not consider the behavioral information existing in the network. However, we managed to get better results compared to the literature with an influence propagation of 70.3% of the network. The results improved when using the modified LTM that we called OpSimTM. Starting with 32 seed nodes, we were able to influence 83% of the network.

The time complexity of [5] is really high since it is a greedy algorithms. This explains the delayed execution time. Based on sentiment analysis, the user profile information and their activity [20] determine the influence scores. However, the results were low in our dataset because they selected specific topics. The model is not robust when using other topics. We were able to get results for three of our comparison approaches [5,20,22] on the first dataset. But, when the size of the network increases those approaches take a long time (more than a month) to produce results. We can conclude that those approaches are not scalable.

4.4 Performance on a Large Dataset

Dataset. The data used in those experiments is the public coronavirus dataset provided by [14].

The dataset was supplemented by adding non-duplicated tweets obtained through the process of hydrating the original set of tweets. These tweets are just a sample of all the tweets generated that are provided by Twitter. It does not represent the whole population of tweets at any given point. We extracted the only English tweets from this dataset and we mapped them to the users in the graph [11]. The network used in this dataset is semi-synthetic. This dataset contains 136794 nodes, 33760792 edges and 48315724 tweets.

Table 4. Results on large dataset

r	Number of leaders	# Infected nodes	IP%
1	1529	136794	100
2	1687	136794	100
3	2184	136794	100
4	2589	136794	100
5	2589	136794	100
6	2589	136794	100

(a) OpSimTM and LTM-136794 nodes

r	leaders	Activated % under ICM	Infected % under ICM	Activated % under modified ICM	Infected % under modified ICM
1	1529	91.94	100	53.61	57
2	1687	91.94	100	57.02	59.63
3	2184	91.94	100	63.56	65.46
4	2589	91.94	100	65.09	66.69
5	2589	91.94	100	65.09	66.69
6	2589	91.94	100	65.09	66.69

(b) ICM and Modified ICM-136794 nodes

Results Using Different Diffusion Models and Different Voting Radius. Our proposal AMIR appears to have a higher influence propagation rate than the other approaches. Using the classical LTM as the diffusion model limited our approach since it does not consider the behavioral information existing in the network.

The time complexity of both [5,22] is really high since they are both greedy algorithms. This explains the delayed execution time. Based on sentiment analysis, the profile information of users and their activity, [20] determines the influence scores. However, the results were low in our dataset due to the fact that they selected specific topics. The model is not robust using other topics.

Due to the diffusion model being a probabilistic model, we get better results on the much larger dataset (15k nodes) instead of the smaller dataset. This means that the wider the graph, the better the results.

Our proposed approach presents a powerful tool for detecting influential nodes in the social network. In addition, its distributed nature minimizes both time and space complexity. Our analysis of the large dataset revealed that using our OpSimTM diffusion model, our proposed model maintained a high performance of 100% in terms of influence propagation. The obtained result is higher than its results on the other datasets as shown in Table 3a and Table 3b. The obtained results of AMIR on the two datasets for a radius equal to 2 are respectively 83% and 79.9%. Our model's results show that it is highly scalable since it performs better on large datasets. Also, we can explain this by the fact that the large dataset as described in Table 4.4 is a sparse graph with a density of 0.0018 compared to 0.0023 for dataset1 and 0.0043 for dataset2. So the number of candidates and, therefore, the number of voters is higher than the other datasets. This leads to a meticulous voting process.

We can notice that our model achieved the best results on 130k dataset using OPSimTm and the modified ICM as diffusion models and the modified ICM with an influence propagation of 100% for all the radius values starting from $r = 1$ (see Table 4a and Table 4b). The rapid convergence observed in this case can be attributed to the low average shortest path length of 2.83. This value is notably small, indicating the high clustering degree of our network. The lowest results are obtained by classical ICM and LTM with an influence propagation of 66% and 65% respectively.

5 Conclusion

Due to the NP-hard nature of the influence maximization problem, scalability has been a major challenge for proposed algorithms in the field. Additionally, existing research in this domain often overlooks user semantic interactions. To address this, our approach incorporates user post semantics and social actions in an abstraction of the graph, where links indicate similarity of opinions on common topics, and each node is assigned an influence power, ψ. We leverage this graph to identify candidate nodes using a modified version of PageRank, allowing us to shrink the research area. The primary objective of this step is to initiate the influence propagation process using a minimal cardinality of seed nodes and to achieve maximum influence propagation value.

Furthermore, our proposed multi-agent architecture enables nodes within a radius r to vote for candidates and elect leaders. To evaluate our strategies, we conducted experiments on two datasets, demonstrating improved influence propagation and execution time compared to existing literature. Our use of the OpSimTM diffusion model yielded particularly noteworthy improvements. We also evaluated the performance of our architecture on a large-scale dataset. The outcomes showed promise.

In terms of future research, we propose applying our approach to dynamic networks and Multiplex Social Networks to expand its applicability. Additionally, exploring alternative voting methods for electing leaders and their impact on performance would be a valuable area of study.

References

1. Adineh, M., Nouri-Baygi, M.: Maximum degree based heuristics for influence maximization. In: 2018 8th International Conference on Computer and Knowledge Engineering (ICCKE), pp. 256–261. IEEE (2018)
2. Azaouzi, M., Romdhane, L.B.: An evidential influence-based label propagation algorithm for distributed community detection in social networks. Procedia Comput. Sci. **112**, 407–416 (2017)
3. Banerjee, S., Pal, B.: Budgeted influence and earned benefit maximization with tags in social networks. Soc. Netw. Anal. Min. **12**(1), 21 (2021)
4. Chen, S., He, K.: Influence maximization on signed social networks with integrated pagerank. In: 2015 IEEE International Conference on Smart City/SocialCom/SustainCom (SmartCity), pp. 289–292. IEEE (2015)

5. Chen, W., Wang, Y., Yang, S.: Efficient influence maximization in social networks. In: Proceedings of the 15th ACM SIGKDD International Conference on Knowledge Discovery and Data Mining (KDD 2009), pp. 199–208. Association for Computing Machinery, New York (2009). https://doi.org/10.1145/1557019.1557047
6. Corò, F., Cruciani, E., D'Angelo, G., Ponziani, S.: Exploiting social influence to control elections based on scoring rules. arXiv preprint arXiv:1902.07454 (2019)
7. Franchi, E., Poggi, A.: Multi-agent systems and social networks. In: Handbook of Research on Business Social Networking: Organizational, Managerial, and Technological Dimensions, pp. 84–97. IGI Global (2012)
8. Fu, Y.H., Huang, C.Y., Sun, C.T.: Using global diversity and local topology features to identify influential network spreaders. Physica A Stat. Mech. Appl. **433**, 344–355 (2015)
9. Goyal, A., Lu, W., Lakshmanan, L.V.: CELF++ optimizing the greedy algorithm for influence maximization in social networks. In: Proceedings of the 20th International Conference Companion on World Wide Web, pp. 47–48 (2011)
10. Kempe, D., Kleinberg, J., Tardos, É.: Maximizing the spread of influence through a social network. In: Proceedings of the Ninth ACM SIGKDD International Conference on Knowledge Discovery and Data Mining, pp. 137–146 (2003)
11. Kwak, H., Lee, C., Park, H., Moon, S.: What is Twitter, a social network or a news media? In: Proceedings of the 19th International Conference on World Wide Web, pp. 591–600 (2010)
12. Leskovec, J., Krause, A., Guestrin, C., Faloutsos, C., VanBriesen, J., Glance, N.: Cost-effective outbreak detection in networks. In: Proceedings of the 13th ACM SIGKDD International Conference on Knowledge Discovery and Data Mining, pp. 420–429 (2007)
13. Liang, W., Shen, C., Li, X., Nishide, R., Piumarta, I., Takada, H.: Influence maximization in signed social networks with opinion formation. IEEE Access **7**, 68837–68852 (2019). https://doi.org/10.1109/ACCESS.2019.2918810
14. Lopez, C.E., Gallemore, C.: An augmented multilingual Twitter dataset for studying the COVID-19 infodemic. Soc. Netw. Anal. Min. **11**(1), 1–14 (2021)
15. Messaoudi, C., Guessoum, Z., Ben Romdhane, L.: Opinion mining in online social media: a survey. Int. J. Soc. Netw. Anal. Min. **12**, 5 (2022). https://doi.org/10.1007/s13278-021-00855-8
16. Messaoudi, C., Guessoum, Z., BenRomdhane, L.: Topic extraction in social networks. Comput. Inform. **41**(1), 56–77 (2022)
17. Messaoudi, C., Guessoum, Z., Romdhane, L.B.: Topic extraction in social network. In: International Conference on Applied Data Science and Intelligence (2021)
18. Messaoudi, C., Guessoum, Z., Romdhane, L.B.: A deep learning model for opinion mining in Twitter combining text and emojis. In: 26th International Conference on Knowledge Based and Intelligent Information and Engineering Systems (2022)
19. Narayanam, R., Narahari, Y.: A shapley value-based approach to discover influential nodes in social networks. IEEE Trans. Autom. Sci. Eng. **8**(1), 130–147 (2011). https://doi.org/10.1109/TASE.2010.2052042
20. Sardana, N., Tejwani, D., Thakur, T., Mehrotra, M.: Topic wise influence maximisation based on fuzzy modelling, sentiments, engagement, activity and connectivity indexes (IC3 2021), pp. 443–449. Association for Computing Machinery, New York (2021). https://doi.org/10.1145/3474124.3474192
21. Suri, N.R., Narahari, Y.: Determining the top-k nodes in social networks using the shapley value. In: Proceedings of the 7th International Joint Conference on Autonomous Agents and Multiagent Systems, vol. 3, pp. 1509–1512. Citeseer (2008)

22. Wang, W., Street, W.N.: Modeling and maximizing influence diffusion in social networks for viral marketing. Appl. Netw. Sci. **3**(1), 6 (2018)
23. Wang, Y., Cong, G., Song, G., Xie, K.: Community-based greedy algorithm for mining top-k influential nodes in mobile social networks. In: Proceedings of the 16th ACM SIGKDD International Conference on Knowledge Discovery and Data Mining, pp. 1039–1048 (2010)
24. Yan, B., Song, K., Liu, J., Meng, F., Liu, Y., Su, H.: On the maximization of influence over an unknown social network. In: AAMAS, vol. 19, pp. 13–17 (2019)

Check for updates

Extracting Knowledge from Testaments - An Ontology Learning Approach

Shahzod Yusupov[1], Anabela Barros[2], and Orlando Belo[1](✉)

[1] Algoritmi Research Centre/LASI, University of Minho, 4710-059 Braga, Portugal
a82617@alunos.uminho.pt, obelo@di.uminho.pt
[2] CEHUM, Centre for Humanistic Studies, University of Minho, 4710-059 Braga, Portugal
aldb@elash.uminho.pt

Abstract. The extraction of ontologies from textual data is not a simple task. It is necessary to have specific expertise and knowledge about the ontology application domain to carry it out, and knowing how to use properly a set of sophisticated methods and techniques, requiring the use of advanced ontology learning tools. For some years, ontologies have been instruments of great importance in the process of designing and developing knowledge systems. In this paper, we present and describe the design and development of a semi-automatic system for extracting an ontology of Portuguese testaments from a set of ancient texts of the 18th century, towards the acquisition of the knowledge about the legacies of people of a Portuguese region on that period. This ontology has great interest and relevance for knowing many aspects of that time, namely linguistic, historical, religious, cultural, economic or agricultural, among others.

Keywords: Ontology Learning · Linguistic Ontologies · Knowledge Extraction · Unstructured Textual Data · Natural Language Processing · Linguistic Patterns

1 Introduction

In the last few years, ontology learning processes [1, 2] have gained a vast space for discussion and work, being today essential instruments for knowledge discovery. However, its design and implementation is not simple. By nature, they are quite complicated processes, which require a high level of knowledge and use different types of strategies and methods for extracting knowledge from distinct sources of information. Currently, ontology learning provides us a large diversity of tools especially oriented to extract, develop and maintain ontologies from different types of information sources, which may range from conventional datasets to textual data. Combining machine-learning [3] with natural language processing [4] techniques, these tools simplify ontology extraction processes, requiring less human resources and processing time. The gains are evident when we look at ontology extraction from unstructured texts. These texts raise many problems on ontology extraction processes. From a same text, domain experts may have different interpretations and recognize dissimilar ontological structures, which may cause disparate extraction of terms and, consequently, concepts and relationships.

S. Ossowski et al. (Eds.): DCAI 2023, LNNS 740, pp. 254–263, 2023.
https://doi.org/10.1007/978-3-031-38333-5_26

Approaching automatically the process of extracting an ontology, using machine-learning techniques and natural language processing, the identification and extraction of terms, concepts, properties, and relationships presented in data is simplified and is done systematically independently from domain experts, following strict rules and methods, keeping the same "way of acting" along all the documents that must be processed. However, it is quite difficult to achieve a complete automatic process. There are many aspects related to knowledge identification and extraction that still require human intervention to resolve ambiguities, errors, or inconsistencies found in texts. In a near future, probably some (or all) of these problems can be eliminated based on the experience and software developed in past ontology learning processes.

In this paper, we present the implementation and development of a semiautomatic ontology learning system, with the ability for extracting the information present in a manuscript of the 18th century, containing a large diversity of testaments (wills). Barros and Alves (2019) [5] edited these documents under the title "The Book of Testaments – Picote 1780–1803", from an unpublished codex from the Municipal Archive of Miranda do Douro, Portugal. The manuscript, previously prepared by the notary for use as a notebook of Picote's testaments, written in classical to modern Portuguese by the scribe Manuel Domingues, with wide linguistic variation, allows the study of this language along the years, and of Mirandese, the second official language of Portugal and the mother tongue of the natives of Terra de Miranda. The acquisition of the knowledge contained in these documents is of great interest and relevance, as it allows the systematization of the very abundant and varied types of data present in this textual subgenre: linguistic, historical, genealogical, religious, geographical, economic, agricultural or sociocultural, among others. These documents are written in Portuguese from the 18th century. Some of them are quite long. Their analysis and manual extraction of knowledge about its content is an arduous, time-consuming and not always successful task. Knowing this, we conceived and implemented an ontology learning tool, especially for analyzing the content of testaments, with a simple and intuitive graphical interface, containing mechanisms that facilitate the acquisition of knowledge. The remaining part of this paper is organized as follows. Section 2 exposes the domain of ontology learning, Sect. 3 discusses the testament ontology learning process we conceived and implemented, revealing how we extracted the ontology from the ancient texts we had available, how we can explore the ontology and a brief results analysis. Finally, Sect. 4 presents some conclusions and points out a few ideas for future work.

2 Ontology Learning

Representing and exchanging information between systems, applications and computational services depends a lot on the design and development of consistent and comprehensive models and structures for representing application domain knowledge. For some years now, we have been witnessing a very significant increase in the size and complexity of computing systems and knowledge bases. On the Web, this increase has been extraordinary, not only in terms of storage resources, but also in terms of the information that circulates in the Web environment, every day, every minute. Such circumstances imposed the creation of sophisticated mechanisms for facilitating the representation, storage and communication of information. Of all the mechanisms proposed and implemented, probably the ontologies stood out the most.

Today, ontologies [6, 7] are recognized as one of the most sophisticated structures we have access to accommodate knowledge in a large range of application domains. When well defined, ontologies are able to support the representation and exploration of the various aspects of a given application domain, incorporating concepts and relationships extracted through the characteristics and semantics of the acquired knowledge. In addition, ontologies' data models increase the sharing of knowledge between systems, or even between systems and people, sustaining a systematized vocabulary of some application domain that allows improving the quality of entities, concepts and relationships established in a given information system, their use and reuse and the way we interpret the information used by that same system. However, the process of idealizing and conceiving an ontology is not simple. It requires time and rigorous planning, expertise and robust knowledge in the field of the ontology, and dexterity in the use of adequate computational tools. All these are essential elements in guaranteeing the success of an ontology, whether applied manually or automatically in the construction process. As we know, manual processes require great attention and time from experts in the field of ontology, which increases almost exponentially, depending on the volume of information we have to deal with. In particular, when we want to apply for a process of this kind to a set of texts written in natural language, the process of extracting an ontology becomes complicated, not only because of the nature of the texts – the language in which it is written, the time of the texts, syntactic and semantic structures, the vocabulary used, etc. –, as well as the number of texts to be processed. Thus, it is recognized easily that, after a certain moment, a manual approach it is not feasible, or even possible, to enrich or maintain ontology structures that were gradually extracted and documented. The option for a manual or automatic process in the definition and development of an ontology must obviously be evaluated at the beginning of the process, taking into account aspects such as the size of the textual information sources to be used, the number and availability of experts in the field, the tools available or the time to spend on carrying it out. Today, in these cases, the adoption of a (semi)automatic approach in an ontology extraction process is a very common step, strongly encouraged by the availability of a large and diversified number of methods and tools specially oriented to the implementation of learning ontology processes. Ontology Learning [7, 8] provide today a large variety of automatic tools for performing tasks in which the application of natural language processing, and machine learning programs are used for discovering valuable knowledge about specific domains, identifying concepts and their relationships and establish ontologies.

3 Learning an Ontology from Testament Texts

3.1 The Book of Testaments

In this work, we present and describe the implementation of a semi-automatic ontology learning system designed especially for extracting information present in texts containing old testaments, from the 18th century, edited by Alves and Barros in 2019 [5] – The Book of Testaments – Picote 1780–1803 –, and represent it in an ontological structure.

The book is a parchment-bound codex, measuring 30.8 by 20 cm, consisting of one hundred folios, of which only seventy-six were filled. During the second half of the 20th century, the book remained at the Museum of Terra de Miranda, and since 2013 has been in the Municipal Archive of Miranda do Douro, Portugal. It was previously prepared for use as the "Book of Notes on the Testaments of Picote", to be written by the clerk Manuel Domingues [5]. The content of these documents is based on the testaments that the inhabitants of Picote, a parish in Miranda do Douro, ordered to be drawn up, usually when they were weak and bedridden. These testaments were drawn up by the clerk Manuel Domingues, in the presence of seven witnesses who, normally, were also inhabitants of Picote. As these documents were written in Portuguese from the 18th century, some of which are quite long. Thus, the manual extraction of relevant information is not a viable process, since it involves some arduous and time-consuming tasks. In Fig. 1, we can see two small fragments of a testament (semi diplomatic version) written in classical Portuguese, corresponding respectively to an original text (a) and its corresponding preprocessed version (b).

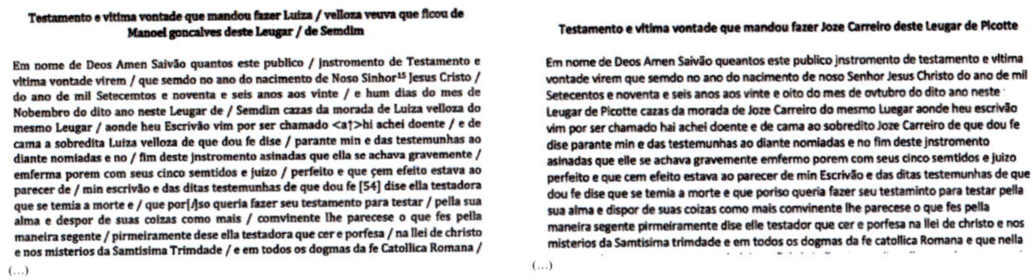

Testamento e vltima vontade que mandou fazer Luiza / velloza veuva que ficou de Manoel goncalves deste Leugar / de Semdim

Em nome de Deos Amen Saivão quantos este publico / jnstromento de Testamento e vltima vontade virem / que semdo no ano do nacimento de Noso Sinhor[15] Jesus Cristo / do ano de mil Setecemtos e noventa e seis anos aos vinte / e hum dias do mes de Nobembro do dito ano neste Leugar de / Semdim cazas da morada de Luiza velloza do mesmo Leugar / aonde heu Escrivão vim por ser chamado <a†>hi achei doente / e de cama a sobredita Luiza velloza de que dou fe dise / parante min e das testemunhas ao diante nomiadas e no / fim deste jnstromento asinadas que ella se achava gravemente / emferma porem com seus cinco semtidos e juizo / perfeito e que çem efeito estava ao parecer de / min escrivão e das ditas testemunhas de que dou fe [54] dise ella testadora que se temia a morte e / que por[/]so queria fazer seu testamento para testar / pella sua alma e despor de suas coizas como mais / comvinente lhe parecese o que fes pella maneira segente / pirmeiramente dese ella testadora que cer e porfesa / na llei de christo e nos misterios da Samtisima Trimdade / e em todos os dogmas da fe Catollica Romana /

(...)

a) Original text

Testamento e vltima vontade que mandou fazer Joze Carreiro deste Leugar de Picotte

Em nome de Deos Amen Saivão queantos este publico jnstromento de testamento e vltima vontade virem que semdo no ano do nacimento de noso Senhor Jesus Christo do ano de mil Setecentos e noventa e seis anos aos vinte e oito do mes de ovtubro do dito ano neste Leugar de Picotte cazas da morada de Joze Carreiro do mesmo Luegar aonde heu escrivão vim por ser chamado hai achei doente e de cama ao sobredito Joze Carreiro de que dou fe dise parante min e das testemunhas ao diante nomiadas e no fim deste jnstromento asinadas que elle se achava gravemente emfermo porem com seus cinco semtidos e juizo perfeito e que cem efeito estava ao parecer de min Escrivão e das ditas testemunhas de que dou fe dise que se temia a morte e que poriso queria fazer seu testaminto para testar pella sua alma e dispor de suas coizas como mais comvinente lhe parecese o que fes pella maneira segente pirmeiramente dise elle testador que cer e porfesa na llei de christo e nos misterios da Samtisima trimdade e em todos os dogmas da fe catollica Romana e que nella

(...)

b) Preprocessed text

Fig. 1. Fragments of a text of a testament extracted from [5].

3.2 The Ontology Learning Process

The ontology learning process we implemented (Fig. 2) follows the model proposed by Asim et al. [9]. However, we adjusted some of the tasks indicated by the authors in order to fulfill the specificities of the testament ontology extraction. The first task of the process is the selection of the testament texts we want to process. Then, we proceed to the text preprocessing task, which is responsible to clean and prepare the testament texts for ontology extraction, removing invalid or superfluous characters, ambiguities, or structural or semantic inconsistency, among other things that could constraint or avoid the success of the ontology learning process [10].

Having texts prepared, we start the third task, applying syntactic lexical patterns [10] for discovering key data elements, concepts and relationships of the future testament ontology. All these elements are analyzed and combined based on the dependencies between other discovered elements [11]. In the following task, we reinforced the knowledge extraction process applying dependency parsing [12] for discovering other ontological elements not identified by the lexical patterns established and applied in previous task. Finally, we created and populated knowledge structures that received all

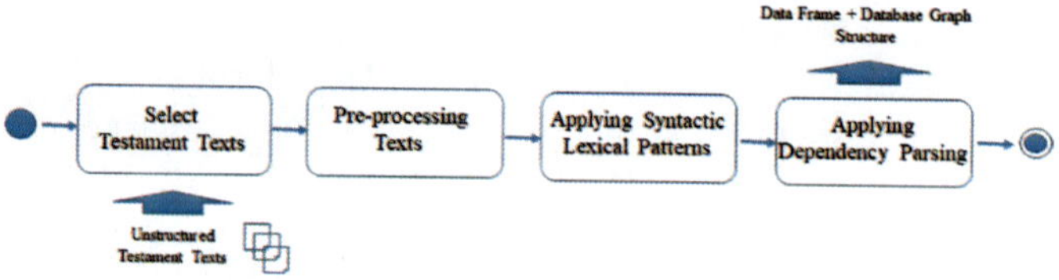

Fig. 2. The main tasks of the ontology learning process.

the ontological elements extracted in previous tasks. In the next section, we will describe in detail each one of these tasks.

Preprocessing Texts. Before applying any ontology learning technique, it is necessary to ensure that the available data are in conditions that can be used, having the required quality to carry out the process of extracting concepts, relationships and properties from a given application domain. Therefore, it is essential to carry out a pre-processing stage, to analyze the content of the texts to process, proceed with their cleaning or transforming textual elements in accordance with the requirements of the ontology learning process. Data cleaning involves the removal of data elements considered incorrect or irrelevant, while data transformation implies the normalizing and restructuring data elements, in order to facilitate knowledge extraction and the ontology definition. Previously, in Fig. 1a, we present a small fragment of an original text of a testament, before undergoing some kind of preprocessing operation. We can see that the interpretation of the text or the extraction of any piece of data by the computer is not easy. Through the analysis the text, which resulted from editing the original text of the manuscript, we verified that it does not present punctuation and contains some characters that make its interpretation difficult, such as, for example, the character '/', representing a change of line, or 'the character =' indicating the beginning of a sentence. Other characters also appear, such as '[', ']', or '+', which add nothing to the content of the testament. As such, all these characters have been deleted. However, we added '.' characters to all sentence endings. After carrying out these cleaning tasks, we proceeded to the standardization of certain words, which throughout the testaments appeared written in different ways, which, later, could have different meanings in grammatical analysis. At the end of carrying out all the pre-processing tasks, we obtained a text ready to be processed (Fig. 1b).

Applying Syntactic Lexical Patterns. Similarly to what happened in one of our previous works [13], to improve the process of extracting concepts through the discovery of semantic relationships between the words contained in the texts of testaments, we decided to use lexical-syntactic patterns. Their application allowed us to improve the process of extracting an ontology from a set of old texts about remedies. These patterns were the result of a pioneering work done by Hearst [10]. Its definition and application makes it possible to extract hyperonym and hyponym relationships from textual sources. Thus, after a detailed analysis of the texts, we applied pattern-based extraction techniques and identified a specific set of concepts that appeared in many of the texts. Out of this set, stood out terms relating to the testator, the person who had his will draw up, the clerk, the person who wrote the will, the place where the document was drawn up, the date on which the will was drawn up, the witnesses present at the time of the writing of the will, to the heirs and family members of the testator and also to the legatees, the individuals to whom the testator left a legacy. For example, to extract a testator's child we used the pattern: '*seu filho {nome} + de? {nome}**', which we can translate to English as 'your son {name} + of? {name}'. With this pattern, it was possible to extract from the texts expressions such as "Deixa a seu filho Manuel..." or "Deixa a seu filho Manuel Garcia". Furthermore, this pattern was used also for extracting all members of a family. Many other patterns were developed in order to cover all the cases that we considered important for the creation of the ontology. In particular, we have developed lexical patterns especially for extracting heirs, testators' representatives and legatees.

Applying Dependency Parsing. In the dependency parsing [12] task we extracted the information related to the inheritances and legacies left by the testators. Based on the analysis of the testaments, we also verified the existence of patterns in expressions such as 'Item dise que deixa uma camisa...' or 'Item dise que deixa cinco alquires de pão cozido...', in which several verbs appear, such as, for example, *leave, want,* or *name*, which are used to indicate things that the testator left as an inheritance or something that he wants to be done to him after his death. However, the technique for identifying patterns used in the previous task cannot be used here, since it is not possible to know exactly the number of attributes that may appear after each verb, nor the order in which they may appear in the sentence under analysis. To overcome this difficulty, we chose to use dependency parsing for analyzing the syntactic dependencies that could exist between words. The dependency analysis allows for discovering relationships between terms, using information from dependencies present in the parsing trees that were performed from the sentences. A dependency relation is an asymmetric binary relation established between a word designated by "head" and a word designated by "modifier" or "dependent" [14]. Before applying this technique, after extracting the name of the testator, in the previous task, we replaced expressions such as "Item disse que..." or "Item disse o testador que..." by the respective name of the testator. Doing this, we can obtain, for example, "José Folgado deixa vinte missas..." instead of "Item disse que deixa vinte missas...". With this substitution we were able to divide each sentence into three parts, as with a triple in an ontology (subject, predicate, object), in which the predicate is the verb to be captured and the object the inheritance or legacy left by the testator. Concerning the subject, there are several distinct cases, such as: "José" or "José Castro" or even "José de Castro". In Fig. 3a and 3b we can see two parsing trees, showing the dependencies we found, respectively, for a simple and a composite subject.

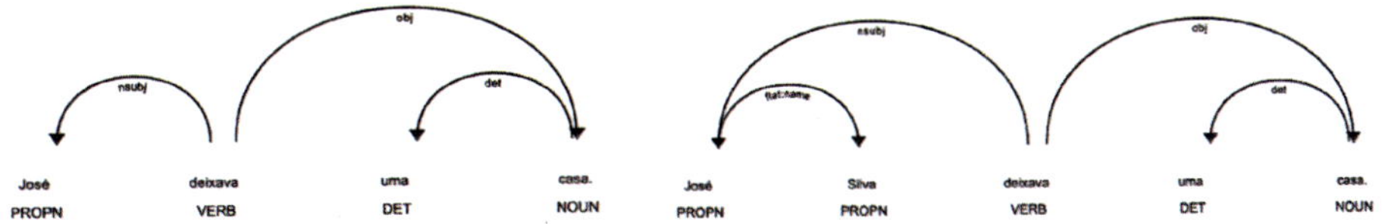

a) b)

Fig. 3. Parsing trees for a subject, simple, and compound.

The analysis and treatment of the inheritance (or legacies) field implied some extra work, not only because of the existence of a large number of cases, but also due to the fact that the model we used from spaCy [15] is not prepared for classical Portuguese. In some cases, this led to an incorrect part of speech assignment or syntactic dependency. In other cases, we verified alterations in the type of dependencies between words. In the process of extracting the inheritances and legacies left by the testator, we identified around thirty different cases.

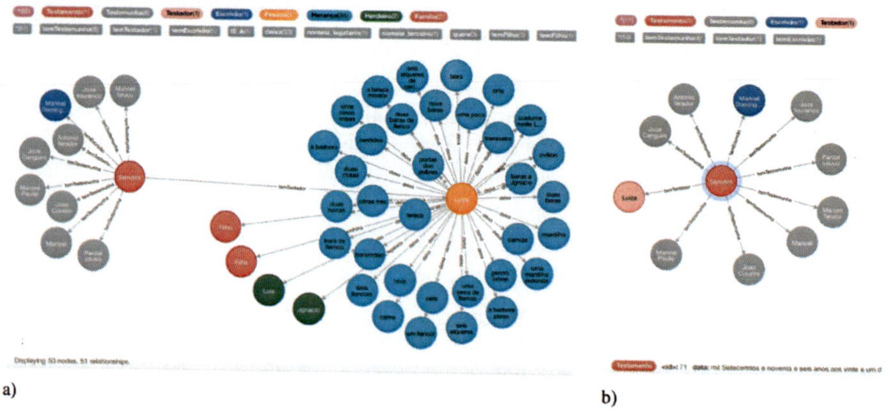

Fig. 4. Distinct views of the structure of the ontology.

3.3 Exploring the Ontology and Results

In order to explore the ontology, we exported to Neo4J the ontological structure that we created earlier during the ontology learning process. Neo4j [16] is a graph-based database management system, which is part of a new generation of software for NoSQL database systems. This system provides several tools and libraries that were built especially for the creation, exploration and maintenance of databases based on graphs. In Fig. 4a, we can see a small excerpt of the ontology graph that was extracted from the texts of the testaments we processed. A brief analysis of the figure allows us to see the existence of various types of relationships. For example, the 'Testamento' (testament) node is related to the 'Testemunha' (witness), 'Escrivão' (scribe) and 'Testador' (testator) nodes through the relationships 'temTestemunha' (hasWitness), 'temEscrivão' (hasScribe) and 'temTestador' (hasTestator), respectively. Thus, we can know that during the writing of this testament, in particular, eight witnesses were present, in addition to the clerk 'Manoel Domingues' and that 'Luiza' was the person who had the testament drawn up. All the properties of a node can be easily seen, just by positioning the system cursor on the referred node (Fig. 4b).

To evaluate the effectiveness and efficiency of the system, we evaluated the results of applying the various techniques we used to process testaments. In Table 1, we can see the results of the evaluation made. Analyzing the values of the table, we can see that the

Table 1. Text processing evaluation results.

Processed Words	Pattern Precision	Pattern and Dependency Parsing Precision	Processing Time (s)
4130	1.0	0.90	83.8
4715	1.0	0.82	81.9
5020	0.9	0.82	85.7
5206	1.0	0.83	81.7
6138	0–9	0.72	84.1
6832	1.0	0.80	85.4

application of lexical-syntactic patterns got an average accuracy of 96.7% in the data extraction process and the combination of the patterns technique with the dependency parsing technique resulted in an average accuracy of around 82%, an average less than of the patterns. This difference can be explained by the fact that the lexical-syntactic patterns are, in most cases, fixed structures, which allow us to know exactly what we want to extract, which means that there is not much room for the extraction of incorrect data. On the other hand, the parsing dependency depends a lot on the grammatical classes of the words and the dependencies between them. As the texts are written in classical to modern Portuguese of the 18th/early 19th century, sometimes occurred errors in the definition of the grammatical class made by spaCy to the words of the processed texts and, consequently, in the type of dependencies that the words have between themselves. Furthermore, the values of Table 1 reveal that the processing time for each testament is almost identical, even when the number of words processed per document varies between 4130 and 6832.

4 Conclusions and Future Work

In this paper, we presented the development of a semiautomatic ontology learning system, which was conceived for extracting information containing in ancient testament texts, mostly from the 18th century, edited by Alves and Barros [5]. The acquisition of the knowledge contained in these documents allowed the construction of a specific domain ontology, in a semiautomatic approach, which is capable of characterizing and describing the knowledge expressed in the testaments we processed. This ontology, it will be a valuable instrument for all that are interested to study linguistic, genealogical, geographical, theological or economic aspects of the region of Miranda do Douro, Portugal, in the 18th century. As in other ontology learning works, the knowledge acquisition task involved natural language text processing, and the application of lexical-syntactic patterns and dependency parsing improved significantly the quality of the knowledge extracted. It was possible to identify a very significant set of linguistic patterns in the testaments, which allowed a (semi)automatic extraction, with a good level of confidence, of the main concepts, such as "testador" or "herdeiro", and relationships such as "temTestator" or "temTestemunha", contained in testaments. Later, we exported the ontology to a graph-oriented database in Neo4J. Based on the environment and services used by this tool, we can analyze, correlate and visualize the different elements of the ontology in a very effective and practical way. For example, how many masses were asked to be said by the testators of Picote between 1780 and 1803 or which types of lands were named and their respective microtoponyms are questions whose answer we can obtain very quickly with a simple query to the system's database. As in similar previous works, the greatest difficulties we faced during the development process were with the preprocessing of the texts, given the nature of the texts (testaments), the type of language in which they were written, classical Portuguese, and to the vocabulary used at the time the texts were written.

Although we consider that the results obtained are positive, we can still improve them, reducing the level of noise that results from incorrectly classified concepts. For this, we intend to design and develop a specific model in spaCy, especially aimed at the Portuguese used in the texts of the testaments we processed. We are convinced that this new model will allow us to obtain better results by applying dependency parsing, assigning the correct grammatical class to each word and creating valid semantic triples.

Acknowledgements. This work has been supported by FCT – Fundação para a Ciência e Tecnologia within the R&D Units Project Scope: UIDB/00319/2020, and the PhD grant: 2022.12728.BD.

References

1. Keet, M.: An Introduction to Ontology Engineering, University of Cape Town (2018). https://people.cs.uct.ac.za/~mkeet/OEbook/. Accessed 24 Apr 2023
2. El Kadiri, S., Terkaj, W., Urwin, E.N., Palmer, C., Kiritsis, D., Young, R.: Ontology in engineering applications. In: Cuel, R., Young, R. (eds.) FOMI 2015. LNBIP, vol. 225, pp. 126–137. Springer, Cham (2015). https://doi.org/10.1007/978-3-319-21545-7_11
3. Sharma, A.: Natural language processing and sentiment analysis, in international research. J. Comput. Sci. **8**(10), 237 (2021). https://doi.org/10.26562/irjcs.2021.v0810.001
4. Zhang, L., Wang, S., Liu, B.: Deep learning for sentiment analysis: a survey. WIREs Data Min. Knowl. Disc. **8**(4) (2018). https://doi.org/10.1002/widm.1253
5. Alves, A., Barros, A.: O Livro dos Testamentos - Picote 1780–1803. Traços do português e do mirandês setecentistas na língua jurídica. Frauga, Picote (2019). ISBN 978–989–99411–8–2
6. Guarino, N., Oberle, D., Staab, S.: What is an ontology? In: Staab, S., Studer, R. (eds.) Handbook on Ontologies. IHIS, pp. 1–17. Springer, Heidelberg (2009). https://doi.org/10.1007/978-3-540-92673-3_0
7. Cimiano, P., Mädche, A., Staab, S., Völker, J.: Ontology learning. In: Staab, S., Studer, R. (eds.) Handbook on Ontologies. IHIS, pp. 245–267. Springer, Heidelberg (2009). https://doi.org/10.1007/978-3-540-92673-3_11
8. Asim, M., Wasim, M., Khan, M., Mahmood, W., Abbasi, H.: A survey of ontology learning techniques and applications. Database 2018 (2018). https://doi.org/10.1093/database/bay101
9. El Ghosh, M., Naja, H., Abdulrab, H., Khalil, M.: Ontology learning process as a bottom up strategy for building domain-specific ontology from legal texts. In: Proceedings of the 9th International Conference on Agents and Artificial Intelligence, pp. 473–480. SCITEPRESS - Science and Technology Publications (2017). https://doi.org/10.5220/0006188004730480

10. Hearst, M.: Automatic acquisition of hyponyms from large text corpora. In: Proceedings of the Fourteenth International Conference on Computational Linguistics, Nantes France (1992). https://doi.org/10.3115/992133.992154
11. Wong, W., Liu, W., Bennamoun, M.: Ontology learning from text. ACM Comput. Surv. **44**(4), 1–36 (2012). https://doi.org/10.1145/2333112.2333115
12. Jaiswal, S.: Natural Language Processing — Dependency Parsing. Towards Data Science. Industrial Data (2021). https://towardsdatascience.com/natural-language-processing-dependency-parsing-cf094bbbe3f7
13. Nunes, J., Belo, O., Barros, A.: Mining ancient medicine texts towards an ontology of remedies - a semi-automatic approach. In: Proceedings of the 1st International Conference on Intelligent systems and Machine Learning (ICISML 2022), Hyderabad, India, 16–17 December (2022)
14. Hazman, M., El-Beltagy, S.R., Rafea, A.: A survey of ontology learning approaches. In: CEUR Workshop Proceedings, pp. 36–43 (2008)
15. Honnibal, M., Montani, I.: spaCy Industrial-strength Natural Language Processing in Python (2017)
16. Neo4J: Neo4J Graph Data Platform (2023). https://neo4j.com/. Accessed 24 Apr 2023

AI-Based Platform for Smart Cities Urban Infrastructure Monitorization

Francisco Pinto-Santos[1(✉)], Juan Antonio González-Ramos[2], Javier Curto[1], Ricardo S. Alonso[1], and Juan M. Corchado[1,3]

[1] Air Institute, IoT Digital Innovation Hub (Spain), Carbajosa de la Sagrada, 37188 Salamanca, Spain
{franpintosantos,jcurto}@air-institute.com, {corchado,juanan}@usal.es
[2] Servicios Informáticos, Universidad de Salamanca, Salamanca, Spain
[3] BISITE Research Group, University of Salamanca, Salamanca, Spain

Abstract. This paper presents a case study of using an AI-based platform to monitor and manage urban infrastructure elements, such as waste containers and traffic lights, in a smart city. The platform leverages Internet of Things (IoT) technologies and MQTT protocol for data ingestion, preprocessing, analysis, and visualization. The case study demonstrates the ease of creating interactive visualizations and dashboards using the platform, enabling city officials and other stakeholders to monitor the status of various urban elements effectively, without the need for extensive technical knowledge or programming skills.

Keywords: Smart cities · Internet of Things · Software platform

1 Introduction

Smart cities are rapidly becoming a reality worldwide, driven by the increasing adoption of Internet of Things (IoT) technologies and artificial intelligence (AI) to enhance urban environments' efficiency, sustainability, and livability. In a smart city, various interconnected systems and infrastructure elements, such as waste management, traffic control, public transportation, and energy grids, are monitored and managed in real-time to optimize resource utilization and provide better services to citizens. The growing complexity of these interconnected systems and the vast amounts of data generated by IoT devices present significant challenges for city officials and other stakeholders responsible for the planning and management of urban environments.

One of the key challenges faced by these stakeholders is the ability to effectively monitor and manage various urban infrastructure elements, such as waste containers and traffic lights, which play a crucial role in maintaining a clean, safe, and well-functioning city. Traditional methods for monitoring and managing these elements often rely on manual inspections, data collection, and paper-based reporting systems, which can be time-consuming, labor-intensive, and prone to errors. Moreover, these methods often fail to provide real-time insights into the

S. Ossowski et al. (Eds.): DCAI 2023, LNNS 740, pp. 264–274, 2023.
https://doi.org/10.1007/978-3-031-38333-5_27

performance and status of urban infrastructure elements, leading to inefficiencies and suboptimal decision-making.

IoT technologies offer a promising solution to these challenges by providing real-time, continuous data on the status of various urban infrastructure elements. IoT devices, such as sensors and actuators, can be installed on waste containers and traffic lights to collect data on parameters such as container fill levels, battery status, and traffic light status. The data collected by these devices can be used to monitor the performance of urban infrastructure elements, detect issues and anomalies, and inform decision-making in areas such as maintenance scheduling, resource allocation, and infrastructure planning.

However, the sheer volume and complexity of the data generated by IoT devices can be overwhelming for city officials and other stakeholders, making it difficult for them to derive meaningful insights and make informed decisions. This is where AI-based platforms, such as the one described in a previous paper, can play a crucial role in simplifying the process of data ingestion, preprocessing, analysis, and visualization. By providing an intuitive, user-friendly interface and an AI-driven assistant, these platforms can help bridge the gap between complex IoT data and its practical applications in urban planning and management.

The AI-based platform presented in this paper is designed to democratize the use of AI in smart cities by making it accessible to a wide range of stakeholders, regardless of their technical expertise. The platform integrates an AI-driven assistant that guides users through the process of importing data, applying machine learning algorithms, creating visualizations, and designing dashboards using drag-and-drop techniques. The platform supports data ingestion from IoT devices using the MQTT protocol, ensuring seamless integration with the smart city infrastructure.

This case study aims to demonstrate the potential of the AI-based platform in monitoring and managing urban infrastructure elements, such as waste containers and traffic lights, in a smart city context. The paper will focus on the process of using the platform to import data from IoT devices, preprocess and analyze the data, and create interactive visualizations and dashboards for monitoring the status of different urban elements. By showcasing a practical application of the platform, the paper seeks to highlight the benefits of integrating AI and IoT technologies in the planning and management of smart cities and to encourage further research and development in this area.

The remainder of this paper is organized as follows: Sect. 2 provides background information on the challenges faced in monitoring urban infrastructure elements and the need for user-friendly solutions. Section 3 describes the process of using the AI-based platform to import data, preprocess and analyze it, and create visualizations and dashboards for monitoring the status of waste containers and traffic lights. Section 4 presents the results of the case study, focusing on the effectiveness of the platform in addressing the challenges faced by city officials and other stakeholders in monitoring and managing urban infrastructure elements. Section 5 concludes the paper and discusses future work in the development of AI-based solutions for smart cities.

By presenting this case study, we hope to provide a comprehensive understanding of the potential applications and benefits of the AI-based platform in the context of smart cities. Furthermore, we aim to contribute to the growing body of research on integrating AI and IoT technologies in urban planning and management and to inspire further innovation in this area. Ultimately, the goal is to create more sustainable, efficient, and livable urban environments through the effective use of AI and IoT technologies.

2 Background

2.1 Smart Cities and Urban Infrastructure Challenges

In recent years, the concept of smart cities has gained significant traction as a means to address the various challenges associated with rapid urbanization and population growth [6]. Smart cities leverage advanced technologies, such as the Internet of Things (IoT) [8], artificial intelligence (AI) [7], and big data analytics [9], to optimize the efficiency, sustainability, and livability of urban environments [10]. The overarching goal of a smart city is to enhance the quality of life for its residents by providing better services, improving resource utilization, and fostering economic growth [1].

A key component of smart cities is the efficient management of urban infrastructure elements [4], such as waste containers [2], traffic lights [3], public transportation systems [14], and energy grids [13]. The proper monitoring and management of these elements are crucial for maintaining a clean, safe, and well-functioning city [5]. However, city officials and other stakeholders responsible for the planning and management of urban environments face several challenges in this regard.

Firstly, the sheer scale and complexity of urban infrastructure systems make it difficult to monitor and manage individual elements effectively [11]. Traditional methods for monitoring and managing urban infrastructure elements often rely on manual inspections, data collection, and paper-based reporting systems, which can be time-consuming, labor-intensive, and prone to errors. These methods also lack the ability to provide real-time insights into the performance and status of infrastructure elements, leading to inefficiencies and suboptimal decision-making [12].

Secondly, the rapid proliferation of IoT devices in urban and industrial environments generates vast amounts of data, which can be overwhelming for city officials and other stakeholders [15]. The effective analysis and utilization of this data are essential for deriving actionable insights and informing decision-making processes. However, the volume, variety, and velocity of IoT data pose significant challenges in terms of data storage, processing, and analysis [16].

2.2 The Role of AI and IoT in Smart Cities

The integration of AI and IoT technologies offers a promising solution to these challenges by providing the tools and capabilities needed to effectively monitor, manage, and analyze urban infrastructure elements in real-time. IoT devices, such as sensors and actuators, can be installed on various infrastructure elements to collect data on parameters such as container fill levels, battery status, and traffic light status. This data can be used to monitor the performance of infrastructure elements, detect issues and anomalies, and inform decision-making in areas such as maintenance scheduling, resource allocation, and infrastructure planning.

AI techniques, such as machine learning and data analytics, can be applied to the data collected by IoT devices to uncover hidden patterns, trends, and relationships. By analyzing the data, AI algorithms can generate actionable insights and predictions that can help city officials and other stakeholders make more informed decisions and optimize the management of urban infrastructure elements. For example, AI algorithms can be used to predict container fill levels and identify containers that are likely to overflow or require maintenance. Similarly, AI algorithms can be applied to traffic data to optimize traffic light schedules and improve traffic efficiency.

2.3 The Need for User-Friendly AI-Based Platforms

While the potential benefits of integrating AI and IoT technologies in smart cities are evident, there remains a significant gap between the complexity of these technologies and the ability of city officials and other stakeholders to effectively harness their capabilities. The lack of technical expertise and the steep learning curve associated with AI and IoT technologies can present significant barriers to their adoption and use in urban planning and management.

To address this issue, there is a growing need for user-friendly AI-based platforms that can simplify the process of data ingestion, preprocessing, analysis, and visualization, making it accessible to a wide range of stakeholders regardless of their technical expertise. Such platforms should provide an intuitive, user-friendly interface that guides users through the process of importing data, applying machine learning algorithms, creating visualizations, and designing dashboards. Additionally, the platform should support seamless integration with IoT devices and data sources, ensuring compatibility with the diverse array of technologies used in smart cities.

Moreover, AI-based platforms should incorporate features that facilitate collaboration and communication among stakeholders. This includes the ability to securely share visualizations and dashboards with other users, as well as the integration of user and role management capabilities to control access to data and analytics resources. By providing a centralized platform for data analysis and visualization, stakeholders can work together more effectively, promoting a more holistic and data-driven approach to urban planning and management.

2.4 Existing Solutions and Limitations

Several existing solutions aim to address the challenges associated with monitoring and managing urban infrastructure elements in smart cities. These solutions typically leverage IoT and AI technologies to collect, analyze, and visualize data from various infrastructure elements, such as waste containers and traffic lights. However, many of these solutions are limited in their scope and functionality, often requiring a high degree of technical expertise to use and customize.

Furthermore, existing solutions may not offer the level of integration and interoperability needed to support the diverse range of technologies and data sources used in smart cities. This can result in fragmented and siloed data, hindering the ability of city officials and other stakeholders to derive meaningful insights and make informed decisions. Additionally, many existing solutions lack the collaboration and communication features needed to promote a more data-driven and transparent approach to urban planning and management.

2.5 The Value of AI-Driven Assistants and Drag-and-drop Techniques

One of the key innovations in AI-based platforms for smart cities is the integration of AI-driven assistants that guide users through the process of importing data, applying machine learning algorithms, creating visualizations, and designing dashboards. These AI-driven assistants leverage natural language processing, machine learning, and other AI techniques to understand user inputs, recommend appropriate actions, and automate various tasks, making it easier for users to navigate and interact with the platform.

Another important innovation is the use of drag-and-drop techniques to simplify the process of creating visualizations and dashboards. Drag-and-drop interfaces allow users to easily select, arrange, and customize various components, such as charts, maps, and tables, without the need for programming or technical expertise. This enables users to quickly create visually appealing and informative dashboards that can be shared with other stakeholders to facilitate collaboration and communication.

2.6 Integration with IoT Technologies and Data Sources

To support the diverse range of technologies and data sources used in smart cities, AI-based platforms must be capable of seamless integration with IoT devices and other data sources. This includes support for popular IoT communication protocols, such as MQTT, as well as the ability to ingest data from various formats, such as CSV, JSON, and APIs. By providing flexible and extensible data ingestion capabilities, AI-based platforms can help city officials and other stakeholders consolidate and analyze data from multiple sources, enabling a more comprehensive understanding of urban infrastructure elements and their performance.

2.7 The Importance of Scalability, Security, and Privacy

As smart cities continue to grow and evolve, the volume and complexity of data generated by IoT devices will only increase. AI-based platforms must be capable of scaling to accommodate this growth, ensuring that city officials and other stakeholders can continue to derive meaningful insights and make informed decisions. This includes the ability to handle large volumes of data, support real-time data processing, and maintain high levels of performance and reliability.

In addition to scalability, security and privacy are critical considerations for AI-based platforms in smart cities. Sensitive data, such as personal information and infrastructure status, must be protected from unauthorized access and tampering to ensure the safety and privacy of city residents. AI-based platforms should incorporate robust security features, such as secure data storage, encryption, and user authentication, to safeguard sensitive data and maintain the trust of stakeholders.

3 Using the AI-Based Platform for Monitoring Urban Infrastructure Elements

This section describes the process of using the AI-based platform to monitor the status of waste containers and traffic lights in a smart city. The process involves the following steps: data ingestion from IoT devices, data preprocessing and analysis, and the creation of visualizations and dashboards.

A schematic of the deployment of the solution in the designed architecture is presented in Fig. 1.

3.1 Data Ingestion from IoT Devices

The first step in the process involves ingesting data from IoT devices installed on waste containers and traffic lights. These devices transmit data using the MQTT protocol, a lightweight messaging protocol that is widely used in IoT applications due to its low bandwidth requirements and efficient resource utilization. The AI-based platform supports seamless integration with MQTT-enabled devices, allowing users to easily import real-time data from various urban infrastructure elements.

3.2 Data Preprocessing and Analysis

Once the data has been ingested, the platform's AI-driven assistant guides users through the process of preprocessing and analyzing the data. This may involve tasks such as data cleaning, aggregation, and transformation to ensure that the data is in a suitable format for analysis. The AI-driven assistant also helps users select appropriate machine learning algorithms from the platform's comprehensive algorithm library to analyze the data and uncover hidden patterns, trends, and relationships.

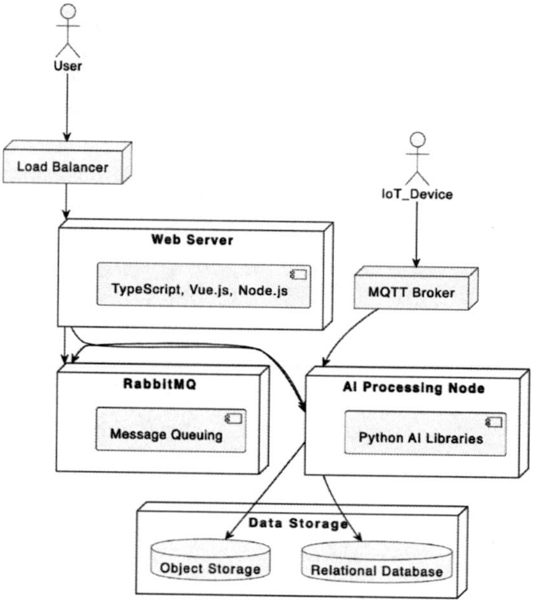

Fig. 1. Architecture schema

In the case of waste containers, the analysis might involve predicting container fill levels and identifying containers that are likely to overflow or require maintenance. For traffic lights, the analysis could focus on identifying patterns in traffic flow, detecting malfunctioning lights, or optimizing traffic light schedules to improve traffic efficiency.

3.3 Creation of Visualizations and Dashboards

After the data has been preprocessed and analyzed, users can create interactive visualizations and dashboards using the platform's visualization engine and dashboard designer. These tools allow users to effectively communicate their findings and explore their data in a visually engaging and immersive manner.

For waste containers, visualizations might include maps showing the location of containers and their current fill levels, bar charts comparing the fill levels of different containers, or line charts displaying trends in container fill levels over time. For traffic lights, visualizations could include maps showing the status of traffic lights across the city, heatmaps displaying traffic congestion levels, or bar charts comparing the efficiency of different traffic light schedules.

The dashboard designer enables users to create visually appealing and informative dashboards using drag-and-drop techniques, with a variety of layout options and widgets to customize the appearance and functionality of the dashboard. These dashboards can be easily shared and accessed by city officials, urban planners, and other stakeholders, promoting collaboration and informed decision-making in the management of urban infrastructure elements.

3.4 Security and Data Privacy Issues

In addition to the benefits of the AI-based platform and its capabilities for data ingestion, preprocessing, analysis, and visualization, it is crucial to consider the security and privacy aspects of the data being used. Given the sensitive nature of the data collected by IoT devices in smart cities, such as personal information and infrastructure status, it is essential to ensure that the platform provides robust mechanisms to protect the data from unauthorized access and tampering.

One of the strengths of MQTT, the IoT communication protocol used in this case study, is its built-in support for security features such as Transport Layer Security (TLS) encryption and user authentication. TLS encryption ensures that data transmitted between IoT devices and the AI-based platform is secure and protected from eavesdropping and man-in-the-middle attacks. User authentication, on the other hand, verifies the identity of IoT devices and users connecting to the platform, ensuring that only authorized entities can access the data.

The AI-based platform also incorporates additional security and privacy measures, such as secure data storage and role-based access control (RBAC). Secure data storage ensures that the data collected from IoT devices is stored in an encrypted format, protecting it from unauthorized access and potential data breaches. RBAC allows the platform administrators to define user roles and permissions, controlling access to data and analytics resources based on the user's role within the organization. This not only helps to maintain the privacy of sensitive data but also facilitates collaboration among different stakeholders without compromising security.

Furthermore, the platform's dashboard sharing capabilities, which enable stakeholders to share visualizations and dashboards through secure URLs or dynamic user and role management, ensure that sensitive information is only accessible to authorized users. This enhances data security and privacy while promoting a more data-driven and transparent approach to urban planning and management.

4 Results

The case study demonstrated the effectiveness of the AI-based platform in addressing the challenges faced by city officials and other stakeholders in monitoring and managing urban infrastructure elements, such as waste containers and traffic lights. The platform's seamless integration with MQTT-enabled IoT devices, AI-driven assistant, and user-friendly visualization tools enabled users to easily import data, preprocess and analyze it, and create interactive visualizations and dashboards for monitoring the status of various urban elements.

The visualizations and dashboards created using the platform provided valuable insights into the performance and status of waste containers and traffic lights, allowing city officials and other stakeholders to make informed decisions and take timely actions to maintain efficient operations and ensure a sustainable and livable environment. Furthermore, the platform facilitated collaboration and communication among stakeholders, promoting a more holistic approach to urban planning and management.

An important feature of the platform is the ability to securely share dashboards with other stakeholders, either through a secure URL or by leveraging the platform's dynamic user and role management capabilities. This enables a wide range of stakeholders, including city officials, urban planners, maintenance teams, and even citizens, to access real-time information on the status of connected urban elements and the predictions made by the system. By providing a centralized and accessible source of information, the platform promotes data-driven decision-making and enhances transparency in urban management processes.

For example, maintenance teams can use the shared dashboards to quickly identify waste containers that are nearing their full capacity or traffic lights that are malfunctioning, allowing them to prioritize their tasks and allocate resources more efficiently. Urban planners can use dashboards to analyze trends in waste generation and traffic patterns, informing the development of data-driven strategies for infrastructure improvement and resource allocation. Citizens can also access the dashboards to stay informed about the status of their local waste containers and traffic lights, fostering a sense of community engagement and ownership in the management of their urban environment.

So the results of the case study highlight the potential of the AI-based platform to transform the way urban infrastructure elements are monitored and managed in smart cities. By providing a user-friendly solution for data ingestion, preprocessing, analysis, and visualization, the platform enables city officials and other stakeholders to effectively harness the power of IoT and AI technologies to improve the efficiency, sustainability, and livability of urban environments. The platform's secure dashboard-sharing capabilities further enhance collaboration and communication among stakeholders, promoting a more data-driven and transparent approach to urban planning and management.

5 Conclusion and Future Work

This case study has demonstrated the potential of using an AI-based platform to monitor and manage urban infrastructure elements in a smart city, with a focus on waste containers and traffic lights. The platform's integration with IoT technologies, AI-driven assistant, and user-friendly visualization tools make it an effective and accessible solution for city officials and other stakeholders, regardless of their technical expertise.

Future work in the development of AI-based solutions for smart cities could involve expanding the scope of the platform to include additional urban infrastructure elements, such as public transportation systems, energy grids, and water supply networks. This would provide a more comprehensive understanding of the city's overall performance and facilitate the development of integrated strategies for urban planning and management.

Additionally, the integration of advanced machine learning and AI techniques, such as deep learning and reinforcement learning, could further enhance the platform's capabilities in analyzing complex data and generating actionable insights. This could lead to the development of more sophisticated predictive models and optimization algorithms, enabling city officials and other stakeholders to proactively address potential issues and improve the overall efficiency and sustainability of urban environments.

Another area of potential improvement is the development of more advanced visualization techniques and tools, such as augmented reality (AR) and virtual reality (VR) applications, that could provide users with even more immersive and interactive experiences when exploring and analyzing their data. These technologies could also facilitate remote collaboration and real-time decision-making among stakeholders, further enhancing the platform's potential in supporting smart city initiatives.

In conclusion, the AI-based platform presented in this case study represents a significant step towards democratizing the use of AI in smart cities and making it accessible to a wide range of stakeholders. By continuing to develop and refine this platform, researchers and developers can help unlock the full potential of AI and IoT technologies in creating more sustainable, efficient, and livable urban environments.

Acknowledgment. This work has been partially supported by the project TED2021-132339B-C43 (idrECO), funded by MCIN/AEI/10.13039/501100011033 and by the European Union "NextGenerationEU"/PRTR.

References

1. Bellini, P., Nesi, P., Pantaleo, G.: IoT-enabled smart cities: a review of concepts, frameworks and key technologies. Appl. Sci. **12**(3), 1607 (2022)
2. Bokhari, S.A.A., Myeong, S.: Use of artificial intelligence in smart cities for smart decision-making: a social innovation perspective. Sustainability **14**(2), 620 (2022)
3. Canizes, B., Pinto, T., Soares, J., Vale, Z., Chamoso, P., Santos, D.: Smart city: a GECAD-BISITE energy management case study. In: De la Prieta, F., et al. (eds.) PAAMS 2017. AISC, vol. 619, pp. 92–100. Springer, Cham (2018). https://doi.org/10.1007/978-3-319-61578-3_9
4. Casado-Vara, R., Martin-del Rey, A., Affes, S., Prieto, J., Corchado, J.M.: IoT network slicing on virtual layers of homogeneous data for improved algorithm operation in smart buildings. Futur. Gener. Comput. Syst. **102**, 965–977 (2020)
5. Casado-Vara, R., Prieto-Castrillo, F., Corchado, J.M.: A game theory approach for cooperative control to improve data quality and false data detection in WSN. Int. J. Robust Nonlinear Control **28**(16), 5087–5102 (2018)
6. Chamoso, P., González-Briones, A., Rodríguez, S., Corchado, J.M.: Tendencies of technologies and platforms in smart cities: a state-of-the-art review. Wirel. Commun. Mob. Comput. **2018** (2018)
7. Chamoso, P., González-Briones, A., Rivas, A., De La Prieta, F., Corchado, J.M.: Social computing in currency exchange. Knowl. Inf. Syst. **61**(2), 733–753 (2019). https://doi.org/10.1007/s10115-018-1289-4
8. Corchado, J.M., et al.: Deepint. net: a rapid deployment platform for smart territories. Sensors **21**(1), 236 (2021)

9. Corchado, J.M., Pinto-Santos, F., Aghmou, O., Trabelsi, S.: Intelligent development of smart cities: deepint.net case studies. In: Corchado, J.M., Trabelsi, S. (eds.) SSCTIC 2021. LNNS, vol. 253, pp. 211–225. Springer, Cham (2022). https://doi.org/10.1007/978-3-030-78901-5_19
10. Dash, B., Sharma, P.: Role of artificial intelligence in smart cities for information gathering and dissemination (a review). Acad. J. Res. Sci. Publ. **4**(39) (2022)
11. García-García, L., Jiménez, J.M., Abdullah, M.T.A., Lloret, J.: Wireless technologies for IoT in smart cities. Netw. Protoc. Algorithms **10**(1), 23–64 (2018)
12. Garcia-Retuerta, D., Chamoso, P., Hernández, G., Guzmán, A.S.R., Yigitcanlar, T., Corchado, J.M.: An efficient management platform for developing smart cities: solution for real-time and future crowd detection. Electronics **10**(7), 765 (2021)
13. González-Briones, A., Chamoso, P., de Alba, F.L., Corchado, J.M.: Smart cities energy trading platform based on a multi-agent approach. Artif. Intell. Environ. Sustain. Challenges Solut. Era Ind. **4**, 131–146 (2022)
14. Martí, P., Jordán, J., Chamoso, P., Julian, V.: Taxi services and the carsharing alternative: a case study of Valencia city (2022)
15. Rivas, A., Fraile, J.M., Chamoso, P., González-Briones, A., Sittón, I., Corchado, J.M.: A predictive maintenance model using recurrent neural networks. In: Martínez Álvarez, F., Troncoso Lora, A., Sáez Muñoz, J.A., Quintián, H., Corchado, E. (eds.) SOCO 2019. AISC, vol. 950, pp. 261–270. Springer, Cham (2020). https://doi.org/10.1007/978-3-030-20055-8_25
16. Shoeibi, N., Shoeibi, N., Julian, V., Ossowski, S., Arrieta, A.G., Chamoso, P.: Smart cyber victimization discovery on twitter. In: Corchado, J.M., Trabelsi, S. (eds.) SSCTIC 2021. LNNS, vol. 253, pp. 289–299. Springer, Cham (2022). https://doi.org/10.1007/978-3-030-78901-5_25

Bitcoin Price Prediction Using Machine Learning and Technical Indicators

Abdelatif Hafid[1(✉)], Abdelhakim Senhaji Hafid[2], and Dimitrios Makrakis[1]

[1] University of Ottawa, Ottawa, ON K1N 6N5, Canada
{ahafid2,dmakraki}@uottawa.ca
[2] University of Montreal, Montreal, QC H3T 1J4, Canada
ahafid@iro.umontreal.ca

Abstract. With the rise of Blockchain technology, the cryptocurrency market has been gaining significant interest. In particular, the number of cryptocurrency traders and market capitalization has grown tremendously. However, predicting cryptocurrency prices is very challenging and difficult due to the high price volatility. In this paper, we propose a classification machine learning approach in order to predict the direction of the market (i.e., if the market is going up or down). We identify key features such as Relative Strength Index (RSI) and Moving Average Convergence Divergence (MACD) to feed the machine learning model. We illustrate our approach through the analysis of Bitcoin's close price. We evaluate the proposed approach via different simulations. Particularly, we provide a backtesting strategy. The evaluation results show that the proposed machine learning approach provides buy and sell signals with more than 86% accuracy.

Keywords: Bitcoin price movement · classification models · market predictions · random forest · technical indicators

1 Introduction

The cryptocurrency market is transforming the world of money and finance [7], and has seen significant growth in the last years [7,15]. In particular, the number of cryptocurrencies reached more than 7000 in 2021 [6], and the crypto market capitalization hit $3 trillion the same year [6].

The banking and financial industry has taken notice of Blockchain benefits. The underlying technology behind every cryptocurrency is Blockchain technology. Blockchain is a distributed/decentralized database that is organized as a list of blocks, where the committed blocks are immutable. It has many attractive properties including transparency and security [15].

The crypto market has many good characteristics including high market data availability and no closed trading periods. However, it suffers from its high price

Supported by organization x.

S. Ossowski et al. (Eds.): DCAI 2023, LNNS 740, pp. 275–284, 2023.
https://doi.org/10.1007/978-3-031-38333-5_28

volatility and relatively smaller capitalization. In crypto financial trading, data can be available to all traders. However, the analysis and the selection of this data makes the difference between executing good trades and bad trades. Therefore, one of the main challenges in financial trading is to develop methods/approaches to extract meaningful knowledge and insights from the data. Furthermore, due to the high price volatility of cryptocurrencies price, forecasting the price becomes more challenging.

Up to now, there are few studies that have attempted to create profitable trading strategies in the cryptocurrency market. In 2018, Saad et al. [12] provided a machine learning model to predict Bitcoin price. In particular, they made use of a regression model and involved many factors that impact the price of Bitcoin. However, they did not provide *Buy* and *Sell* signals, which are the most important in building a trading strategy/approach. Furthermore, they did not consider any kind of technical indicators. The use of technical indicators as features to feed machine learning models for financial trading has been successfully employed by many researchers [11,13]. McNally et al. [10] proposed a machine learning model that makes a recurrent neural network, called Long Short Term Memory Model (*LSTM*). *LSTM* achieves an accuracy of 52%, for classification. However, this is not acceptable for building a trading strategy. Our approach achieves 86%, which is quite acceptable.

Recently, Jay et al. [8] proposed a stochastic neural network model for cryptocurrency price prediction. Precisely, they made use of random walk theory, and Multi-Layer Perceptron (*MLP*) and *LSTM* machine learning models [8]. The approach achieves good mean absolute percentage error (*MAPE*). However, they did not consider *Buy* and *Sell* signals. They also did not consider any kind of technical indicators to feed their machine learning models.

In a more recent work, Singh et al. [14] proposed three machine learning models to predict the price of cryptocurrency. They reported that Gated Recurrent Unit (*GRU*) provides a good accuracy compared to others with a *MAPE* of 0.2454% for Bitcoin. However, the authors did not provide *Buy* and *Sell* signals, and did not consider technical indicators to feed this machine learning model.

In this paper, we contribute to the development of profitable trading strategies by proposing a new approach that integrates various features including technical indicators and historical data. The key contribution of the proposed approach is providing *buy* and *sell* signals with high accuracy. We evaluate the proposed approach through the analysis of Bitcoin cryptocurrency.

The remaining of the paper is organized as follows. Section 2 presents the mathematical modelling of the proposed approach. Section 3 compares the machine learning models and evaluates the proposed approach through a back-testing strategy and confusion matrices. Finally, Sect. 4 concludes the paper.

2 Mathematical Modeling

In this section, we provide a mathematical modelling of the proposed approach.

2.1 Notations and Definitions

Table 1 below provides the definitions of parameters and abbreviation.

Table 1. Notations & Abbreviations

Notation	Description
m	Total number of observations/samples
$n_{training}$	Number of samples in the training set
n_{test}	Number of samples in the test set
n_x	Total number of input features
$\mathcal{C}_p^{(i)}$	Close price at time t_i
$\mathcal{O}_p^{(i)}$	Opening price at time t_i
$\mathcal{H}_p^{(i)}$	High price at time t_i
$\mathcal{L}_p^{(i)}$	Low price at time t_i
$\mathcal{V}^{(i)}$	Volume of the cryptocurrency that is being in trade at time t_i
s	Span ($s \geq 1$)
$RSI_\alpha^{(i)}$	Relative strength index at time t_i within a time period α
$MACD^{(i)}$	Moving average convergence divergence at time t_i
$EMA_\alpha^{(i)}$	Exponential moving average at time t_i within a period of time α
$PROC_\alpha^{(i)}$	Price rate of change at time t_i within a period of time α
$\%K_\alpha^{(i)}$	Stochastic oscillator at time t_i within a period of time α
$MOM_\alpha^{(i)}$	Momentum at time t_i within a period of time α
$\mathbb{R}$	Set of real numbers
$\mathcal{T}$	Set of targets, $\mathcal{T} = \{1, -1\}$

Definition 1 (Backtesting). *Backtesting is a technique that allows the evaluation and the assessment of a trading strategy through data-driven simulations using historical data.*

2.2 Mathematical Model

In this section, we present a mathematical modeling of our approach.

Let $(x^{(i)}, y^{(i)})$ denotes a single sample/observation. The set of samples is represented by:

$$\mathcal{S} = \{(x^{(1)}, y^{(1)}), (x^{(2)}, y^{(2)}), \ldots, (x^{(m)}, y^{(m)})\} \tag{1}$$

where $x^{(i)} \in \mathbb{R}^{n_x}$ and $y^{(i)} \in \mathcal{T}$.

Since we consider both technical indicators and Blockchain historical data to predict the price, we need to combine/merge different data sets. Specifically, technical indicators and historical data are input to our model. Our feature vector in a given time t can be expressed as follows:

$$x^{(i)} = \begin{pmatrix} \mathcal{C}_p^{(i)} \\ \mathcal{V}^{(i)} \\ RSI_{14}^{(i)} \\ RSI_{30}^{(i)} \\ RSI_{200}^{(i)} \\ MOM_{10}^{(i)} \\ MOM_{30}^{(i)} \\ MACD^{(i)} \\ PROC_{9}^{(i)} \\ EMA_{10}^{(i)} \\ EMA_{30}^{(i)} \\ EMA_{200}^{(i)} \\ \%K_{10}^{(i)} \\ \%K_{30}^{(i)} \\ \%K_{200}^{(i)} \end{pmatrix}, \quad x^{(i)} \in \mathbb{R}^{n_x} \tag{2}$$

Let us generalize our model by stacking all features vectors in one matrix X. This matrix can be expressed as follows:

$$X = \begin{matrix} & x^{(1)} \cdots\cdots\cdots x^{(m)} \\ \begin{matrix} \mathcal{C}_p \\ \mathcal{V} \\ RSI_{14} \\ \vdots \\ \%K_{30} \\ \%K_{200} \end{matrix} & \begin{pmatrix} \mathcal{C}_p^{(1)} & \mathcal{C}_p^{(2)} \cdots & \mathcal{C}_p^{(m)} \\ \mathcal{V}^{(1)} & \mathcal{V}^{(2)} \cdots & \mathcal{V}^{(m)} \\ RSI_{14}^{(1)} & RSI_{14}^{(2)} \cdot & RSI_{14}^{(m)} \\ \vdots & \vdots \ddots & \vdots \\ \%K_{30}^{(1)} & \%K_{30}^{(2)} \cdot & \%K_{30}^{(m)} \\ \%K_{200}^{(1)} & \%K_{200}^{(2)} \cdot & \%K_{200}^{(m)} \end{pmatrix} \end{matrix} \tag{3}$$

where

$$X \in \mathbb{R}^{n_x \times m}$$
$$\mathcal{C}_p = \left(\mathcal{C}_p^{(1)}, \mathcal{C}_p^{(2)}, \ldots, \mathcal{C}_p^{(m)}\right)$$
$$\mathcal{V} = \left(\mathcal{V}^{(1)}, \mathcal{V}^{(2)}, \ldots, \mathcal{V}^{(m)}\right)$$
$$RSI_{14} = \left(RSI_{14}^{(1)}, RSI_{14}^{(2)}, \ldots, RSI_{14}^{(m)}\right)$$
$$\cdots\cdots\cdots$$
$$\%K_{200} = \left(\%K_{200}^{(1)}, \%K_{200}^{(2)}, \ldots, \%K_{200}^{(m)}\right)$$

The output matrix can be expressed as follows:

$$Y = \left(y^{(1)}, y^{(2)}, \ldots, y^{(m)}\right) \tag{4}$$

where $y^{(i)} \in \mathcal{T}$ and $i \in \{1, 2, \ldots, m\}$.

2.3 Features

In this section, we present different features. In particular, we make use of historical market data and technical analysis indicators. All these features are evaluated in Sect. 3 by using the random forest model; most of these features show good importance to contribute to the accuracy of our approach (see Fig. 1d).

Historical Data. Regarding the historical data, we consider close price and volume.

Close Price. Close price refers to the price at which a cryptocurrency closes at a given time period.

Volume. Volume is the number of units (e.g., number of Bitcoins) traded in the market during a given time period.

Technical Analysis Indicators. Technical analysis is a trading discipline employed to evaluate investments and identify trading opportunities by analyzing statistical trends gathered from trading activity, such as price movement and volume [1]. In this work, we consider the exponential moving average, moving average convergence divergence, relative strength index, momentum, price rate of change, and stochastic oscillator.

Exponential Moving Average. The exponential moving average (EMA) was first introduced by Roberts (1959) [9]. It is a type of moving average (MA) that places a greater weight and significance on the most recent data points. EMA is also referred to as the exponentially weighted moving average.

$C_p^{(1)}$ represents the closing price at time t_1, $C_p^{(2)}$ represents the close price at time t_2, and gradually $C_p^{(m)}$ represents the close price at time t_m. Knowing that $t_1 < t_2 < \dots t_{m-1} < t_m$ $\gamma = t_k - t_{k-1}$ measures the step time (e.g., 1 min, 15 min, 1 day). EMA can be expressed recursively as follows:

$$\begin{aligned} EMA_1 &= C_p^{(1)} \\ EMA_t &= (1-\alpha)EMA_{t-1} + \alpha C_p^{(t)}, \end{aligned} \tag{5}$$

where α is smoothing factor ($\alpha \in [0,1]$) and is expressed as $\alpha = 2/(s+1)$.

Moving Average Convergence Divergence. Moving Average Convergence Divergence ($MACD$) is a technical indicator created by Gerald Appel in 1970 [2]. $MACD$ helps investors understand the movement of the price (i.e., the market will be in bullish or bearish movement) [2]. Usually, $MACD$ is calculated by subtracting the 26-period EMA from the 12-period EMA. Formally, it is expressed as follows:

$$MACD = EMA_n - EMA_{n+j} \tag{6}$$

where EMA_i is the i-period EMA and j equals 14 ($j = 26 - 12$).

The Relative Strength Index. The relative strength index (RSI) is a technical indicator used to chart the current and historical strength or weakness of a stock/market based on the closing price of a recent trading period. It was originally developed by J. Welles Wilder [16]. RSI is classified as a momentum oscillator, which is an indicator that varies over time within a band. Technically, RSI is typically used on a 14-days period and is measured on a scale from 0 to 100. RSI takes the values 70 and 30 with high and low levels of the market [16]. RSI within a band α (α usually equals 14) can be mathematically expressed as follows:

$$RSI_\alpha = 100 - \frac{100}{1 + RS} \tag{7}$$

$$RS = \frac{A^g_\alpha}{A^l_\alpha}$$

where A^g_α and A^l_α represent average gain over α-days and average loss over α-days, respectively.

Momentum. Momentum (MOM) measures the velocity of a stock price over a period of time, which means the speed at which the price is moving; typically we use the close price [5]. MOM helps investors identify the strength of a trend [5]. Formally, the momentum can be expressed as follows:

$$MOM_\zeta = \mathcal{C}_p^{(i-(\zeta-1))} - \mathcal{C}_p^{(i)} \tag{8}$$

where ζ is the number of days.

Price Rate of Change. The Price Rate Of Change (PROC) measures the most recent change in price. It can be expressed as follows:

$$PROC_t = \frac{C_p^t - C_p^{t-n}}{C_p^{t-n}} \tag{9}$$

where $PROC_t$ is the price rate of change at time t and n is the number of periods to look back.

Stochastic Osillator. A stochastic oscillator is a popular technical indicator for generating overbought and oversold signals. Usually the current value of the stochastic indicator is denoted by $\%K$ and it is computed as follows:

$$\%K = \frac{\mathcal{C} - \mathcal{L}_d}{\mathcal{H}_d - \mathcal{L}_d} * 100 \tag{10}$$

where $\mathcal{C}$ represents the most recent closing price, $\mathcal{L}_d$ represents the lowest price traded during the d previous periods, and $\mathcal{H}_d$ represents the highest price traded during the d previous periods.

Signal. Let $\mathcal{Y}$ be a random variable that takes the values of 1 or -1 (*Buy* and *Sell*, respectively).

To generate *Buy* and *Sell* signals, we employ a technical indicator called Moving Average (MA). MA identifies the trend of the market. The MA rule that generates *Buy* and *Sell* signals at time t consists of comparing two moving averages. Formally, the rule is expressed as follows:

$$\mathcal{Y}(t) = \begin{cases} 1 & \text{if } \mathrm{MA}_{s,t} \geq \mathrm{MA}_{l,t}, \\ -1 & \text{if } \mathrm{MA}_{s,t} < \mathrm{MA}_{l,t} \end{cases} \tag{11}$$

where

$$\mathrm{MA}_{j,t} = (1/j) \sum_{i=0}^{j-1} \mathcal{C}_p^{(t-i)}, \quad \text{for } j = s \text{ or } l; \tag{12}$$

s (l) is the length of the short (long) MA ($s < l$). We denote the MA indicator with MA lengths s and l by MA(s, l). In this paper, we consider the MA(10, 60) because of it high accuracy.

3 Results and Evaluation

In this section, we compare the most common and popular machine learning models for classification: logistic regression, support vector machine, random forest, and voting classifier. We also present a simulation-based evaluation of our approach.

3.1 Simulation Setup

We make use of *sklearn* Python package to simulate the proposed approach. In particular, we take advantage of *sklearn.preprocessing* module to scale the data, and *sklearn.metrics* module to calculate the accuracy, classification report, and confusion matrix. We make use of *sklearn.ensemble* module to import random forest classifier and voting classifier, *sklearn.linear_model* to import logistic regression classifier, and *sklearn.svm* to import support vector machine classifier.

For the data, we stream real-time historical market data directly from Binance via Binance API [4]. The data is from 01 February, 2021 to 01 February, 2022 and with a time step of 15 min (i.e., $\gamma = 15$ min). We choose 15 min because it gives us good accuracy. We split the data into 80% for training set and 20% for testing set.

3.2 Results and Analysis

Table 2 provides a K-fold cross-validation comparison among the different proposed machine learning models. This comparison is based on the score (accuracy) presented in Table 2. This score is calculated as the average of the accuracy of 10 folds. Table 2 shows that the four models provide approximately the same score. We choose RF (an ensemble model) to forecast crypto market since it has

Table 2. K-fold Comparison

Model	LR	SVM	RF	VC
Score	0.863	0.854	0.867	0.864

the ability to deal with very larger sizes of data, a large number of features, and an expected non-linear relationship between the predicted variable and the features [3].

Table 3. Classification Report

	Model	Accuracy	Precision		Recall	
			1	−1	1	−1
Bitcoin	RF	0.884	0.885	0.884	0.891	0.876

Table 3 shows the classification report of the proposed approach. It shows the accuracy, precision, and the recall.

Figure 1a shows the distribution of the predicted variable for Bitcoin. Particularly, Fig. 1a shows that the predicted variable's class 1 is slightly more bigger than 50% of the time, meaning there are more *buy* signals than *sell* signals. The predicted variable is relatively balanced.

Figures 1b, 1c, and 1d show an evaluation of the proposed approach (which makes use of the RF classifier), starting by a simple backtesting strategy, then the calculation of the confusion matrix, and eventually the evaluation of the feature's importance. The proposed backtesting strategy consists of calculating the predicted returns (aka, strategy returns) and comparing it with the actual returns.

Figure 1b shows that the predicted returns are very close to the actual returns. This means that the proposed approach performs well for predicting Bitcoin.

Figure 1c shows the confusion matrices corresponding to Bitcoin. In particular, Fig. 1c shows that (in the first column of the matrix) the model predicts that we should execute 2953 + 388 *Buy* operations. However, the truth is that we should only execute 2953 Buy operations and the rest (388) should be executed as a *Sell* operations. Moreover, in the second column of the matrix, the model predicts that we should execute 416 + 3198 *Sell* operations. However, the reality is that we should only execute 3198 *Sell* operations and the rest (416) should be executed as *Buy* operations.

Figure 1d shows that $MACD$, RSI_{30}, and MOM_{30} are the features that contribute highly in improving the performance of the proposed approach. $\mathcal{C}_p$ (Close price), EMA_{10}, EMA_{30}, EMA_{200}, and the $\mathcal{V}$ (Volume), are the features that contribute less in improving the performance of our approach.

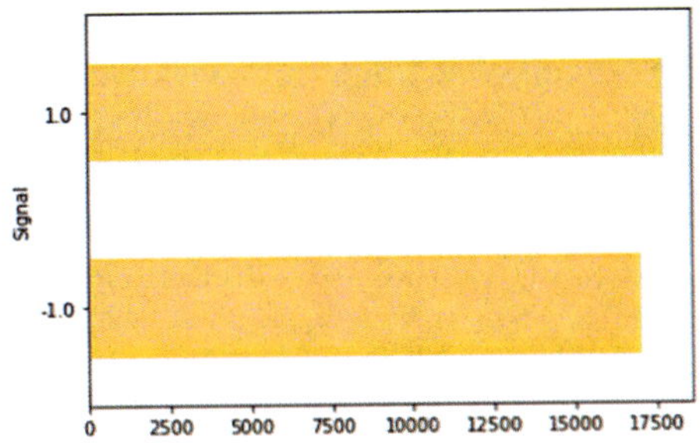

(a) The predicted variable of Bitcoin

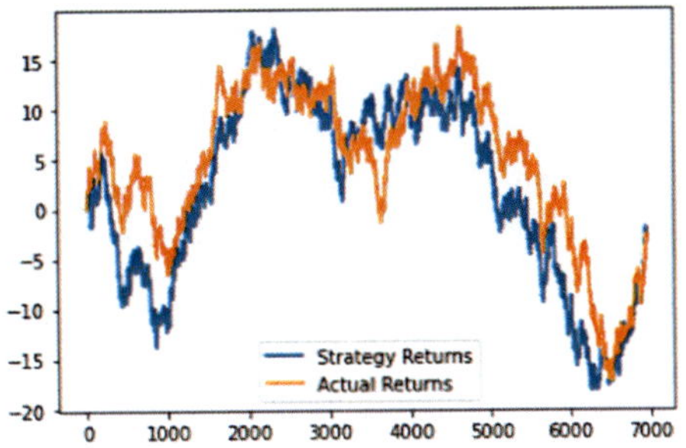

(b) Bitcoin backtesting

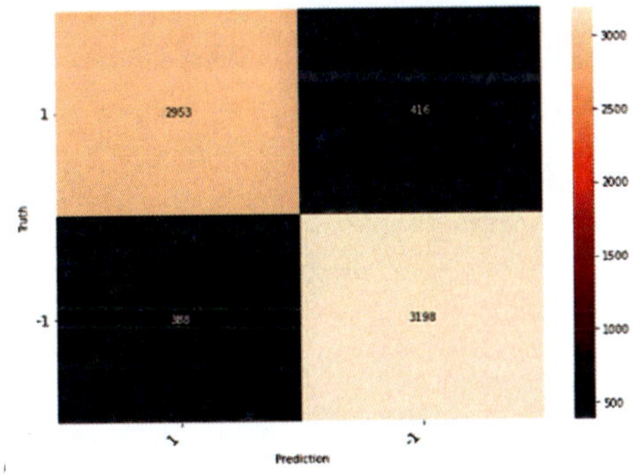

(c) Bitcoin confusion matrix

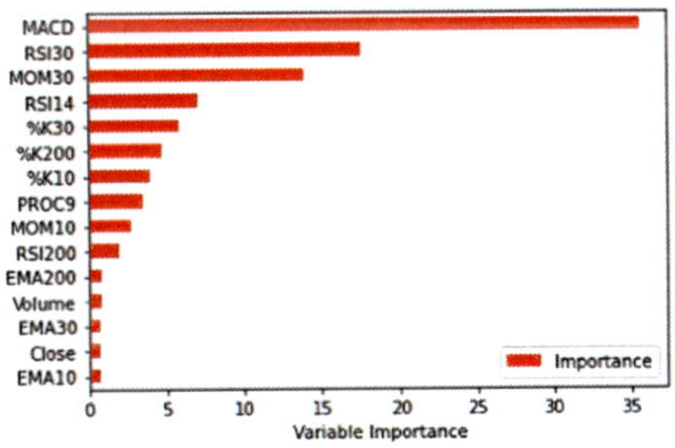

(d) Bitcoin feature importance

Fig. 1. The predicted variable, backtesting, confusion matrix, and feature importance of Bitcoin close price.

To conclude, we collect crypto market data and employ the most significant features that influence the price including volume and technical indicators. Technical indicator features appear to be most influential (Fig. 1d) compared to close price and volume. Our approach achieves a good accuracy, precision, and recall. However, this approach still needs improvement (for making actual/real returns as close as possible to the returns of our approach/strategy returns).

It is worth noting that predicting cryptocurrency price is very challenging since unconventional factors such as social media, investor psychology have great influence on the crypto market. Future research directions could focus on building upon a forecasting approach that involves more features (e.g., cash flow, mining rate, number of transactions) as well as employing more machine learning models.

4 Conclusion

In this paper, we analyze cryptocurrency market price by using the most common technical indicators and compare four well-known classification machine learning models. We get historical market data from Binance to compute the technical indicators. We also add the volume and the close price as features. We predict the direction of market by providing *buy* and *sell* signals. Compared to the existing models that predict the future price based on the past price or even models that use other features, our approach is highly accurate. In the future, we aim to identify more key features by adding more technical indicators and compare more classification models, including *XGBoost*, for the purpose of enhancing the speed and accuracy of the proposed approach.

References

1. Achelis, S.B.: Technical analysis from A to Z (2001)
2. Appel, G.: Technical Analysis: Power Tools for Active Investors. FT Press (2005)
3. Biau, G., Scornet, E.: A random forest guided tour. TEST **25**(2), 197–227 (2016). https://doi.org/10.1007/s11749-016-0481-7
4. Binance: Binance API. https://www.binance.com/en/binance-api. Accessed 16 July 2022
5. Chan, L.K., Jegadeesh, N., Lakonishok, J.: Momentum strategies. J. Financ. **51**(5), 1681–1713 (1996)
6. CoinGecko: Cryptocurrency prices, charts, and crypto market cap. https://www.coingecko.com/. Accessed 02 July 2022
7. Hileman, G., Rauchs, M.: Global cryptocurrency benchmarking study. Cambridge Centre Alternative Financ. **33**(1), 33–113 (2017)
8. Jay, P., Kalariya, V., Parmar, P., Tanwar, S., Kumar, N., Alazab, M.: Stochastic neural networks for cryptocurrency price prediction. IEEE Access **8**, 82804–82818 (2020)
9. Lucas, J.M., Saccucci, M.S.: Exponentially weighted moving average control schemes: properties and enhancements. Technometrics **32**(1), 1–12 (1990)
10. McNally, S., Roche, J., Caton, S.: Predicting the price of bitcoin using machine learning. In: 2018 26th Euromicro International Conference on Parallel, Distributed and Network-Based Processing (PDP), pp. 339–343. IEEE (2018)
11. Oncharoen, P., Vateekul, P.: Deep learning for stock market prediction using event embedding and technical indicators. In: 2018 5th International Conference on Advanced Informatics: Concept Theory and Applications (ICAICTA), pp. 19–24. IEEE (2018)
12. Saad, M., Choi, J., Nyang, D., Kim, J., Mohaisen, A.: Toward characterizing blockchain-based cryptocurrencies for highly accurate predictions. IEEE Syst. J. **14**(1), 321–332 (2019)
13. Shynkevich, Y., McGinnity, T.M., Coleman, S.A., Belatreche, A., Li, Y.: Forecasting price movements using technical indicators: investigating the impact of varying input window length. Neurocomputing **264**, 71–88 (2017)
14. Singh, H., Agarwal, P.: Empirical analysis of bitcoin market volatility using supervised learning approach. In: 2018 Eleventh International Conference on Contemporary Computing (IC3), pp. 1–5. IEEE (2018)
15. Treleaven, P., Brown, R.G., Yang, D.: Blockchain technology in finance. Computer **50**(9), 14–17 (2017)
16. Wilder, J.W.: New concepts in technical trading systems. Trend Research (1978)

RKTUP Framework: Enhancing Recommender Systems with Compositional Relations in Knowledge Graphs

Lama Khalil(✉) and Ziad Kobti

School of Computer Science, University of Windsor, Windsor, ON N9B 3P4, Canada
{khali121,kobti}@uwindsor.ca
https://www.uwindsor.ca/science/computerscience/

Abstract. Advances in joint recommendation and knowledge graph completion (KGC) learning have enhanced the performance and explainability of recommendations. Recent studies have established that taking the incomplete nature of knowledge graphs (KG) into consideration can lead to enhancements in recommender systems' (RS) performance. The existing models depend on translation-based knowledge graph embedding (KGE) methods for KGC. They cannot capture various relation patterns, including composition relations, even though the composition relationships are prevalent in real-world KG. This study proposes a simple and effective approach to enhance the KGC task while training it with the RS. Our approach, rotational knowledge-enhanced translation-based user preference (RKTUP), is an advanced variant of the knowledge-enhanced translation-based user preference model (KTUP), an existing MTL model. To enhance KTUP, we use the rotational-based KGE techniques (RotatE or HRotatE) to model and infer various relation patterns, such as symmetry/asymmetry, composition, and inversion. Unlike earlier MTL models, RKTUP can model and infer diverse relation patterns while learning more robust representations of the entities and relations in the KGC task, leading to improved recommendations for users. Using RotatE improved the recommender system's performance while using HRotatE enhanced the model's efficiency. The experimental results reveal that RKTUP outperforms existing methods and achieves state-of-the-art performance on the recommendation and KGC tasks. Specifically, it shows a 13.7% and 11.6% improvement in the F1 score, as well as a 12.8% and 13.6% increase in the hit ratio on DBbook2014 and MovieLens-1m, respectively.

Keywords: Recommender System · Knowledge Graph Completion · Multi-Task Learning Model

1 Introduction

Knowledge graphs (KGs) are powerful tools for storing and managing complex, nonlinear information in a structured and interpretable manner [8]. KGs have

S. Ossowski et al. (Eds.): DCAI 2023, LNNS 740, pp. 285–295, 2023.
https://doi.org/10.1007/978-3-031-38333-5_29

been integrated into recommender systems (RS) to enhance performance and explainability, particularly in addressing data sparsity and cold start problems [3,19]. Most KG-based RSs assume that KGs are complete. In reality, they are inherently incomplete [8]. To address this issue, multi-task learning (MTL) models have been proposed to jointly learn KG completion (KGC) and recommendation tasks [2,7]. However, previous models fail to capture various relation patterns, such as composition relations, which are common in real-world KGs [5]. As evident in our research, the impact of such missing relations on the RS's performance has not been fully explored.

Capturing composition relations is critical for personalized recommendations, as illustrated in Fig. 1. The red dotted arrow between *Dan Aykroyd* and *Canada* represents the missing relation *isFrom*. Assuming that the reason for the user's choices is his preference for movies starring Canadian actors, we can infer that the relation *isFrom* represents the user's preference. Although the user's preference was determined, the RS did not recommend *Ghostbusters II*, one of the user's preferred items, due to a missing relation in the KG (red dotted arrow) that led to overlooking the fact that *Dan Aykroyd* is Canadian.

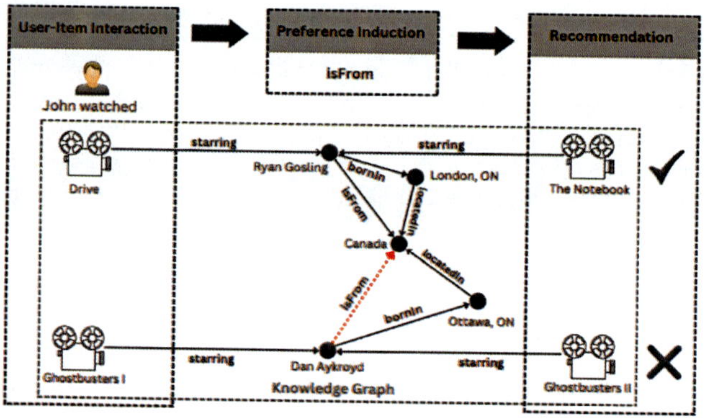

Fig. 1. An illustration of the significance of predicting missing composition relations for better recommendations.

Composition relations, which are defined as a composition of two or more relations, are prevalent in real-world KGs [5,17]. Utilizing KGC methods capable of capturing a wider range of relation patterns may provide a more accurate representation of user preferences, which could ultimately lead to improved RS performance. Rotational KGE methods, such as HRotatE [14] and RotatE [17], can infer and capture a wide range of relationship patterns including composition.

In this study, our hypothesis is that integrating RotatE into a new MTL model called RKTUP can improve RS and KGC performance by incorporating diverse relationship patterns captured by RotatE during KGE training into the RS model. Additionally, we hypothesize that using HRotatE could enhance the RKTUP's efficiency.

To summarize, our contributions are as follows:

1. In Sect. 4, we introduce RKTUP, a new MTL model that incorporates diverse relations, including composition relations, from the KG into the RS during knowledge enhancement.
2. In Sect. 5, we evaluate RKTUP and show that incorporating composition relations from the KG into the RS improves its performance.

2 Problem Formulation

Given a knowledge graph $G = (E, R, T)$, where E is the set of entities, R is the set of relations and T is the set of triples in G.

Additionally, we have a list of user-item interactions denoted by $Y = \{(u, i)\}$. Here $u \in U$ denotes the set of all users and $i \in I$ denotes the set of all items.We know that $I \cap E \neq \emptyset$, which means that there exists at least one item that is common to both sets I and E. The value of $Y_{ui} \in \{0, 1\}$, depending on whether the user u engaged with item i. Specifically, if $Y_{ui} = 1$, it means that the user u interacted with item i, whereas if $Y_{ui} = 0$, it means that the user did not interact with the item [19].

Our goal is to learn more expressive vector representations of entities E and relations R in a KG that is part of a Multi-task Learning (MTL) model. The MTL model jointly learns to recommend the top-N items for a user and complete G through finding the missing set of triplets $T' = \{(e_s, r, e_t) \mid e_s \in E,\ r \in R,\ e_t \in E,\ (e_s, r, e_t) \notin T\}$ [15], where e_s, r and e_t denote the source entity, the relation and the head entity, respectively.

3 Related Work

Classical RS typically use a rating and similarity structure. Collaborative recommendations rely on the assumption that users with similar preferences tend to like similar items. Collaborative filtering (CF) [13], factorization machines (FM) [11] and Bayesian personalized ranking matrix factorization (BPRMF) [12] are popular approaches in this category, but their performance can be hindered by data sparsity due to their reliance on feature extraction for numerical similarity computation.

The content-based approach gathers user information and alleviates data sparsity. The co-factorization model (CoFM) [10] leads this approach and uses two methods: Share and Regularize. Share assumes that linked items and entities' latent factors are the same, while Regularize reduces the distance between latent item and entity representations. However, enforcing similar representations may lead to information loss.

With classical methods many issues arise. Some of the challenges are the cold start problem, sparsity in ratings, scalability of the approach, accuracy of the prediction and explainability of the recommendations [3]. In an attempt to solve these issues, KGs have been used as a source of side information [3]. These attempts are classified into three groups: connection-based models, propagation-based models and graph embedding-based models [3]. In this study, we focus on graph embedding-based models. The purpose of these models is to enhance the users' and items' vector representation. There are three approaches in this category:

Two-Stage Learning Models (TSL): In this approach, the embedding model learns the vector representations of the users and items first. Then, the pretrained KG embeddings are fed to the RS to predict the best recommendations [3]. Some of the leading models in this category are the deep knowledge-aware network (DKN) [18] and the knowledge-enhanced sequential recommender (KSR) [4].

Joint Learning Models (JL): The joint training of embedding model and RS is a technique in which the KG stores users' side information like social media profiles. Collaborative knowledge base embedding (CKE) [21] and collaborative filtering with KG model (CFKG) [22] are common methods in this category. Both use TransE [1] for graph embedding and cannot handle N-to-N relations commonly found in practice.

Multi-Task Learning Models (MTL): Recent research combines recommendation and knowledge graph completion tasks using multi-task learning (MTL) [3]. KTUP [2] is an example of such models that transfer low-level features between items and entities to enhance recommendation performance. Unlike other MTL models, KTUP represents user preferences with KG relations, making its recommendations more explainable. Furthermore, KTUP utilizes TransH [20] for entity and relation embedding. However, KTUP's limitation is its use of TransH, which has limited performance in KGC, as it cannot infer all relation patterns, such as composition.

4 Methodology

This section presents our proposed approach, RKTUP, which integrates TUP for the recommendation task and RotatE (or HRotatE) for KGC task. Initially, we introduce TUP, followed by an explanation of the KGE methods. Finally, we provide a detailed explanation of the RKTUP approach.

- Translation-Based User Preference (TUP) for Item Recommendation: In this work, we use translation-based user Preference (TUP) for item recommendation, which is an RS proposed in [2]. TUP aims to model user preferences as translational relationships between users and items by projecting the user and item vector representations to the preference hyperplane and employing a score function similar to that used in TransH [20].

TUP takes user, item and preference as inputs and outputs a relevance score for a user-item pair. It learns embeddings of preference, user and item such that $u + p \approx i$. The set of preferences is a hyperparameter and each user-item pair has an induced preference that represents the relationship between the user and item.

TUP has two components: preference prediction and hyperplane-based translation. The preference prediction task predicts the user's preference from a set of latent features P given a user-item pair (u, i). Two strategies handle the user's implicit and varied preferences: a hard approach selecting one preference from P and a soft approach combining all preferences with attention. The hard strategy uses the Straight-Through Gumbel SoftMax technique to sample a user's preference, while the soft strategy combines all preferences using an attention mechanism.

TUP addresses the N-to-N problem by using a hyperplane-based translation strategy. Each preference is assigned its own hyperplane in TUP, allowing for preference-specific information to be preserved for each user-item pair. TUP's scoring function is defined as follows:

$$g(u, i; p) = ||(u^{\perp} + p) - i^{\perp}|| \tag{1}$$

where $u^{\perp}$ and $i^{\perp}$ are the projections of the user and item embeddings onto the preference hyperplane defined by the preference p.

To train the model, the loss function aims to minimize the difference between the predicted score of a triplet (u, p, i) and the actual truth value of that triplet. The loss is calculated as follows:

$$L_{RS} = \sum_{(u,i)\in Y} \sum_{(u,i')\in Y'} -\log \sigma \left[g(u, i'; p') - g(u, i; p)\right] \tag{2}$$

where Y is the set of positive triplets and Y' is the set of corrupted triples constructed by randomly corrupting an interacted item to a non-interacted one for each user.

- RotatE (or HRotatE) for Knowledge Graph Completion: RotatE maps entities and relations into a complex space, inspired by Euler's identity [16]. This approach overcomes limitations of existing KGC models like TransE and TransH. The model uses the real and imaginary parts of $e^{(i\theta)}$ to represent head and tail entities, where θ is the rotation angle for relation r.

The scoring function measures the proximity between rotated entity vectors and calculates relation scores to determine the likelihood of a triple being true or false as follows:

$$||h \circ r = t||^2 \tag{3}$$

where h, r, $t \in C^d$, $||.||$ denotes the L1-norm distance function and $\circ$ represents the element-wise (Hadamard) product of two complex vectors.

During training, RotatE uses a self-adversarial negative sampling loss to optimize the model. The loss function is defined as:

$$L_{KGC} = -\log \sigma(\gamma - d_r(h, t)) - \sum_{i=1}^{n} p(h'_i, r, t'_i) \log \sigma(d_r(h'_i, t'_i) - \gamma) \tag{4}$$

where γ is a fixed margin, σ is the sigmoid function and (h'_i, r, t'_i) is the i-th negative triplet. The training goal is to minimize the distance between the modelled triples and the actual triples in the dataset.

HRotatE [14] improves upon RotatE's high computational cost by combining elements of SimplE [6] and RotatE. It inherits the loss function and self-adversarial negative sampling technique from RotatE and uses the principle of inverse embedding from SimplE to improve prediction scores. HRotatE uses different embedding-generation classes to generate embedding vectors for the head and tail entities which allows for more efficient learning. The scoring function of HRotatE is:

$$d_r(h,t) = \frac{1}{2}(||(h_{e_i} \circ v_r) - t_{e_j}|| + ||(h_{e_j} \circ v_r^{-1}) - t_{e_i}||) \quad (5)$$

In HRotatE, each entity, e, is represented by the two vectors, $h_e, t_e \in C^d$ and each relation, r, is also represented by two vectors $v_r, v_r^{-1} \in C^d$.

RotatE and HRotatE use complex space rotations to capture diverse entity relationships in a KG, including one-to-one, many-to-one, one-to-many, many-to-many, symmetric/antisymmetric, inversion and composition relationships.

- Rotational Knowledge-Enhanced Translation-Based User Preference (RKTUP): RKTUP takes as inputs a KG, a user-item interaction list Y and a set of item-entity alignments A. It outputs $g(u, i; p)$, which is the score that measures the likelihood of a user u interacting with an item i with a preference p and also a score $f(e_h, r, e_t)$ that indicates the plausibility of the given fact. This is based on the jointly learned embeddings of users, items, preferences, entities and relations.

After completing the alternating training process between TUP and KGC models, the knowledge enhancement of TUP with the guidance of the embeddings generated by the rotational-based KGC methods is done as follows:

Step 1. Calculate the enhanced item embedding $\hat{i}$ as $i + e$ where $(i, e) \in A$.

$$\hat{i} = i + e, (i, e) \in A \quad (6)$$

Step 2. Calculate the translation vector $\hat{p}$ and the projection vector $\hat{w}_p$ enhanced by the corresponding relation embedding.

$$\hat{p} = p + r \qquad \hat{w}_p = w_p + w_r \quad (7)$$

Step 3. Project the enhanced item embedding $\hat{i}$ onto the enhanced preference hyperplane.

$$\hat{i}^{\perp} = \hat{i} - \hat{w}_p^T \hat{i} \hat{w}_p \quad (8)$$

Step 4. Calculate the enhanced score of recommendation $g(u, i; p)$.

$$g(u, i; p) = ||(u^{\perp} + \hat{p} - \hat{i}^{\perp})|| \quad (9)$$

Step 5. Calculate the final loss as a weighted sum of the RS and KGC losses with a hyperparameter λ used to adjust their hyperparameters.

$$L = \lambda L_{RS} + (1 - \lambda) L_{KGC} \quad (10)$$

5 Experimental Evaluation

Here, we describe our experimental setup, including the datasets and hyperparameters used for training. Then, we evaluate our RKTUP model against several state-of-the-art models for recommendation and knowledge graph completion tasks and report the results. We used RotatE and HRotatE codes by [14,17] for KGC and TUP code by [2] for recommendation.

- Material and Data: In our experiments, we used two widely-used datasets in the field: MovieLens-1m and DBbook2014 datasets which were refined for LODRecSys by mapping entities to DBpedia KG, where available. To ensure fair comparison, we followed the same pre-processing steps as [2]. The datasets were randomly split into training, validation and testing subsets using a 70:10:20 ratio and each user had at least one item in the test set. Table 1 summarizes the dataset statistics after pre-processing.

Table 1. Datasets characteristics

Attribute	MovieLens-1m	DBbook2014
Users	6,040	5,576
Items	3,240	2,680
Ratings	998,539	65,961
Avg. ratings	165	12
Positive ratings	56 %	45.8%
Sparsity	94.9 %	99.6 %
Entity	14,708	13,882
Relation	20	13
Triple	434,189	334,511
I-E Mapping	2,934	2,534

- Evaluation Metrics: We used widely-utilized standard evaluation metrics from the literature. For the recommendation task, we measured the performance of our approach using five metrics. These are Precision (Precis.), Recall, F1 Score (F1), Hit Ratio (Hit) and Normalized Discounted Cumulative Gain (nDCG). As for the KGC task, we used the Hit Ratio (Hit) and Mean Rank (MR).

- Hyperparameters: We performed a grid search to determine the optimal hyperparameters for both the recommendation and KGC tasks. For the learning rate (η), we searched in values of 0.0005, 0.005, 0.001, 0.05, 0.01 and set it to

0.001. The L2 coefficient was searched in values of $10^{-5}, 10^{-4}, 10^{-3}, 10^{-2}, 10^{-1}, 0$, and set to 10^{-5}. The self-adversarial sampling temperature (α) was searched in values of 0.5, 1.0 and set to 1.0. The fixed margin (γ) was searched in values of 3, 6, 9, 12, 18, 24, 30 and set to 24. The joint hyperparameter (λ) was set to 0.5 in MovieLens and 0.7 in DBook after searching in 0.7, 0.5, 0.3. The embedding and batch sizes were set to 100 and 256, respectively. The number of preferences was predefined as 20 in MovieLens-1m and 13 in DBook2014. Adam was selected as the optimizer. The maximum number of training steps was set to 140,000 for RotatE and 70,000 for HRotatE to demonstrate its effectiveness. To ensure a fair comparison, we trained RKTUP and all baseline models in the same environment and with the same hyperparameters.

- Result Analysis: Here we evaluates our model's performance by comparing it to several state-of-the-art models for both the RS and KGC tasks. Specifically, for the RS, we compare our model against classical models like FM [11] and BPRMF [12], as well as three joint-learning models (CFKG [22], CKE [21] and CoFM [10]) and a multi-task learning model (KTUP [2]). For the KGC task, we compare our model against TransE [1], TransH [20] and TransR [9].

Table 2. Recommender systems' experimental results on two datasets.

Models	MovieLens-1M (@10, %)					DBook2014 (@10, %)				
	Percis.	Recall	F1	Hit	NDCG	Percis.	Recall	F1	Hit	NDCG
FM	28.91	12.07	12.34	80.01	58.74	4.19	20.89	4.99	31.28	19.87
BPRMF	31.07	13.10	12.98	82.76	60.56	4.59	21.03	6.14	30.67	22.38
CKE	37.54	15.87	18.68	87.32	66.98	4.01	22.73	6.60	34.09	27.35
CFKG	30.14	11.82	13.76	81.93	57.86	4.50	20.85	4.28	30.01	18.94
CoFM(share)	31.78	12.53	15.62	82.98	57.19	3.27	21.04	6.31	29.80	20.87
CoFM(reg)	30.87	11.98	13.05	81.32	57.45	4.05	20.97	5.13	29.32	20.98
TUP(hard)	36.67	16.71	18.83	88.12	66.81	3.98	22.08	4.99	30.14	20.35
TUP(soft)	36.03	15.94	18.41	87.95	66.17	3.86	22.15	7.08	30.59	22.67
KTUP(hard)	40.90	17.11	18.72	87.88	69.04	4.86	25.06	7.50	34.38	28.16
KTUP(soft)	41.06	17.64	18.75	87.94	69.31	4.89	25.12	7.53	34.50	28.25
RKTUP$(hard)_R$	46.85	19.24	21.15	89.72	74.96	5.78	29.61	8.62	38.20	30.79
RKTUP$(soft)_R$	47.26	19.84	21.18	90.34	75.08	5.81	29.80	8.70	38.64	31.07
RKTUP$(hard)_H$	46.91	19.31	21.20	89.87	75.12	5.81	29.73	8.63	38.29	30.86
RKTUP$(soft)_H$	**47.28**	**19.89**	**21.23**	**90.42**	**75.23**	**5.85**	**29.82**	**8.73**	**38.70**	**31.16**

Table 2 and 3 use abbreviations for specific models and strategies. For example, KTUP(hard) and KTUP(soft) indicate the hard or soft strategies used for preference induction. RKTUP$(hard)_R$ and RKTUP$(soft)_R$ indicate the use of RotatE for the KGC task, while RKTUP$(hard)_H$ and RKTUP$(soft)_H$ indicate the use of HRotatE.

RKTUP outperformed state-of-the-art models on both the RS and KGC tasks (see Tables 2 and 3). The performance of RKTUP$(soft)_H$ RKTUP$(soft)_R$ was combarable on both tasks, with RKTUP$(soft)_H$ requiring half the training

Table 3. Knowledge graph completion experimental results on two datasets.

Model	MovieLens-1M		DBook2014	
	Hit @10 (%)	MR	Hit @10(%)	MR
TransE	47.50	536	59.87	532
TransH	46.85	536	61.02	554
TransR	40.29	608	55.23	565
CKE	35.26	584	55.02	592
CFKG	42.16	522	58.93	547
CoFM(share)	47.65	513	58.01	529
CoFM(reg)	47.38	505	61.09	521
KTUP(hard)	49.00	525	59.94	502
KTUP(soft)	49.68	526	60.37	499
RKTUP$(hard)_R$	56.19	499	67.66	474
RKTUP$(soft)_R$	56.42	494	68.07	466
RKTUP$(hard)_H$	56.30	498	67.71	473
RKTUP$(soft)_H$	**56.47**	**493**	**68.13**	**465**

steps (i.e. 70,000 vs. 140,000). RKTUP$(soft)_H$ demonstrated larger improvements on DBbook2014 than MovieLens-1m (i.e., 13.7% vs. 11.6% gains in F1), indicating the effectiveness of integrating more knowledge for sparse data, but it performed better on MovieLens-1m compared to DBbook2014. In the KGC task, RKTUP$(soft)_H$ outperformed all other models on both datasets, achieving a higher Hit Ratio for MovieLens-1m than for DBbook2014 (13.6% vs. 12.8%). This could be due to the fact that MovieLens-1m has more connectivities between users and items that can help in modeling structural knowledge between entities. These findings support our hypothesis.

6 Conclusion

Our proposed RKTUP MTL model enhances the performance of RS by using rotational-based KGE approaches to learn better representations of users, items, entities and relations. It incorporates most relation patterns from the KG, including composition relations. Experiments on MovieLens-1m and DBbook2014 showed that RKTUP outperforms several state-of-the-art methods in both recommendation and KGC tasks, due to the use of RotatE (or HRotatE) for KGC. Future research could explore advanced dynamic KGE models to capture changes in the graph over time.

Acknowledgement. We acknowledge the support of the Natural Sciences and Engineering Research Council of Canada (NSERC) [funding reference number 03181].

References

1. Bordes, A., Usunier, N., Garcia-Duran, A., Weston, J., Yakhnenko, O.: Translating embeddings for modeling multi-relational data. In: Advances in Neural Information Processing Systems, vol. 26 (2013)
2. Cao, Y., Wang, X., He, X., Hu, Z., Chua, T.S.: Unifying knowledge graph learning and recommendation: towards a better understanding of user preferences. In: The World Wide Web Conference, pp. 151–161 (2019)
3. Guo, Q., et al.: A survey on knowledge graph-based recommender systems. IEEE Trans. Knowl. Data Eng. **34**(8), 3549–3568 (2022). https://doi.org/10.1109/TKDE.2020.3028705
4. Huang, J., Zhao, W.X., Dou, H., Wen, J.R., Chang, E.Y.: Improving sequential recommendation with knowledge-enhanced memory networks. In: The 41st International ACM SIGIR Conference on Research & Development in Information Retrieval, pp. 505–514 (2018)
5. Huang, X., Tang, J., Tan, Z., Zeng, W., Wang, J., Zhao, X.: Knowledge graph embedding by relational and entity rotation. Knowl.-Based Syst. **229**, 107310 (2021)
6. Kazemi, S.M., Poole, D.: Simple embedding for link prediction in knowledge graphs. In: Advances in Neural Information Processing Systems, vol. 31 (2018)
7. Li, Q., Tang, X., Wang, T., Yang, H., Song, H.: Unifying task-oriented knowledge graph learning and recommendation. IEEE Access **7**, 115816–115828 (2019)
8. Lin, X.V., Socher, R., Xiong, C.: Multi-hop knowledge graph reasoning with reward shaping. arXiv preprint arXiv:1808.10568 (2018)
9. Lin, Y., Liu, Z., Luan, H., Sun, M., Rao, S., Liu, S.: Modeling relation paths for representation learning of knowledge bases. arXiv preprint arXiv:1506.00379 (2015)
10. Piao, G., Breslin, J.G.: Transfer learning for item recommendations and knowledge graph completion in item related domains via a co-factorization model. In: Gangemi, A., et al. (eds.) ESWC 2018. LNCS, vol. 10843, pp. 496–511. Springer, Cham (2018). https://doi.org/10.1007/978-3-319-93417-4_32
11. Rendle, S.: Factorization machines. In: 2010 IEEE International Conference on Data Mining, pp. 995–1000. IEEE (2010)
12. Rendle, S., Freudenthaler, C., Gantner, Z., Schmidt-Thieme, L.: BPR: Bayesian personalized ranking from implicit feedback. arXiv preprint arXiv:1205.2618 (2012)
13. Sarwar, B., Karypis, G., Konstan, J., Riedl, J.: Item-based collaborative filtering recommendation algorithms. In: Proceedings of the 10th International Conference on World Wide Web, pp. 285–295 (2001)
14. Shah, A., Molokwu, B., Kobti, Z.: Hrotate: hybrid relational rotation embedding for knowledge graph. In: 2021 International Joint Conference on Neural Networks (IJCNN), pp. 1–8. IEEE (2021)
15. Shi, B., Weninger, T.: Open-world knowledge graph completion. In: Proceedings of the AAAI Conference on Artificial Intelligence, vol. 32 (2018)
16. Stipp, D.: A Most Elegant Equation: Euler's Formula and the Beauty of Mathematics. Hachette UK (2017)
17. Sun, Z., Deng, Z.H., Nie, J.Y., Tang, J.: Rotate: knowledge graph embedding by relational rotation in complex space. arXiv preprint arXiv:1902.10197 (2019)
18. Wang, H., Zhang, F., Xie, X., Guo, M.: DKN: deep knowledge-aware network for news recommendation. In: Proceedings of the 2018 World Wide Web Conference, pp. 1835–1844 (2018)

19. Wang, H., et al.: Knowledge-aware graph neural networks with label smoothness regularization for recommender systems. In: Proceedings of the 25th ACM SIGKDD International Conference on Knowledge Discovery & Data Mining, pp. 968–977 (2019)
20. Wang, Z., Zhang, J., Feng, J., Chen, Z.: Knowledge graph embedding by translating on hyperplanes. In: Proceedings of the AAAI Conference on Artificial Intelligence, vol. 28 (2014)
21. Zhang, F., Yuan, N.J., Lian, D., Xie, X., Ma, W.Y.: Collaborative knowledge base embedding for recommender systems. In: Proceedings of the 22nd ACM SIGKDD International Conference on Knowledge Discovery and Data Mining, pp. 353–362 (2016)
22. Zhang, Y., Ai, Q., Chen, X., Wang, P.: Learning over knowledge-base embeddings for recommendation. arXiv preprint arXiv:1803.06540 (2018)

A Classification Method for "Kawaii" Images Using Four Feature Filters

Daiki Komiya(✉) and Masanori Akiyoshi

Kanagawa University, 3-27-1, Rokkakubashi, Kanagawa-ku, Yokohama 221-8686, Japan
r202370133nq@jindai.jp, akiyoshi@kanagawa-u.ac.jp

Abstract. The Japanese word "kawaii", similar to but slightly different from "cute", has many varied words. Images expressed by these words can be classified based on human intuitive judgements. However, in computational image classification, conventional image classification methods like CNN (Convolutional Neural Network) can classify ordinary images like human faces with high accuracy, but it is considered to be quite difficult to classify images that do not have common objects and concepts like "kawaii". In this paper, we propose a method for classifying "kawaii" images by extracting color, shape, and other latent features of "kawaii" images using four proposed feature filters, quantitatively representing them, and then comparing classification accuracy using various classifiers based on machine learning. In the experiments, the RGB color, SIFT (Scale Invariant Feature Transform), and line features of the images are extracted using such filters and compared using NN (Neural Network), Random Forests, AdaBoost, and SVM (Support Vector Machine). The experimental results show the effectiveness of the proposed feature filters and the predominance of "Random Forests" as a classifier for "kawaii" images.

Keywords: Kawaii · Image Classification · Feature Value · Machine Learning · Sentient Expression Images

1 Introduction

Machine learning-based image classification method is currently being extensively studied, and its classification accuracy is improving. This technology is used in a wide range of fields in society that require complex and accurate classification, such as immigration screening at airports, search engines that use images as input, and disease detection assistance in the medical field. The development of such image classification technology has demonstrated classification accuracy equal to or better than that of the human, and has contributed to solving the recent labor shortage.

However, while artificial intelligence has been able to classify images with common objects and concepts shared among people such as dog images and

S. Ossowski et al. (Eds.): DCAI 2023, LNNS 740, pp. 296–305, 2023.
https://doi.org/10.1007/978-3-031-38333-5_30

human faces with very high accuracy, there has been few research on the classification of ambiguous images represented by "sentient words". Sentient words are adjectives that express human emotions, such as "cute" and "cool". These words do not have quantitative values, there exists no clear criteria. The content of images expressed by these "sentient words" varies widely, ranging from animal images, landscapes, illustrations to photographs. Such images can only be classified according to human subjective judgement and such expression handling is one of the challenges in artificial intelligence. Our research target is concerning such unestablished image classification fields and the proposed method aims to evoke future discussions on this target by indicating experimental results.

Among the "sentient words", the word "kawaii," which means "cute" in Japanese, is considered to be quite complex. The objects are not only people and animals, but also colors, designs, food, and many other things. And due to the young female pop culture of coined words, many varied words as to "kawaii" exist in the Japanese language. Table 1 shows examples of varied words in "kawaii".

Table 1. Examples of varied words in "kawaii"

varied words in "kawaii"	meaning
KIMOKAWA	Disgusting, but cute.
BUSAKAWA	Ugly but cute.
YUMEKAWA	Cute as a dream.
YURUKAWA	Soothing cuteness.
YAMIKAWA	Mentally and physically ill, but cute

Figure 1 is also examples of images as to varied words in "kawaii".

KIMOKAWA

BUSAKAWA

Fig. 1. Examples of "kawaii images"

In addition, there have been many psychological studies on "cute", one of which pointed out that the typical characteristic of "cute" is a baby schemer [1], and the experiment using visual narratives indicated common characteristics of "cute" images [2].

In this paper, we propose an image classification method using a machine learning-based classifier that extracts and quantifies image features using filters in the classification of images represented by the varied word "kawaii".

2 Related Works and Problem

2.1 Related Works

Various methods have been used in image classification experiments using CNNs. Some experiments have improved classification accuracy by investigating the structure of the CNN model [3], while others have improved accuracy by solving the problems of overfitting and gradient loss with parameter reduction [4]. Some experiments not only use CNNs as classifiers, but also extract features from CNN models and fuse them with other classifiers to enable classification with accuracy better than simply use of CNNs [5].

In addition, some comparison experiments with multiple machine learning classifiers, including CNNs, to set the best classifier for a data set [6].

Experiments have also been conducted on the relationship between images and human impressions. Some experiments have investigated the relationships between natural images and human impressions [7], while others have extracted sentient features based on artistic principles and classified the impressions of images [8].

2.2 Problem to be Tackled

CNN (Convolutional Neural Network) is the main method for image classification. This method can classify images with high accuracy by finding common objects in the image.

However, it is believed that image classification using CNNs cannot classify "kawaii" images with sufficient accuracy. This is because there are no common objects within the these images. Machine learning classifiers, which can classify with high accuracy by finding commonalities, are not suited to handle "kawaii" images as input as they are. In order to classify these "kawaii" images, we believe it is necessary to quantify the characteristics of these images and find latent commonalities.

It is also necessary to choose an appropriate classifier that can take maximum advantage of the quantified features.

3 Methods

There is an experiment that attempts to classify "kawaii" images [9]. In the experiment, features were extracted from images using three filters, which were then used for classification by NN (Neural Network).

To improve the classification accuracy of the previous work, we propose optimized classification scheme for "kawaii" images. Figure 2 shows the proposed method.

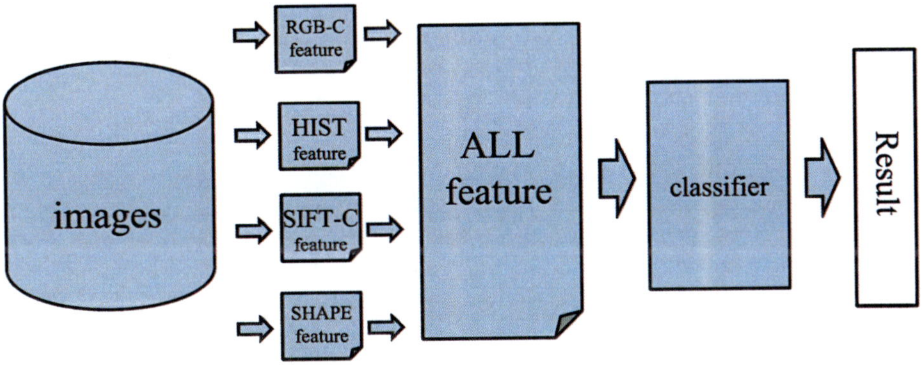

Fig. 2. Configuration of design image classification method

Four features are extracted. After the extracted features are synthesized into a "total feature data", experiments are conducted to compare the classification accuracy using multiple classifiers. Based on the results, we propose a classification method suitable for "kawaii" images.

3.1 Feature Filters

Our proposed image feature filters are as follows; "RGB histogram feature filter", "RGB convolutional feature filter", "SIFT convolutional feature filter" and "Shape feature filter".

RGB Histogram Feature Filter. Each pixel in an image represents a color using a combination of RGB numbers, each RGB having an integer value ranging from 0 to 225. By making a histogram of these numerical values of all pixels in the image for each RGB, we can obtain a feature of 768-dimensional vectors corresponding to three colors.

RGB Convolutional Feature Filter. This filter performs convolution on the image to obtain the characterized RGB information. This process uses CNN. Since CNN convolve images in the process of the learning process, we can directly extract the values of the middle layer after convolution to obtain the features of the image where only the convolution process has been performed.

However, since the learning process of the CNN is used, the convolution process is affected by the model and the order of image input. Therefore, this problem is solved by preparing the same initial model for all images at the time of feature extraction, and inputting them to the CNN under the same conditions for convolution processing. The feature set is a 512-dimensional vector equal to the number of intermediate layers to be extracted.

SIFT Convolutional Feature Filter. SIFT (Scale Invariant Feature Transform) [10] features are extracted local features of an image that are invariant to image "scaling" and "rotation" and their feature values. The extracted feature values for the grayscale of the image do not include RGB information. But, simply extracting "SIFT features" is difficult to handle because the number of such feature points varies from image to image. Therefore, pseudo-images are generated and SIFT features are converted to a standardized data format.

Figure 3 shows the method. For pixels in the image from which SIFT feature points are extracted, the RGB information that the original image has per pixel is replaced with 129-dimensional SIFT features. For pixels from which no feature points were extracted, the 129-dimensional zero element vector is used. By executing this process for all pixels, a pseudo-image is generated. Once the pseudo-image is created, the values of the middle layer of the CNN are extracted and the convolutional features are obtained in the same way as in the "RGB convolutional feature filter" described above. The feature set is a 512-dimensional vector equal to the number of intermediate layers to be extracted.

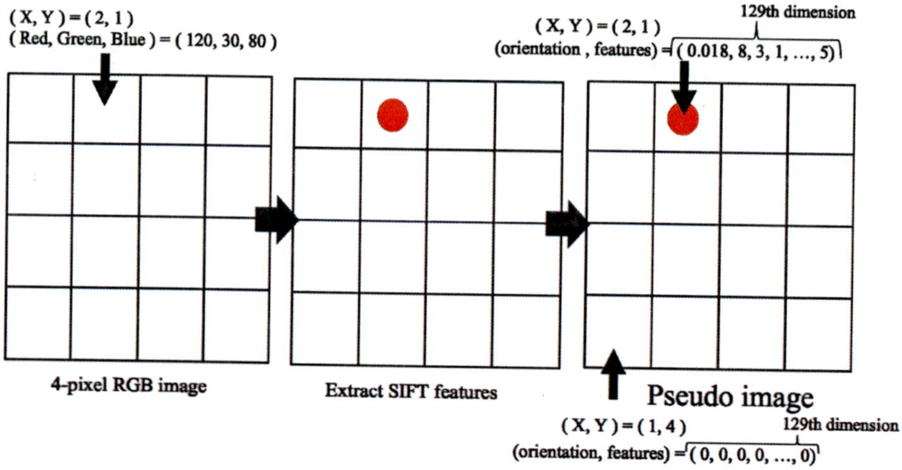

Fig. 3. Pseudo-image generation of SIFT features

Shape Feature Filter. The above three filters extract color and local features. However, they do not extract information about the shape of the image. Although "kawaii" images do not have a fixed shape or object, there are commonalities among the contours and the lines that make up the image due to the shape of the partial lines, and we believe that these features are useful information in classification.

In this filter, convolution processing is performed on the line drawings in the image to create a feature value as a contour image. In some cases, the contour of an image is considered in the process of highlighting the shape of an image, but by using line drawings, the line information inside the contour can be extracted in addition to the contour information of the image. In addition, there is a study that the classification accuracy using line drawings was superior to that using grayscale images as the result of a classification comparison experiment using a CNN of grayscale images and line drawings [11]. It was considered that line drawings may have been more effective for classification due to the abstraction of information. The feature set is a vector of 512 dimensions, which is the same as the number of intermediate layers to be extracted.

3.2 Classifiers

The machine learning classifiers used in the experiments are CNN, NN, Random Forests, Ada Boost, and SVM (Support-Vector Machine). The CNN does not use any feature filters, but uses the pixels of the original image itself as input, and the experiments will be conducted as a benchmark against the proposed method.

4 Experiment

4.1 Data Sets

All images used in the experiment are typical varied words derived from "kawaii" as described in the Sect. 1. The images were extracted from the top-ranked images in the search engines using the word in question. Although the criteria for "kawaii" images vary depending on the human evaluator, we believe that the search engine results reflect the average criteria in society, and we used this as the criterion for genre in this experiment.

Table 2 shows the five "kawaii" images used in this experiment. In addition, the image size is converted to 252×252 pixel for all of them.

In addition, an experiment was conducted to confirm whether the genres of "kawaii" images collected were valid as genre data. 187 sampling images were randomly presented to testers with a total of 50 images, 10 for each genre, and each was classified into 5 genres. Table 2 also shows the results.

The results will also be used as a measure of classification accuracy as the accuracy of manually classifying "kawaii" images.

Furthermore, we conducted a questionnaire survey to determine the points of interest in distinguishing between datasets during the aforementioned sampling image labeling experiments. The top three responses are summarized below. The number of testers to the questionnaire was 122, and multiple points of interest were allowed.

The results confirmed the most common opinion that color is important. On the other hand, there were also many opinions that could not be explained in words, such as "impression".

Table 2. Experiment data on "kawaii" images and label assignment statistics

Genre name	Number of data	Coincidence rate (%)
KIMOKAWA	570	82.7
BUSAKAWA	464	69.4
YUMEKAWA	702	82.1
YURUKAWA	830	81.5
YAMIKAWA	559	76.8
Average	–	78.5

Table 3. Points of interest in classifying "kawaii" Images

Answer	Number of responses
Color	62
Feeling	29
Contents of image	27

4.2 Results

The experiment is conducted through cross-validated five times with 80% of these data for each genre as training data and 20% as test data. The classification results are evaluated using the average of the five correct responses.

Table 4 below summarizes the classification results for each classifier.

Table 4. Comparison of classification accuracy (%)

	CNN	NN	Random Forests	Ada Boost	SVM
KIMOKAWA	33.4	38.2	64.4	44.2	63.5
BUSAKAWA	14.5	22.4	53.4	40.2	48.2
YUMEKAWA	43.5	74.1	89.3	65.3	22.3
YURUKAWA	41.2	62.6	72.2	79.4	34.2
YAMIKAWA	31.3	53.3	71.5	45.5	10.2
Average	32.8	50.1	70.2	54.9	35.7

The accuracy of all classifiers using the proposed method resulted in a higher average rate compared to the CNN.

In an experiment comparing the four classifiers, the average classification for NN was 50.1%, compared to 70.2% for Random Forest, 54.9% for Ada Boost, and 35.7% for SVM. Random forests, which had the highest accuracy, were classified with high accuracy when depth restrictions are set to the decision trees themselves. In addition, classification accuracy increased monotonically as the number of decision trees increased.

It was also confirmed that there were differences in classification accuracy among genres. The classification accuracy of "Busakawa" tended to be low for all classifiers.

The results in Table 3 indicate that the information held by color is important in the classification of "kawaii" images by hand. From this, it can be assumed that the extraction of color features is also quite effective in classification by machine learning. We conducted an additional experiment to confirm this by modifying the two feature filters using RGB information as follows. In the "RGB histogram feature filter", the conversion to the grayscale was used as a 256-valued histogram instead of RGB, and the feature was a vector of 256 dimensions. In the "RGB convolution feature filter", the image to be convolved was converted to grayscale, and the features were extracted. In the CNN used as a benchmark, the input was also a grayscale image. The results are summarized in Table 5 below.

Table 5. Comparison of classification accuracy excluding RGB information (%)

	CNN	NN	Random Forests	Ada Boost	SVM
KIMOKAWA	21.4	42.9	43.8	50.3	48.4
BUSAKAWA	27.14	36.9	33.8	30.2	38.8
YUMEKAWA	22.8	31.9	37.8	37.8	25.8
YURUKAWA	34.4	38.9	42.2	42.5	33.5
YAMIKAWA	28.8	24.6	20.0	38.8	28.0
Average	26.9	35.0	35.5	39.9	34.9

Compared to the results shown in Table 4, a significant decrease in classification accuracy was observed.

4.3 Discussion

Comparison with CNN indicates that the feature filter of the proposed method is able to extract latent features of the image and convert them into information that is useful classify. In addition, the shape features newly added to the previous study are effective in improving the classification accuracy. This is thought to be because the shape features were able to compensate for the lack of information that is useful in the conventional features.

The difference in classification accuracy by genre tended to be lower for the "Busakawa" category in all classifiers, but this was also true for the actual human classification accuracy shown in Table 2. From this result, we believe that "Busakawa" images are particularly difficult to discriminate even within the "kawaii" image group, and that this is because the sentient word "Busakawa" is weakly recognized in common. In image processing, many of the "Busakawa" images are of mammals such as dogs and cats, and it is difficult to capture their characteristics in acquiring color and line information from the images.

Based on the comparison of different classifiers, we believe that training with a strong learner such as NN on the training data may result in lower classification accuracy for the test data when the images are ambiguous and do not have a quantitative index, such as the "kawaii" images. On the other hand, classification using a weak learner is said to be insufficient for classifying images with fixed objects or elements, but in this experiment, in which ambiguous images were the input, classification was achieved with higher accuracy than with a learner using a rigid algorithm.

The classification with Random Forests, which was the most accurate in this study, is based on ensemble learning using weak learners in parallel. Each weak learner performs classification using partial features extracted randomly from a total of 2304 dimensional features. Although the images are "kawaii" images with no quantitative indicators or fixed objects or elements, we believe that there are partial commonalities among the filtered features, and that some partial features can be used to classify images that could not be classified by the total feature set.

We also believe that the method in which each weak learner predicts the genre and the final classification result is obtained by majority vote was suitable in terms of the standard of average sensitivity of people in the world, which is used as reference to evaluations of input images for this experiment.

Experiments in which RGB information was removed showed a significant decrease in accuracy compared to the proposed method. These results confirm that color information is as important in computational "kawaii" image classification as it is in human classification.

5 Conclusions

In this study, we introduced new features in addition to the features used in the previous study for "kawaii" images, which are expressed by sentient words and do not have quantitative index, and conducted comparative experiments on image classification using four different classifiers. The classification accuracy was improved by adding shape features, which are not covered by the features used in the previous study, and by enhancing the information in the image. In comparison experiments using different classifiers, Random Forests, which is based on ensemble learning with a weak learner, was able to classify with high accuracy compared to the classifier using a strong algorithm.

For classification of "kawaii" images, which do not have fixed shapes or elements and do not have quantitative indicators, we believe that the data extracted features for each element the image has, and that the majority voting structure with partial feature extraction and weak learner of Random Forests is suitable for the classifier. It is also confirmed that color information is very important in the classification of "kawaii" images.

Our future work is concerning the design of sentient filters beside the proposed filters and automatic selection of such filters.

References

1. Lorenz, K.: Die angeborenen formen möglicher erfahrung. Z. Tierpsychol. **5**(2), 235–409 (1943)
2. Yamada, Y., Kido, A.: Why do we feel "kawaii"?: a diverse joint method for visual narratives. Jpn. Assoc. Qual. Psychol. **16**(1), 7–24 (2017). (in Japanese)
3. Litao Liu: Improved Image Classification Accuracy by Convolutional Neural Networks, ICITEE '21, Proceedings of the 4th International Conference on Information Technologies and Electrical Engineering, Article No:50, pp.1-5(2021)
4. Shuo, H., Kang, H.: Deep CNN for classification of image contents. In: Proceedings of the 2021 3rd International Conference on Image Processing and Machine Vision (IPMV 2021), pp. 60–65 (2021)
5. Yan, H., Liu, X., Hong, R.: Image classification via fusing the latent deep CNN feature. In: Proceedings of the International Conference on Internet Multimedia Computing and Service (ICIMCS 2016), pp. 110–113 (2016)
6. Tropea, M., Fedele, G.: Classifiers comparison for convolutional neural networks (CNNs) in image classification. In: IEEE/ACM 23rd International Symposium on Distributed Simulation and Real Time Applications (DS-RT), pp. 1–4 (2019)
7. Dellagiacoma, M., Zontone, P., Boato, G., Albertazzi, L.: Emotion based classification of natural images. In: Proceedings of the 2011 International Workshop on DETecting and Exploiting Cultural diversiTy on the Social Web (DETECT 2011), pp. 17–22 (2011)
8. Zhao, S., Gao, Y., Jiang, X., Yao, H., Chua, T.-S., Sun, X.: Exploring principles-of-art features for image emotion recognition. In: Proceedings of the 22nd ACM International Conference on Multimedia (MM 2014), pp. 47–56 (2014)
9. Kowatari, S., Akiyoshi, M.: A method of classifying images containing "cute" factor elements using feature filters. In: IEEJ, IS-22, IS22004, pp. 15–18 (2022). (in Japanese)
10. Fujiyoshi, H.: SIFT and recent approaches in image local features. Trans. Jpn. Soc. Artif. Intell. **25**(6), 753–760 (2010). (in Japanese)
11. Asahi, S., Narumatsu, H., Yamato, J., Taira, H.: Image type classification for object recognition of illustration images. In: JSAI2021, 2Yin5-20 (2021). (in Japanese)

Exploring Dataset Patterns for New Demand Response Participants Classification

Cátia Silva[1], Pedro Campos[2,3,4], Pedro Faria[1], and Zita Vale[1](✉)

[1] Intelligent Systems Associated Laboratory (LASI), Research Group on Intelligent Engineering and Computing for Advanced Innovation and Development (GECAD), Polytechnic of Porto (P.PORTO), Porto, Portugal
{cvcds,pnf,zav}@isep.ipp.pt

[2] Laboratory of Artificial Intelligence and Decision Support, LIAAD INESC TEC, Porto, Portugal
pcampos@fep.up.pt

[3] Faculty of Economics, University of Porto R. Dr. Roberto Frias, 4200-464 Porto, Portugal

[4] Statistics Portugal, Porto, Portugal

Abstract. There is a growing trend towards consumer-focused approaches integrating distributed generation resources in the power and energy sector. This, however, adds complexity to community management as new players are introduced. The authors have designed a trustworthy rate (TR) system to address this issue of selecting participants for demand response events based on their previous performance. The aim is to avoid discomfort for consumers and reduce aggregator costs by selecting participants fairly. However, this poses a challenge for new players with a performance history. This study aims to develop a method to assign the TR to new players. For this purpose, the authors used supervised clustering and subgroup discovery to identify the relevant features of the dataset without compromising privacy after employing techniques such as Decision Trees, Random Forest, and Extreme Gradient boosting to assign the appropriate TR to each player. The performance of the methods has been evaluated using metrics such as accuracy, precision, recall, and f1 score.

Keywords: Classification · Demand Response · Subgroup Discovery · Supervised Clustering · Uncertainty

1 Introduction

In the power and energy sector, all the players are working toward decarbonizing the system by replacing fossil fuels with renewable-based Distributed Generation (DG) technologies, such as wind and solar. However, these renewable sources have uncertain and non-programmable behavior, leading to a new paradigm where generation no longer follows demand needs [1]. Consumers must provide flexibility to avoid negative impacts on grid management, and Demand Response (DR) programs are an effective solution for this. The definition of consumers is changing as they become new market players in the energy system [2]. But they need proper knowledge to participate effectively. While

S. Ossowski et al. (Eds.): DCAI 2023, LNNS 740, pp. 306–317, 2023.
https://doi.org/10.1007/978-3-031-38333-5_31

many consider consumers as rational and economic agents, the reality may differ, and they require time, education, and resources to make better judgments [3].

So, despite the need for consumer flexibility through DR programs, business models do not include or cannot deal with the uncertainty associated with these new resources. Consumer participation in the market is indirect due to the small load flexibility provided and the high response uncertainty [4]. Aggregators manage active communities by gathering all the load flexibility from community members and entering the market on their behalf, but their response is only guaranteed if participation is voluntary [5]. The volatile behavior from consumers and DG units increases the network's complexity. To deal with this problem, the authors have developed a trustworthy rate (TR) to classify consumers based on the context in which DR events are triggered. The trustworthy rate is based on the participant's performance in previous events, and if their performance is good, they receive a high rate and better rewards. Otherwise, if their performance is poor, penalties are applied. The aggregator then uses the trustworthy rate to select the appropriate participants for the DR events triggered. However, this approach relies on historical information on the consumer. As innovation from previous works, such as [6] and [7], for new participants, the authors want to discover the patterns for specific combinations of characteristics that will help classify each consumer using the TR. Based on prior knowledge, supervised clustering and subgroup discovery are used to similar group objects into these predefined categories, assuming a target variable *a priori*. Decision trees, random forests, and Extreme Gradient Boosting (XGBoost) will be used and compared for classification tasks. In general, XGBoost performs better than random forests on large datasets, while random forests may perform better on smaller datasets with fewer features. Decision trees remain a useful tool for understanding the underlying relationships in data by learning simple decision rules inferred from the data features and can be a good starting point for more complex machine learning models.

The paper is structured into five main sections. The first section contains the introduction, which sets the context and outlines the paper's main objectives and research questions. In the second section, the proposed methodology is described in detail. The third section presents the case study, defining the different scenarios. The fourth section discusses the results of the study. Finally, the conclusions are drawn in the last section of the paper.

2 Proposed Methodology

In the present section, the authors define the proposed methodology to manage and select DR resources considering the triggered event context optimally and fairly. Following Fig. 1, DR participants identification, Optimal Scheduling, and Remuneration are the three main steps, where the first is the focus of this study.

As soon as the DR event is triggered, the entity responsible for active community management – the aggregator, should decide which resources are needed to achieve the reduction target. Reducing the uncertainty response and enhancing DR participants' performance are the two key goals behind the proposed rate for selecting the proper DR participants – the Trustworthy Rate (TR). For the authors, this is an important step to avoid discomfort for the participants and increase the profit from the aggregator – preventing calling unnecessary participants. For instance, avoiding turning off the air conditioning or shifting the washing machine program to a different schedule at times of need might increase the motivation to participate in DR events. To do this, it is important to understand the contexts in which the players have more availability.

The idea is to characterize the DR participants with a contextual rate that evaluates the performance for the given context, easing the task of deciding which ones should participate in future events. So, in an early stage, the TR is defined as a preliminary TR (PTR) that depends on three independent rates: Context Rate (CR), Historic Rate (HR), and Last Event Rate (LER).

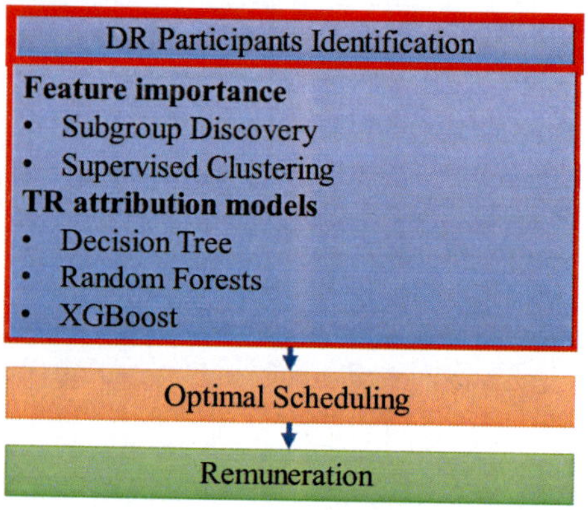

Fig. 1. Proposed methodology.

HR is the average TR from previous events that occurred in similar contexts. The LER alludes just to the last event to avoid misleading. Furthermore, considering the recorded weather and the time the event is triggered, CR depends on the active consumer's availability and willingness to participate.

The study for this paper focuses on this step and the fact that aggregator needs to have historical information on new players. In previous works, for this case, the TRassigned was the lowest, and the participant must earn the trust of the aggregator. However, understanding the behavior of similar participants might lead to a pattern, which can be used to attribute a more accurate TR to this new player.

Following this assumption, the authors intend to find patterns in the dataset to discover which features have more impact on the rate and then classify the participants. Finally, the authors will apply useful techniques such as supervised clustering and subgroup discovery for exploratory data analysis and identifying complex relationships.

According to [8], using an unsupervised clustering algorithm does not necessarily guarantee that objects of the same class or type will be grouped. Some form of supervision or labeling is required to ensure that objects with the same label are grouped into the same cluster. This procedure helps the algorithm identify which attributes or features are important for determining the similarity between objects and which objects should be grouped based on their shared labels or classes [9]. In this way, supervised clustering can be useful in applications with prior knowledge about the classes or labels of the data and where the goal is to group similar objects into these predefined categories, as in the present study. Subgroup discovery, explored by [10], aims to find interesting subsets of data that exhibit a certain property or behavior. It involves identifying subgroups or subsets of data significantly different from the rest in terms of their characteristics or attributes [11].

For the TR attribution, three classification methods are compared: Decision Trees, Random Forests, and XGBoost. Decision trees work by recursively splitting the input data into smaller subsets based on the value of a certain feature. Each split creates a new decision node, and the process continues until a stopping criterion is met, such as reaching a certain depth or minimum number of samples in a node. The final nodes of the tree are called leaves and represent the predicted value for a given input. Decision trees are easy to interpret and visualize but can be prone to overfitting. Random forests are an extension of decision trees that create a collection of decision trees and aggregate their results to make a final prediction [12, 13]. Each tree is trained on a random subset of the data and a random subset of the features, which helps reduce overfitting and improve performance. Random forests are often used for classification tasks and can handle categorical and continuous features. XGBoost is a powerful ensemble method that builds upon the concept of decision trees [14]. It trains a sequence of decision trees that each try to correct the errors made by the previous tree. XGBoost is known for its speed and scalability. While all three algorithms are based on decision trees, random forests, and XGBoost are designed to reduce overfitting and improve performance. Random forests use multiple trees and random sampling to create a more robust model, while XGBoost uses a gradient descent optimization approach to iteratively improve the performance of individual trees.

3 Case Study

The historical information used as input for the case study was taken from previous work by the authors [6] and [7]. An active community where the aggregator triggered several DR events throughout the month. Information such as availability, period, day of the week, day of the month, and temperature were withdrawn. Table 1 shows an initial dataset summary. For the current case study and considering the number of samples from each rate – where rate 1 and rate 5 have fewer observations, the authors created a new range group. So, from now on, the participants with TR lower than three will be in group -1; TR equal to 3 will be in group 0; and TR higher than three will be in group 1. In addition, the authors added new information for this paper based on Portuguese Statistics National Institute (INE - Instituto Nacional de Estatística) data for the North of Portugal, such as the building year, age group, and education level [15]. Table 2 presents the labeling used for the mentioned data. The authors widely discussed the chosen information, considering privacy issues from the participant's perspective. So, the aggregator has some knowledge regarding the participant, which could lead to distinguishing them from others without jeopardizing their privacy. With the information from INE, the authors gathered the percentage of participants from each label and adapted it for the initial dataset. This makes it possible to find patterns and withdraw knowledge using supervised clustering and subgroup discovery. Ultimately, the classification can be performed after finding the important features.

Table 1. Characterization of the number of occurrences for each rate [7].

Day of Month	Day of Week	Temperature	Period	Rate 1	Rate 2	Rate 3	Rate 4	Rate 5
1	1	16	61	17	77	102	179	25
3	3	13	253	19	119	130	132	0
5	5	11	445	7	132	130	131	0
8	1	12	733	14	120	131	135	0
10	3	14	925	2	123	151	124	0
12	5	14	1117	15	77	98	172	38
15	1	17	1405	12	66	147	156	19
17	3	17	1597	5	120	141	134	0
19	5	16	1789	13	133	114	140	0
22	1	19	2077	24	101	56	187	32
24	3	11	2269	8	126	143	123	0
26	5	14	2461	9	135	132	124	0
			Total	145	1329	1475	1737	114

Table 2. Labeling for the new input information – categorical variables.

Building year	Label	Age group		Label	Education level		Label
>1945	A	>25	Men	HJ	None		NN
1946–1960	B		Women	MJ	Basic education	1°	BU
1961–1970	C					2°	BD
1971–1980	D	25–64	Men	HM		3°	BT
1981–1990	E		Women	MM	Secondary education	BS	
1991–1995	F						
1996–2000	G	>65	Men	HI			
2001–2005	H		Women	MI			
2006 –2021	I				Higher education	ES	

4 Results and Discussion

The results from the case study were achieved using libraries written using R language. The dataset was previously handled regarding missing values and categorical variables. Missing values can occur due to various reasons. For instance, the malfunctioning of a smart plug can lead to measurement errors. Categorical variables represent qualitative data that fall into distinct categories, such as those selected in Table 2. Although tree-based algorithms can directly handle these variables, for XGBoost, the authors used one-hot encoding, creating a binary column for each category. After performing both pattern definition methods and rules for TR attribution, the results are analyzed and discussed.

4.1 Dataset Patterns

Firstly, the dataset was tested using subgroup discovery. This technique process involves selecting a target variable of interest and searching for subsets of the data that exhibit significant differences concerning that variable. The library used was *rsubgroup* developed by Martin Atzmueller [16]. As input, the authors provided the dataset, the target, and the set of configuration settings, including the mining method, the quality function, the maximum number of patterns to be discovered, and the parameter to control whether irrelevant patterns are filtered during pattern mining. The quality function chosen was adjusted residuals, where the difference between the observed and expected samples is divided by an estimate of the standard error, according to Eq. (1).

$$\text{Adjusted residual} = (\text{Observed value} - \text{Expected value})/\text{Standard error of the residuals} \quad (1)$$

Table 3 shows the results of subgroup discovery, showing the rules for each TR group and the quality value. According to the rules created, none of the new features was considered important for this technique. The most used were availability, temperature, day of the week, period, and by order.

All the quality values are positive and below 20, presenting the lower ones for TR group 0. A positive value for this quality function indicates that the observed value is higher than predicted by the model, adjusted for sample size. This suggests the presence of an outlier or unusual observation that is not captured by the model and may indicate the need to include additional factors in the model or investigate further. Furthermore, for the goal of the case study, the authors need features that do not rely on historical information.

Table 3. Subgroup Discovery Results.

Trustworthy Rate Group	Rules	Quality
<3	"Availability = 0.000" "DayofWeek = 1.000"	18.17
	"temperature[16.5;∞[" "Availability = 0.000" "DayofWeek = 1.000"	14.65
	"Availability = 0.000"	14.17
3	"temperature]-∞;13.5[" "Availability = 1.000"	5.92
	"temperature]-∞;13.5[" "Period]-∞;829[" "Availability = 1.000"	5.44
	"temperature]-∞;13.5["	5.26
>3	"temperature[16.5;∞[" "Availability = 1.000"	18.53
	"Availability = 1.000"	15.23
	"DayofWeek = 1.000"	12.01

The dataset was tested using a supervised clustering technique, which resorted to a *supclust* library developed by Marcel Dettling and Martin Maechler [17]. This library can perform both "PELORA" and "WILMA" algorithms. The authors have chosen the first one since it performs the selection and supervised grouping of predictor variables in large datasets, where most parameters were considered the default. Only *noc*, the number of clusters that should be searched for on the data, was revised. Equation (2) shows the default quality function used to evaluate this technique – the within-cluster sum of squares (WSS), where *i* ranges over the observations in the cluster, *j* ranges over the clusters, x_i is the i-th observation, c_j is the centroid of the j-th cluster, and d(x_i, c_j) is the distance between the i-th observation and the j-th cluster centroid.

$$\mathrm{WSS} = \sum\nolimits_i \sum\nolimits_j d(x_i, c_i)^2 \tag{2}$$

Results from the supervised clustering algorithm can be seen in Table 4. Regarding the criterion used to evaluate the clusters, the results from Table 4 are relatively high. A high WSS indicates that the observations within the cluster are relatively spread out. However, the results from *pelora* method show the features that have a stronger association with the labels assigned to the data points, which is the purpose of this section. For the first group, where TR 1 and TR 2 are considered, the availability, day of the week, and building year were impactful features. In the next group, with only TR 3 samples, most of the features were included, excluding the day of the month. Regarding the final group, availability was selected as the representative entry. From this, the authors can move to the next phase, considering all the features except the day of the month that was not mentioned in any of the performed techniques. The following sub-section compares the three selected classification methods used for TR rule definition.

Table 4. Supervised Clustering Results.

Trustworthy Rate Group	Features	Quality
<3	Availability	2884.79
	DayofWeek and Availability	2864.62
	Availability and Building Year and DayofWeek	2859.15
3	temperature	2937.35
	Temperature and Availability and Building Year and DayofWeek and Age Group and Period	2921.81
	Temperature and Availability and Building Year and DayofWeek and Age Group and Education level	2916.97
>3	Availability and temperature	3076.53
	Availability and DayofWeek	3038.97
	Availability and temperature and Building Year and Age Group and Education Level	3027.37

4.2 TR Attribution Models

The first method to be tested is decision trees resorting to the *rpart* package [18]. The dependent variable in the model is "TR", and the independent variables are all other variables in the "train" dataset. The "method" parameter is set to "class" indicating a classification problem, and the "maxdepth", "minsplit", and "minbucket" parameters are used to control the size and complexity of the tree. The "cp" parameter is set to a small value, which controls the tree's complexity and helps avoid overfitting.

According to the results in Table 5, for TR less than 3, the following conditions must be met: Availability is 1, Temperature is between 13 and 15 °C, and Building Year is either A, B, C, or F. Additionally, when Availability is 0, the TR will be less than 3. For TR equal to 3, the participant must have Availability equals 1 and Temperature is less than 12 °C, or Availability is 1 and Temperature is between 13 and 15 °C, and Building Year is either D, E, G, H, or I. For TR greater than 3, Availability is 1 and Temperature is between 12 and 13 °C, or Availability is 1 and Temperature is greater than or equal to 15 °C.

Table 5. Decision tree results.

Trustworthy Rate	Rules
<3	when Availability is 1 & Temperature is 13 to 15 & Building Year is A or B or C or F
	when Availability is 0
3	when Availability is 1 & Temperature < 12
	when Availability is 1 & Temperature is 13 to 15 & Building Year is D or E or G or H or I
>3	when Availability is 1 & Temperature is 12 to 13
	when Availability is 1 & Temperature >= 15

Table 6 presents the confusion matrix for the decision tree model – where from the dataset, 80% were used for training and 20% used for testing. The matrix shows the predicted versus actual values for the three groups: "TR < 3", "TR = 3", and "TR > 3". The model correctly predicted 70 instances and misclassified 121 instances as equal to 3 and 103 as superior to 3.

For the group TR equal to 3, the model correctly predicted 110 instances and misclassified 33 instances as inferior and 147 as superior. For the last group, the model correctly predicted 264 instances, misclassified 3 instances as "TR < 3" and 109 instances as "TR = 3". Moving on to the random forest technique, where the R library has the same name, and the default parameters were considered, only using the data training predictors and the dependent variable TR as input [19]. Table 7 presents the resulting confusion matrix for the Random Forest. The first row of the matrix shows that out of 303 instances with the actual class value of TR inferior to 3, 143 instances were correctly classified, 95 instances were misclassified as TR equal to 3, and 65 instances were misclassified as TR

Table 6. Confusion matrix for Decision tree.

	Predicted results			
		<3	3	>3
Actual	<3	70	33	3
	3	121	110	109
	>3	103	147	264

superior to 3. Similarly, the second row of the matrix shows that out of 286 instances with the actual class value of TR equal to 3, 110 instances were misclassified as inferior, 93 instances were correctly classified, and 83 instances were misclassified as TR superior 3. The third row shows that out of 371 instances with the actual class value of TR superior to 3, 78 instances were misclassified as inferior, 85 instances were misclassified as TR equal to 3, and 208 instances were correctly classified.

Table 7. Confusion matrix for Random Forest.

	Predicted results			
		<3	3	>3
Actual	<3	143	95	65
	3	110	93	83
	>3	78	85	208

The final classification model is XGBoost, similarly to the random forest, the library used in R has the same name [20]. The authors adjusted the algorithm's parameters and, besides the input data, limited the maximum number of boosting iterations to 1000; the evaluation metric used to measure the performance of the model during training was multiclass classification error rate; the objective function used to optimize the model was multiclass classification problems and the number of classes in the target variable. The minimum train error obtained was 0.19 at iteration 803 – after this value was not modified. Table 8 presents the confusion matrix for the XGBoost model. By analyzing the results, from the 283 instances with the actual class value of TR inferior to 3, the model correctly classified 114 instances, misclassified 91 instances as TR equal to 3 and misclassified 78 instances as superior.

Table 8. Confusion matrix for XGBoost.

	Predicted results			
		<3	3	>3
Actual	<3	114	91	78
	3	98	100	99
	>3	74	98	208

Similarly, the second row shows that out of 297 instances with the actual class value of TR equal to 3, the model misclassified 98 instances as inferior, correctly classified 100 instances, and misclassified 99 instances as superior. The third row of the matrix indicates that out of 380 instances with the actual class value of TR superior to 3, the model misclassified 74 instances as TR inferior to 3, 98 instances as 3, and correctly classified 208 instances. Resorting to the confusion matrix created in the previous tables, the authors have chosen four commonly used metrics for evaluating the selected classification models: accuracy, precision, recall, and F1 score. Accuracy measures the percentage of correct predictions made by the model, while precision measures the percentage of true positive predictions out of all predicted positives. In other words, a high accuracy score indicates that the model correctly predicts most instances, while high precision can help reduce false positives. Conversely, Recall measures the percentage of true positive predictions out of all actual positives. When false positives are costly, precision is a useful metric, while recall is useful when false negatives are costly. Finally, the F1 score is a metric that balances precision and recall and is particularly useful for imbalanced datasets. By combining these metrics, the authors can gain a more comprehensive understanding of the performance of the selected classification models. Table 9 shows the results. According to the results, there is a small variability in the performance of each method across the different metrics. For example, random forests showed the highest Precision among the three methods, while decision trees had the highest Recall. From the aggregator perspective, using the decision tree model for this dataset might lead to better results since the overall performance evaluation had higher results.

Table 9. Classification models evaluation metrics results.

Metric functions	Method	TR group < 3	3	> 3
Accuracy	Decision Trees	0.59	0.52	0.65
	Random Forest	0.58	0.51	0.65
	XGBoost	0.57	0.53	0.62
Precision	Decision Trees	0.51	0.37	0.49
	Random Forest	0.40	0.31	0.60
	XGBoost	0.40	0.34	0.55
Recall	Decision Trees	0.31	0.28	0.76
	Random Forest	0.47	0.28	0.55
	XGBoost	0.40	0.35	0.54
F1	Decision Trees	0.38	0.32	0.60
	Random Forest	0.43	0.29	0.57
	XGBoost	0.40	0.34	0.54

5 Conclusion

In conclusion, the energy sector is transforming decarbonization with a shift toward renewable-based technologies. DR programs provide a solution by enabling consumers to provide flexibility and avoid negative impacts on grid management. Still, their response can be uncertain. Therefore, the authors developed a trustworthy rate, considering previous performances, to select the optimal participants for the event considering the context. For the study in the present paper, the authors went a step further and wanted to create models for participants without any DR experience. Considering the features selected by the subgroup discovery and supervised clustering techniques, the authors were able to develop three different models and compare their performances. The results prove that availability is an important, crucial feature in defining the TR for a participant. Furthermore, the temperature felt at the event context and the building year are also considered for the rule definition. Regarding overall performance, Decision Trees and Random Forest had similar Accuracy, Recall, and F1 Score values, while XGBoost generally had lower scores across all metrics. However, decision Trees and Random Forest significantly improved Recall and F1 Score for TR superior to 3, while XGBoost performed more consistently across all groups.

Acknowledgements. This work has received funding from FEDER Funds through COMPETE program and from National Funds through (FCT) under the project UIDB/00760/2020. The work has also been done in the scope of projects UIDB/00760/2020, CEECIND/02887/2017, and SFRH/BD/144200/2019, financed by FEDER Funds through COMPETE program and from National Funds through (FCT).

References

1. Prabadevi, B., et al.: Deep learning for intelligent demand response and smart grids: a comprehensive survey, January 2021. https://doi.org/10.48550/arXiv.2101.08013
2. Silva, C., Faria, P., Vale, Z., Corchado, J.M.: Demand response performance and uncertainty: a systematic literature review. Energ. Strat. Rev. **41**, 100857 (2022). https://doi.org/10.1016/j.esr.2022.100857
3. Ilieva, I., Bremdal, B., Puranik, S.: Bringing business and societal impact together in an evolving energy sector. J. Clean Energy Technol. **7**(3), 42–48 (2019). https://doi.org/10.18178/JOCET.2019.7.3.508
4. Ahmed, N., Levorato, M., Li, G.P.: Residential consumer-centric demand side management. IEEE Trans. Smart Grid **9**(5), 4513–4524 (2018). https://doi.org/10.1109/TSG.2017.2661991
5. Khorram, M., Zheiry, M., Faria, P., Vale, Z.: Energy consumption management in buildings in the context of voluntary and mandatory demand response programs in smart grids. In: IEEE PES Innovative Smart Grid Technologies Conference Europe, vol. 2020-October, pp. 275–279, October 2020. https://doi.org/10.1109/ISGT-EUROPE47291.2020.9248750
6. Silva, C., Faria, P., Vale, Z.: Rating consumers participation in demand response programs according to previous events. Energy Rep. **6**, 195–200 (2020). https://doi.org/10.1016/j.egyr.2020.11.101
7. Silva, C., Faria, P., Vale, Z.: Classification of new active consumers performance according to previous events using decision trees. IFAC-PapersOnLine **55**(9), 297–302 (2022). https://doi.org/10.1016/j.ifacol.2022.07.052
8. Al-Harbi, S.H., Rayward-Smith, V.J.: Adapting k-means for supervised clustering. Appl. Intell. **24**(3), 219–226 (2006). https://doi.org/10.1007/s10489-006-8513-8
9. Eick, C.F., Zeidat, N., Zhao, Z.: Supervised clustering - algorithms and benefits. In: Proceedings - International Conference on Tools with Artificial Intelligence, ICTAI, pp. 774–776 (2004). https://doi.org/10.1109/ICTAI.2004.111
10. Atzmueller, M.: Subgroup discovery. Wiley Interdiscip. Rev. Data Min. Knowl. Discov. **5**(1), 35–49 (2015). https://doi.org/10.1002/widm.1144
11. Helal, S.: Subgroup discovery algorithms: a survey and empirical evaluation. J. Comput. Sci. Technol. **31**(3), 561–576 (2016). https://doi.org/10.1007/s11390-016-1647-1
12. Breiman, L., Friedman, J.H., Olshen, R.A., Stone, C.J.: Classification and Regression Trees. Routledge, Abingdon (2017). https://doi.org/10.1201/9781315139470
13. Breiman, L.: Random Forests. Mach. Learn. **45**(1), 5–32 (2001). https://doi.org/10.1023/A:1010933404324
14. Chen, T., Guestrin, C.: XGBoost: a scalable tree boosting system. In: Proceedings of the 22nd ACM SIGKDD International Conference on Knowledge Discovery and Data Mining, in KDD '16. Association for Computing Machinery, New York, NY, USA, 2016, pp. 785–794 (2016). https://doi.org/10.1145/2939672.2939785
15. De Estatística, I.N.: Censos 2021, Censos (2021). https://censos.ine.pt/xportal/xmain?xpgid=censos21_main&xpid=CENSOS21&xlang=pt
16. Atzmueller, M.: rsubgroup: Subgroup Discovery and Analytics (2021). https://cran.r-project.org/web/packages/rsubgroup/index.html
17. Dettling, M., Maechler, M.: supclust: Supervised Clustering of Predictor Variables Such as Genes (2021). https://cran.r-project.org/web/packages/supclust/index.html
18. Therneau, T., Atkinson, B., Ripley, B.: rpart - Recursive Partitioning and Regression Trees (2022). https://cran.r-project.org/web/packages/rpart/rpart.pdf
19. Liaw, A., Wiener, M.: Breiman and Cutler's Random Forests for Classification and Regression (2022). https://cran.r-project.org/web/packages/randomForest/randomForest.pdf
20. Chen, T., et al.: xgboost: Extreme Gradient Boosting (2023). https://cran.r-project.org/web/packages/xgboost/vignettes/xgboost.pdf

Federated Learning of Explainable Artificial Intelligence (FED-XAI): A Review

Raúl López-Blanco[1](✉), Ricardo S. Alonso[2,3], Angélica González-Arrieta[1], Pablo Chamoso[1], and Javier Prieto[1]

[1] BISITE Research Group, University of Salamanca, Edificio Multiusos I+D+i, Calle Espejo 2, 37007 Salamanca, Spain
{raullb,angelica,chamoso,javierp}@usal.es

[2] AIR Institute - Deep Tech Lab, IoT Digital Innovation Hub, Salamanca, Spain
ralonso@air-institute.com

[3] UNIR, International University of La Rioja, Av. de la Paz, 137, 26006 Logroño, La Rioja, Spain
ricardoserafin.alonso@unir.net

https://bisite.usal.es, https://air-institute.com, https://www.unir.net/

Abstract. The arrival of a new wave of popularity in the field of Artificial Intelligence has again highlighted that this is a complex field, with issues to be solved and many approaches involving ethical, moral and even other issues concerning privacy, security or copyright. Some of these issues are being addressed by new approaches to Artificial Intelligence towards explainable and/or trusted AI and new distributed learning architectures such as Federated Learning. Explainable AI provides transparency and understanding in decision-making processes, which is essential to establish trust and acceptance of AI systems in different sectors. Furthermore, Federated Learning enables collaborative training of AI models without compromising data privacy, facilitating cooperation and advancement in sensitive environments. Through this study we aim to conduct a review of a new approach called FED-XAI that brings together explainable AI and Federated Learning and that has emerged as a new integrative approach to AI recently. Thanks to this review, it is concluded that the FED-XAI is a field with recent experimental results and that it is booming thanks to European projects, which are championing the use of this approach.

Keywords: Federated Learning · Explainable Artificial Intelligence · FED-XAI

1 Introduction

Artificial Intelligence is consolidating as an important pillar of many sectors, as this technology is an important complement to many business actions such as automation [21], optimisation [12], prediction [16] and decision support [28].

S. Ossowski et al. (Eds.): DCAI 2023, LNNS 740, pp. 318–326, 2023.
https://doi.org/10.1007/978-3-031-38333-5_32

This is demonstrated by data from the Stanford Institute for Human-Centred Artificial Intelligence, which attest that, from 2017 to 2022, companies that intend to integrate AI into their processes have doubled, thanks in part to the testimony of those that have already done so and have been able to reduce costs and increase revenues [18].

Recently, the Government of Spain has already decided which will be the national headquarters of the Spanish Agency for the Supervision of Artificial Intelligence, in order to comply with its Digital Spain Agenda 2025 [10]. Countries and supranational organisations are legislating in this field because it's going to be very important to predict the future. In fact, the European Commission itself will invest EUR 1 billion a year in AI through programmes such as Digital Europe and Horizon Europe [31]. This is expected to reach 40 billion dollars in the next 20 years [29].

The explosion of this technology is not a coincidence, but has its roots in the progressive maturity of this technology over the decades since its emergence in the 1950s and in the union of some others that, combined, have allowed Artificial Intelligence to flourish. Some of these technologies are the Internet of Things (IoT) capable of producing large amounts of data [1], or Big Data [22], which provides the resources and capabilities needed to process this data.

However, it is not all good prospects for this technology, as a series of criticisms have increased and have reached the general media following the popularisation of Artificial Intelligence with tools such as ChatGPT, Dalle-2 or Midjourney, among others [33]. The major concerns in this field are related to the biases and ethical issues associated with the datasets, training processes and predictions generated by the models.

1.1 The Problems of Today's AI

A recurring problem in AI is the need to use huge amounts of data, which often involves the unification of data from multiple sources to feed Machine Learning models, thus raising privacy and security concerns. In addition, the opacity of AI decision-making processes has generated a problem that has prompted the development of new approaches, such as explainable AI [25] or trusted AI, which seek to improve the transparency and reliability of this technology.

Other approaches have gone in a different direction, seeking to solve the problems of training processes. Among these trends is Federated Learning (FL) [14], which consists of distributing the training processes of the models among the data-producing nodes to avoid the information having to leave them, which often raises privacy and security issues.

It follows from these issues that it is important to consider the implications of AI on society and how the associated ethical and moral issues arising from its use can be effectively addressed. In Europe, ethical guidelines are being developed to guide the future development of AI and ensure its trustworthiness [9].

1.2 New Priorities in Artificial Intelligence

The first of the critical points appears in the explainability of the models, both in the decisions they produce and in the training process. The use of AI models, being so cross-sectoral, can be applied to many processes, which makes the need for explainability crucial. Knowing how these models "think" aims to understand the decisions and output information from the input data, so that biases can be detected, both in the datasets and bias tendencies in training processes [30].

This is why we seek to provide explanability to each of the techniques used to train models. In general, models based on Machine Learning and neural networks complicate interpretability by containing hidden layers necessary in their operation and make it more complex than in other simpler models [30], such as decision trees or linear models.

To undo the complexity in these models, the field of explainability (XAI), has focused on the use of mathematical algorithms applied during training (ante-hoc) and inference (post-hoc) processes [26]. Moreover, explainability does not have a single point of view, but within this field there is global explainability, which seeks to understand how a model works in general, while local explainability focuses on understanding how a model works in a specific situation or in a particular decision.

On the other hand, there is another approach to the organisation and architectures of model training, Federated Learning [15]. This new training approach allows data producer nodes to become aggregators of partial models that through the sharing of parameters will create a joint model. These data-producing nodes can range from devices such as mobile phones to hospital complexes [23] where information must be treated equally and with the same level of privacy.

The distribution of learning between the different nodes means that the information they capture does not have to leave the environment where it was collected, thus avoiding problems related to privacy and data security. After the training phase, the nodes share, among themselves or with a central node, the parameters that characterise the local models they have generated (see Fig. 1).

Considering the impact of these new trends in the field of Artificial Intelligence, this article is proposed as a review, which is divided into the following sections; Sect. 2 which contains works and reviews related to this field of explainability and federated learning. The next section, Sect. 3, presents the review carried out in this work. Section 4 presents the results and conclusions of the work.

2 Related Works

The concept of explainability is not something new in the field of artificial intelligence, as since the 1990s there have already been studies that provided defined rules on which parameters influence the making of a [20] decision. The different types of models (interpretable, black-box and hybrid) and ways of explaining models (pre-modelling and post-modelling) require constant revisions in order not to lose the current status of this technology.

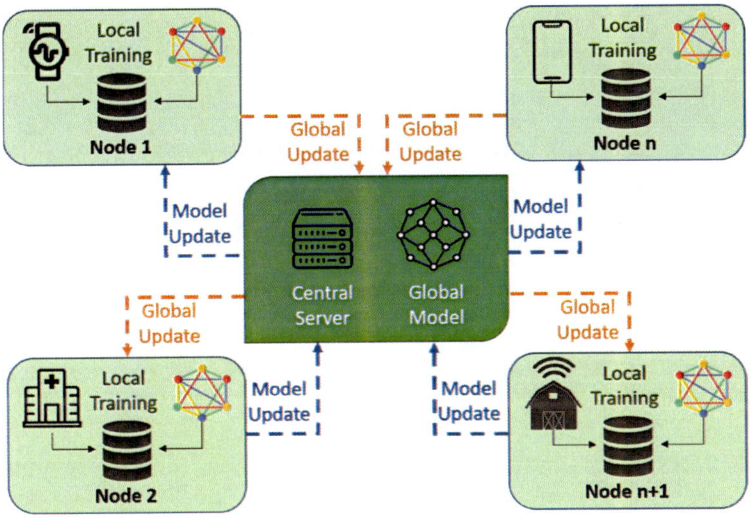

Fig. 1. An example of a distributed learning architecture based on federated learning.

The first of the reviews presented in this sense is the one carried out by [20], which presents different ways of approaching interpretability in models. It lists and explains the processes both prior to the model (centered on the data to be used in the training process), as well as those models that are inherently explainable, and explainability on models already trained, both with methods agnostic to the model and specific to that model.

Another paper with similar reviews focused on the XAI field is Plamen et al. [2] that discusses the challenges of explainability and published methods to explain the predictions of some of the models, some of these techniques are sensitivity analysis, propagation and attribution of feature relevance by layers, local pseudo-explanations by LIME, Shapley additive explanations and gradient-based localization.

The work by Chuang et al. [8] performs a taxonomic review of existing XAI techniques in the literature and examines their limitations and challenges in the field. All these reviews, while dealing in great detail with the field of XAI, do not mention one of the emerging trends in the field of AI such as federated learning.

Because of this, additionally, reviews focusing on the field of federated learning are reviewed. This approach to distributed machine learning allows training AI models using data located on multiple devices without the need to share it centrally. This preserves user privacy and reduces security risks in data processing [19]. Its rapid development in recent years, which can be seen in Fig. 2, is due to the growing concern for data privacy in various fields of application, such as medicine, social networking and mobility [34].

Moreover, federated learning is also efficient in terms of computational resources and communication, especially in scenarios with limited network connections [15]. Among the review studies are some such as Li et al. [14] that review applications of federated learning and some of its most current trends. Other studies, such as the one by Mammen et al. [17] review the opportunities and challenges of this new field within AI.

Advances in recent years in this type of technology bring papers such as the one by Bonawitz et al. [7] that introduced the Federated Averaging (FedAvg) algorithm, which combines local stochastic gradient descent and model averaging, and has been widely adopted in the literature as a benchmark. More efficient communication approaches, such as gradient compression [27] and quantization [13], have been proposed to reduce communication time and overhead in FL.

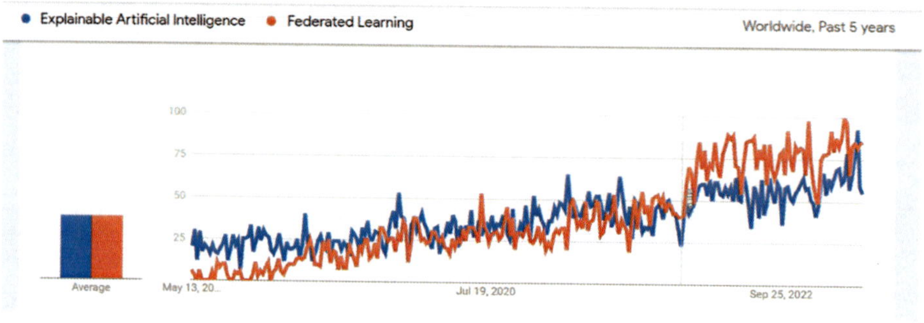

Fig. 2. Search Trends Federated Learning vs Explainable Artificial Intelligence. Source: Google Trends.

From these two trends, a new discipline, FED-XAI, has emerged in the last two years, which aims to bring together these two approaches into one. Building on the reviews seen individually, in the fields of federated learning and XAI, and considering the new approach called FED-XAI, the current review is proposed focusing on articles dealing with this new approach. This review will serve to determine the current status of the field and the articles published in it.

3 Review of Federated Learning and Trustworthy Artificial Intelligence

After reviewing articles on the current state of explainability in AI (XAI) and the current state of the art, there are still few papers on the new trend, but the main contributions have appeared in the years 2022 and 2023, so it is expected that the number of papers related to FED-XAI will continue to grow.

After searching for articles with the text "FED-XAI" in their body, among the main databases and scientific repositories and after eliminating duplicates, 7 noteworthy articles have been compiled, of which the work by Uusitalo et al. [32] stands out. This is one of the first mentions made alluding to this acronym. This article presents the theoretical framework in the Hexa-X project which is based on agents training a global XAI model.

Referring to the same project, the article by Filippou et al. [11] focuses on describing the possibilities of applying AI and ML mechanisms in 6G networks and mentions the term FED-XAI. Explaining the macromodules in the same project is the contribution developed by Bechini et al. [6].

Another series of papers, which are framed as a result of the Hexa-X project, make more comprehensive reviews of how the field is doing and introduce the basic concepts of FL, XAI and FED-XAI, and make a brief survey of interesting works using these concepts [3]. Another paper proposes a review of explainable algorithms, such as decision trees, for predicting QoS in B5G applications [5].

Finally, the most complete works among those reviewed are those by Bácerna et al. [4] and Renda et al. [24]. The former focuses on maintaining explainability in fuzzy regression models while preserving data privacy with federated architectures. The second builds on previous reviews [3,5], to provide insight into 6G networks and their applications in Vehicle to Vehicle (V2V) technology (Table 1).

Table 1. Related works reviewed.

Article	Contributions	Limitations
[3]	Brief but thorough review of the current status of FED-XAI also discusses current limitations.	Some kind of performance indicator of developments in the field could be provided.
[24]	Relationship between 6G networks and V2V technology	It does not provide real applications, nor test the impact of the technology on driving.
[32]	Framework developed for federated learning of eXplainable AI models	No real tests of the performance of the framework or its operation are yet available.
[5]	Provides a large dataset to work with and provides results and lab tests.	The work has been done with an inherently explainable model, the explainability with black box models will have to be seen.
[6]	Mention of the term FED-XAI in the framework of the Hexa-X project.	No contribution to the field
[4]	Explainability framework for backward fuzzy algorithms implemented on a federated architecture.	Dynamic hyperparameter tuning
[11]	Mention of the term FED-XAI as a future feature of B5G and 6G networks	No contribution to the field

4 Conclusions and Future Work Lines

The conclusion of this review of articles is that the Hexa-X project, the flagship of the 6G technology research being carried out by the European Union, mentions the term FED-XAI in one of its demonstrators, which is important for the project and for the future of this technology. In addition, this review finds that the main results in this field come from the research carried out by the Hexa-X project.

Another of the conclusions drawn is that the main problems currently facing the FED-XAI field are related to: achieving results comparable to those achieved by a centralized approach with all the data available in the same silo; finding a balance between the explainability and accuracy of the results, which as can be seen in Fig. 3, the higher the accuracy the lower the explainability; and achieving explainability in black box models containing hidden layers.

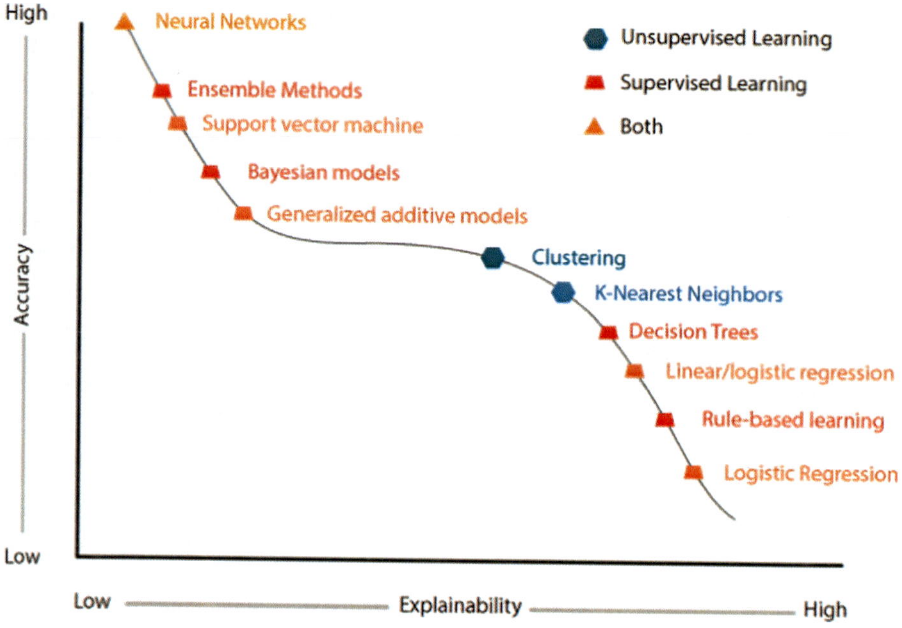

Fig. 3. Accuracy versus explainability in Artificial Intelligence models.

With regard to the lines of future work, as these fields are incipient developments with a great deal of projection within the field of Artificial Intelligence, the following lines are proposed for future work:

- To carry out a Mapping Review that analyses the most relevant articles on explainable Artificial Intelligence and federated learning, in order to know the situation of both fields separately and not only the works in which they complement each other.
- To make a proposal for an architecture that combines the best capabilities of federated learning in terms of security and data privacy without forgetting the explainability of the models.
- Reducing heterogeneity in the Internet of Things field with federated architectures that are device agnostic and adaptive based on the data provided by the device and based on the number of devices in the network.
- To investigate how explainability capabilities evolve in models as they train partial models prior to their aggregation in a general model and the explainability in that joint model.

Acknowledgements. This work has been partially supported by the project TED2021-132339B-C43 (idrECO), funded by MCIN/AEI/10.13039/501100011033 and by the European Union "NextGenerationEU"/PRTR.

References

1. Adli, H.K., et al.: Recent advancements and challenges of AIoT application in smart agriculture: a review. Sensors (Basel) **23**(7) (2023)
2. Angelov, P.P., Soares, E.A., Jiang, R., Arnold, N.I., Atkinson, P.M.: Explainable artificial intelligence: an analytical review. Wiley Interdisc. Rev.: Data Min. Knowl. Discov. **11**(5), e1424 (2021)
3. Bárcena, J.L.C., et al.: Fed-XAI: federated learning of explainable artificial intelligence models (2022)
4. Bárcena, J.L.C., Ducange, P., Ercolani, A., Marcelloni, F., Renda, A.: An approach to federated learning of explainable fuzzy regression models. In: 2022 IEEE International Conference on Fuzzy Systems (FUZZ-IEEE), pp. 1–8. IEEE (2022)
5. Bárcena, J.L.C., et al.: Towards trustworthy AI for QoE prediction in b5g/6g networks (2022)
6. Bechini, A., Bondielli, A., Ducange, P., Marcelloni, F., Renda, A.: Responsible artificial intelligence as a driver of innovation in society and industry
7. Bonawitz, K., et al.: Towards federated learning at scale: system design. Proc. Mach. Learn. Syst. **1**, 374–388 (2019)
8. Chuang, Y.N., et al.: Efficient XAI techniques: a taxonomic survey. arXiv preprint arXiv:2302.03225 (2023)
9. European Commission, Directorate-General for Communications Networks Content and Technology: Ethics guidelines for trustworthy AI. Publications Office (2019). https://doi.org/10.2759/346720
10. de España, G.: Spanish digital agenda 2025 (2020). https://www.lamoncloa.gob.es/presidente/actividades/Documents/2020/230720-Espa%C3%B1aDigital_2025.pdf
11. Filippou, M.C., et al.: Pervasive artificial intelligence in next generation wireless: the Hexa-X project perspective. In: CEUR Workshop Proceedings, vol. 3189 (2022)
12. González-Briones, A., Chamoso, P., De La Prieta, F., Demazeau, Y., Corchado, J.M.: Agreement technologies for energy optimization at home. Sensors **18**(5), 1633 (2018)
13. Konečný, J., McMahan, H.B., Ramage, D., Richtárik, P.: Federated optimization: distributed machine learning for on-device intelligence. arXiv preprint arXiv:1610.02527 (2016)
14. Li, L., Fan, Y., Tse, M., Lin, K.Y.: A review of applications in federated learning. Comput. Ind. Eng. **149**, 106854 (2020)
15. Li, T., Sahu, A.K., Talwalkar, A., Smith, V.: Federated learning: challenges, methods, and future directions. IEEE Signal Process. Mag. **37**(3), 50–60 (2020)
16. López-Blanco, R., Martín, J.H., Alonso, R.S., Prieto, J.: Time series forecasting for improving quality of life and ecosystem services in smart cities. In: Julián, V., Carneiro, J., Alonso, R.S., Chamoso, P., Novais, P. (eds.) ISAmI 2022. LNNS, vol. 603, pp. 74–85. Springer, Cham (2023). https://doi.org/10.1007/978-3-031-22356-3_8
17. Mammen, P.M.: Federated learning: opportunities and challenges. arXiv preprint arXiv:2101.05428 (2021)

18. Maslej, N., et al.: The AI index 2023 annual report (2023). https://aiindex.stanford.edu/wp-content/uploads/2023/04/HAI_AI-Index-Report_2023.pdf
19. McMahan, B., Moore, E., Ramage, D., Hampson, S., Arcas, B.A.: Communication-efficient learning of deep networks from decentralized data. In: Artificial Intelligence and Statistics, pp. 1273–1282. PMLR (2017)
20. Minh, D., Wang, H.X., Li, Y.F., Nguyen, T.N.: Explainable artificial intelligence: a comprehensive review. Artif. Intell. Rev. **55**, 3503–3568 (2021). https://doi.org/10.1007/s10462-021-10088-y
21. Patel, K., Bhatt, C., Corchado, J.M.: Automatic detection of oil spills from SAR images using deep learning. In: Julián, V., Carneiro, J., Alonso, R.S., Chamoso, P., Novais, P. (eds.) ISAmI 2022. LNNS, vol. 603, pp. 54–64. Springer, Cham (2023). https://doi.org/10.1007/978-3-031-22356-3_6
22. Plaza-Hernández, M., Gil-González, A.B., Rodríguez-González, S., Prieto-Tejedor, J., Corchado-Rodríguez, J.M.: Integration of IoT technologies in the maritime industry. In: Rodríguez González, S., et al. (eds.) DCAI 2020. AISC, vol. 1242, pp. 107–115. Springer, Cham (2021). https://doi.org/10.1007/978-3-030-53829-3_10
23. Reisizadeh, A., Tziotis, I., Hassani, H., Mokhtari, A., Pedarsani, R.: Straggler-resilient federated learning: leveraging the interplay between statistical accuracy and system heterogeneity. IEEE J. Sel. Areas Inf. Theory **3**(2), 197–205 (2022)
24. Renda, A., et al.: Federated learning of explainable AI models in 6g systems: towards secure and automated vehicle networking. Information **13**(8), 395 (2022)
25. Rosa, L., Silva, F., Analide, C.: Explainable artificial intelligence on smart human mobility: a comparative study approach. In: Machado, J.M., et al. (eds.) DCAI 2022. LNNS, vol. 585, pp. 93–103. Springer, Cham (2023). https://doi.org/10.1007/978-3-031-23210-7_9
26. Sarkar, A., Vijaykeerthy, D., Sarkar, A., Balasubramanian, V.N.: A framework for learning ante-hoc explainable models via concepts. In: Proceedings of the IEEE/CVF Conference on Computer Vision and Pattern Recognition, pp. 10286–10295 (2022)
27. Sattler, F., Wiedemann, S., Müller, K.R., Samek, W.: Robust and communication-efficient federated learning from non-IID data. IEEE Trans. Neural Netw. Learn. Syst. **31**(9), 3400–3413 (2019)
28. Son, T.H., Weedon, Z., Yigitcanlar, T., Sanchez, T., Corchado, J.M., Mehmood, R.: Algorithmic urban planning for smart and sustainable development: systematic review of the literature. Sustain. Cities Soc. **94**(104562), 104562 (2023)
29. Straus, J.: Artificial intelligence-challenges and chances for Europe. Eur. Rev. **29**(1), 142–158 (2021)
30. Tommasi, T., Patricia, N., Caputo, B., Tuytelaars, T.: A deeper look at dataset bias. In: Csurka, G. (ed.) Domain Adaptation in Computer Vision Applications. ACVPR, pp. 37–55. Springer, Cham (2017). https://doi.org/10.1007/978-3-319-58347-1_2
31. European Union: A European approach to artificial intelligence (2023). https://digital-strategy.ec.europa.eu/en/policies/european-approach-artificial-intelligence
32. Uusitalo, M.A., et al.: Hexa-X the European 6g flagship project. In: 2021 Joint European Conference on Networks and Communications & 6G Summit (EuCNC/6G Summit), pp. 580–585. IEEE (2021)
33. Venkatesh, V.: Adoption and use of AI tools: a research agenda grounded in UTAUT. Ann. Oper. Res. **308**(1), 641–652 (2021). https://doi.org/10.1007/s10479-020-03918-9
34. Yang, Q., Liu, Y., Chen, T., Tong, Y.: Federated machine learning: concept and applications. ACM Trans. Intell. Syst. Technol. (TIST) **10**(2), 1–19 (2019)

Introduction to the Extended Reality as a Teaching Resource in Higher Education

Juan-Carlos de la Cruz-Campos[1], Magdalena Ramos-Navas-Parejo[1], Fernando Lara-Lara[2](✉), and Blanca Berral Ortiz[2]

[1] Department of Didactics and School Organization, Faculty of Education and Sport Sciences of Melilla, University of Granada, Campus Universitario de Melilla, 52071 Melilla, Spain
{juancarlosdelacruz,magdalena}@ugr.es

[2] Department of Didactics and School Organization, Faculty of Education Sciences, University of Granada, Campus Universitario de Cartuja, 18071 Granada, Spain
{fernandolara,blancaberral}@ugr.es

Abstract. Traditional lectures, in which learners play a passive role, have not lived up to the expectations of students and the educational demands of the knowledge society in which we live. Technology has gradually been included in all facets of life, to the point where it has become the core of life. It has been found that this trend has also had a significant impact on the education sector in recent years. Technologies such as virtual and augmented reality offer a series of benefits and potentialities for teaching practice at any educational stage and in any subject. The following paper aims to provide an overview of the benefits of using these methodologies in the classroom, including ensuring that learning is meaningful and contextualised, as it facilitates the understanding of abstract content and enriches educational mathematics. It helps to prevent risks, overcome spatial and temporal obstacles and eliminate barriers to learning. Moreover, they are motivating and innovative resources for students, increasing their interest in learning. For this reason, their implementation in university classrooms should be advocated, regardless of the discipline, as they represent a significant improvement in the quality of teaching.

Keywords: Augmented Reality · Virtual Reality · Higher Education e-learning

1 Introduction

E-learning uses electronic technologies to deliver educational and training content. There are different types of e-learning formats, including online courses, virtual classrooms, webinars and multimedia presentations [1].

This work is linked to the Advanced Teaching Innovation Project entitled "Aula invertida y recursos tecnológicos inmersivos (xr) para el desarrollo de la competencia digital docente en los futuros profesionales de la educación" (Inmer). Reference: 22-06. FIDO Plan. Unidad de Calidad, Innovación Docente y Prospectiva de la Universidad de Granada.

S. Ossowski et al. (Eds.): DCAI 2023, LNNS 740, pp. 327–335, 2023.
https://doi.org/10.1007/978-3-031-38333-5_33

E-learning allows students to access educational materials at any time and from anywhere, following their own learning pace, which makes it a flexible and easy way of acquiring knowledge. It is possible to learn from this methodology by means of a wide variety of electronic devices, such as computers, tablets and smartphones. These can contain multimedia, interactive elements such as videos, animations and simulations [2].

E-learning has multiple uses, which vary according to the wide range of educational and training purposes, such as corporate training, professional development, academic study and personal growth. For people who are short on time, have no possibility to attend face-to-face classes or need to study at a distance, e- learning is very useful [1]. It also allows for individualised learning experiences, as learners can choose the content and pace that best suits them.

The term "augmented reality" (AR) refers to a technology that enriches the real-world environment by adding digital information in real time, such as images, videos and 3D models. Virtual reality (VR), on the other hand, generates a completely fictional environment. AR requires a camera or other devices to capture the participant's environment, then, through an information processing application, digital content is superimposed on the real-world view. The user can manipulate and interact with virtual objects as if they were real thanks to interactive digital content that reacts to their actions [3].

The technology called AR, therefore, adds digital information such as images, videos and 3D models to the real world, allowing participants to interact with digital elements within the physical world, enriching the perception and experience of reality. Applications of AR include gaming, entertainment, education, marketing and healthcare [4]. For example, in education, AR can be applied to design contextualised learning environments, through engaging learning experiences such as virtual field trips and simulations. In healthcare, AR can be very useful for training, surgery and patient care.

With the help of VR technology, users can experience real-time immersion within a three-dimensional digital environment. In contrast to AR, which superimposes digital content on top of the real world, VR generates a completely different virtual world where participants can explore and interact. VR makes use of specialised hardware and software to simulate the physical inclusion of users within the fictional world [5].

The equipment generally consists of virtual reality headsets, motion sensors and controllers that allow users to see, hear and feel the virtual world while interacting with it. VR has a wide range of uses, including gaming, entertainment, education and industrial training [6]. For example, VR is used in the entertainment industry to provide immersive and realistic gaming experiences. In the education sector, it is used to generate immersive learning experiences. It is also notable for its use in training professionals, such as pilots and doctors, to train and develop skills in a safe and controlled environment.

In Higher Education, e-learning and VR are increasingly being put into practice, due to the numerous possibilities they offer for improving the quality of education. They enable interaction and personalisation of the teaching and learning process in a way that is attractive to students. E-learning offers online courses that enable ubiquitous access to educational content, using technological devices such as computers, tablets and smartphones, which is very practical at the Higher Education stage [7]. These online courses are based on the use of tools such as videos, podcasts, online readings, discussion forums, educational games and online quizzes. They allow teachers to interact with students through online chats, e-mails and video conferences, which facilitates synchronous help and guidance.

VR can be used to generate virtual learning environments that can be used to explore and experiment with different contextualised scenarios, leading Higher Education students to develop their skills while avoiding risks and maintaining control of the situation at all times [5]. Examples include simulations of realistic scenarios, such as exploring archaeological sites, performing surgery or simulating natural disasters. Through these simulations, students can hone their skills and acquire knowledge, achieving meaningful learning. Within these virtual learning environments, students can interact with fictitious elements, carry out different activities and experiments, obtain immediate feedback and, in short, receive autonomous and practical learning.

Therefore, immersive technologies are booming in Higher Education, due to their potential to improve the quality of learning and the student experience. These technologies offer a number of advantages, such as the personalisation of learning, the possibility of interaction and practice. These are fundamental aspects for the effective development of competences and skills that could not be acquired through traditional lectures.

2 E-learning and Augmented Reality

Augmented reality (AR) is a technology that combines elements of the real world with digital elements, enabling user interaction with an enriched experience. In augmented reality, graphics, images or virtual information are superimposed on the physical environment to provide a mixed experience in real time.

The use of AR in e-learning brings a number of possibilities to the education sector. These tools make it possible to increase online training opportunities, complementing face-to-face education. AR brought to education offers openness to innovation, through the experimentation of other realities, making possible the assimilation of theoretical contents through experimentation and simulation of contextualised scenarios. It also offers the opportunity to follow a simultaneous continuous assessment [8].

This technology combines virtual and real information at the same time. Generally, it can be accessed from mobile devices, which all higher education students have, thus facilitating the implementation of AR in the classroom, as it does not require specific devices that are expensive and difficult to acquire [9–12]. AR has the following characteristics: it is used in real time, it is perceived in 3 dimensions, which offers more realism in the immersive scenario, it is made up of computer-created content and images superimposed on the field of vision of the participants in the experience [13].

AR has a number of different levels, which vary based on the technological devices used, the level of simplicity or complexity, the parameters and the techniques employed [14, 15]:

- Level 0: This refers to the most basic level of AR, such as image-based information links, like QR codes.
- Level 1: these are virtual images generated in 3D. This level is the most commonly used in education.

- Level 2: Everything related to image recognition or geolocation is considered. It includes parameters such as: orientation, location and tilt of the device.
- Level 3: This level requires the use of special displays to obtain augmented vision, which adds virtual information to the physical or real world.
- Level 4: relates to global positioning, such as GPS.
- Level 5: refers to thermal fingerprints and contact lenses.

AR offers a series of benefits to the teaching work and in the development of students' fundamental competences. These are relevant elements that favour the teaching and learning process in all educational areas and are in line with the demands of students and the active didactic methodologies recommended by the education system as those that effectively develop students' competencies [16, 17] (Table 1).

Table 1. Advantages that Augmented Reality offers to education

Benefits of AR in education	
Advantages in educational practice	- It shows only the most relevant information, eliminating information that may hinder the smooth running of the learning process
	- It enriches the real contents, facilitating their comprehension, especially in the case of the assimilation of abstract concepts
	- It allows students to observe the element of study from any 3-dimensional perspective
	- It adapts to any stage of education
	- Enables ubiquitous learning
	- Offers hands-on, experiential learning
	- Creates motivating learning environments for the learner
	- Enablesexperiencewithinsafeandcontrolled contextualised scenarios
	- Adds extra information displayed in different formats to physical materials
	- It is a very versatile tool that can be used at all educational stages, for learning in all subject areas and disciplines
Aspects of AR that enhance skills development	- Active practice in learning
	- Immersion and presence in the information
	- Placing learning in its real context
	- Verification of information
	- Socialisation of learners

Own elaboration based on Akçayır [16], Barroso and Gallego [17], Madanipour and Cohrs- sen [18] and Gómez-García [19].

Looking at these advantages, it is clear that AR offers a wide range of options for the teaching and learning process [14, 2.]. On the other hand, it also has limitations due to the lack of training in the use of AR in the classroom. The inclusion of AR is still uncharted territory in the education sector. It can be seen that these limitations are exclusively due to lack of experience and contact with AR, which can be overcome through practice (Table 15).

Table 2. Limitations of AR in the educational environment

Limitations of AR in education	- Lack of teacher training
	- Limited number of educational practices and learning objects available
	- There is a need for more intensive theoretical reflection
	- Not enough substantiation of educational models available
	- Students do not have the necessary experience to interact correctly with the elements of AR
	- There are scenarios that are difficult to use

Own elaboration based on Cabero et al. [20].

The use of AR together with mobile technology is booming. Through these resources, applications can be generated, benefiting from the mobility and the possibilities of using information immediately, which these tools allow [21]. In online training, AR offers very useful specific benefits such as [8, 22]:

– In medical education, it offers the opportunity to explore organs in 3D, to observe the functioning of the human body, to analyse the effects of medication or to perform different types of surgery [23].

It develops cognitive, spatial-perceptual-motor and temporal skills.

It improves attention, concentration, short- and long-term memory in visual, auditory and reasoning forms.

– It allows for the improvement of the effectiveness of education, based on the active work of students. Since AR interacts actively and consciously in different scenarios, providing confirmation, refutation or increase of knowledge, generating new ideas, feelings or opinions.
– AR can influence pedagogy and educational didactics, facilitating reflective activities, explaining observable phenomena and offering effective solutions to different concrete problems.
– It provides an ideal learning environment, in any academic area, because of the possibilities of communication at work and the complete knowledge of the elements of learning.
– It increases students' positive attitude towards their learning, developing autonomy and entrepreneurship [24].

3 Higher Education, E-Learning and Virtual Reality

Virtual reality (VR) is a technology that creates a fully digital and immersive environment that simulates reality. VR users can experience and explore this simulated environment through devices such as VR headsets and controllers, and have the feeling of being completely immersed in a computer-generated environment.

Distance learning and virtual reality are two evolving trends that are changing and challenging the ways of teaching in higher education. Both offer interesting advantages and challenges, and when combined, can offer a uniquely effective and engaging educational experience. Distance learning is a teaching model in which students can access course content through an online platform, without the need to attend face-to-face classes at an educational institution. This model has experienced exponential growth in recent years, driven by the evolution of digital technologies and the needs of modern learners who are increasingly looking for flexible and personalised education. In particular, this momentum has been driven by the need to promote the safety of educational actors during the Covid-19 pandemic [24], as well as the need to reduce costs and risks for the delivery of essential internships in various university courses [26].

Distance education offers a wide range of advantages, such as flexible scheduling, the ability to study from anywhere, accessibility to educational materials, reduced transport and accommodation costs, and the ability to adjust the pace of learning according to the individual needs of the student. However, distance learning also presents some challenges, such as the lack of personal interaction between teachers and students, the lack of social and emotional support, and the difficulty in maintaining student motivation and commitment over time.

This is where virtual reality (VR) comes in, offering an innovative and effective way to overcome these challenges [25]. VR is a technology that creates an immersive virtual environment that simulates reality and allows users to interact in this environment in real time. In education, VR is used to create immersive and realistic learning experiences that can help students better understand concepts, and be able to apply them in practical situations [26–28].

In this way, the combination of distance education and virtual reality can offer a unique and effective educational experience. For example, students can access educational materials through an online platform and then use VR to interact with the concepts in a hands-on, immersive way. VR also allows students to connect with other students and teachers more effectively, which can increase their motivation and engagement in learning. It currently supports experiential learning in disciplines such as medicine, geography, construction, nursing, physical education or languages [28]. In addition, VR can also be used to overcome geographical and cultural barriers that often prevent students from accessing higher education. For example, students living in rural or remote areas can use VR to access education without having to move from their city and country. Another advantage of VR is that it can help students develop practical skills in a safe and controlled environment. For example, medical students can use VR to practice complex surgical procedures without putting patients' lives at risk. A good example is the "Valdeci-lla Virtual Hospital" which offers a more complex view on the use of VR than is limited to video games. Its importance in improving student performance has been contractually demonstrated. Along the same lines, significant academic improvements have been perceived in training experiences on the care of traffic emergencies and laboratory accidents [26].

This educational technology will be able to fulfil its function as long as it is planned and implemented from its didactic purpose. It requires that its presence and development be coherent with the curriculum, respectful of the learning outcomes and goals that are aspired to; and finally, it needs a space dedicated to evaluation and feedback actions.

Finally, it allows the democratisation of knowledge and brings scientific culture closer to society. The accessibility that is now possible through the different possibilities offered by this technology can encourage more and better scientific vocations, and encourage traditionally excluded groups to be more motivated towards the formal and non-formal education offered by university institutions [25, 26].

4 Conclusions

We are immersed in the knowledge society, which is characterised by technology as a key resource for the proper functioning of any area of citizens' lives. In the education sector, technological inclusion has occurred in stages, allowing for the evolution of teaching and learning strategies. Traditional lecture-based teaching methods, in which students play a passive role, have been abandoned. It has been proven that these forms of teaching do not meet the demands of the students, nor do they offer an adequate response to current educational needs. For this reason, the implementation of technology in education is on the increase.

Among the technologies that have been included in the education sector, e-learning and immersive technologies, such as augmented reality and virtual reality, stand out. These tools offer a series of benefits and potentialities in educational work at any educational stage and in any discipline. Immersive technologies allow students to experience situations or learning scenarios that would otherwise be difficult or even impossible to interact with. For example, they can simulate experiences in virtual laboratories or virtual trips to historical places or museums.

The incorporation of these technologies in higher education from any discipline profoundly improves the quality of education. These technologies allow learning to be meaningful and contextualised with reality, favour the correct assimilation of abstract content, improve teaching materials, prevent risks, overcome spatial and temporal obstacles and capture the interest of students through the possibilities of motivation and innovation that they offer.

However, the implementation of these resources in university education also poses limitations and challenges that must be addressed. One of the fundamental challenges is to ensure the quality of the information and educational materials presented through these technological tools. It is key that content is provided in a way that is accurate, up-to-date and relevant to the educational subject being taught.

Another key challenge is to ensure effective education and training of teachers who teach classes using these technological media. It is crucial that teachers have the necessary education and training to make appropriate use of these technologies, with the intention of designing and developing quality educational materials that are adapted to the particular needs and characteristics of the students.

Finally, another key challenge is to ensure accessibility and equality in the use of technology by all students. Taking the necessary measures to avoid the so-called digital divide, in order to guarantee equal opportunities in the access and use of these resources, which are so important for students' education, regardless of their socio- economic situation.

In conclusion, it can be affirmed that the incorporation of technologies in Higher Education enables a considerable improvement in the quality of education. However, the challenges and limitations that arise in the inclusion of these resources must be faced in order to ensure their effectiveness and guarantee quality and equity in access and use by all students without exception. In short, the incorporation of technology in Higher Education is a very effective resource for improving the education and training of students who live in a technological and digitalised society.

References

1. Firwana, A., Abu, M., Aqel, M.: Effectiveness of e-learning environments in developing skills for designing E-tivities based on gamification for teachers of technology in Gaza. Educ. Knowl. Soc. EKS **22**, 15–21 (2021)
2. Segovia-García, N., Said-Hung, E.: Factores de satisfacción de los alumnos en e-learning en Colombia. Rev. Mex. Investig. Educ. **26**(89), 595–621 (2021)
3. Tallón, S., Berral, B., Campos, M. N., Martínez, J. A.: La realidad aumentada y sus efectos en el aprendizaje de los escolares. En J. A. Marín-Marín, A. Boffo, M. Ramos-Navas- Parejo y J. C. De La Cruz-Campos (Eds.), Retos de la investigación y la innovación en la sociedad del conocimiento, pp.45–52. Dykinson (2022)
4. Carballo, L., Fernández, Y. F.: La Realidad Aumentada en el enfrentamiento a la COVID-Serie Científica de la Universidad de las Ciencias Informáticas, **13**(11), 1–16 (2020)
5. Chirinos, Y.: La realidad virtual como mediadora de aprendizajes: desarrollo de una aplicación móvil de realidad virtual orientada a niños. Revista Iberoamericana de Tecnología en educación y Educación en Tecnología **27**, 98–99 (2020)
6. Londoño, L.M., Rojas, M.D.: De los juegos a la gamificación: propuesta de un modelo integrado. Educación y educadores **23**(3), 493–512 (2020)
7. Oliveros-Castro, S., Núñez-Chaufleur, C.: Posibilidades educativas de la realidad virtual y la realidad combinada: una mirada desde el conectivismo y la bibliotecología. Revista Saberes Educativos **5**, 46–62 (2020)
8. Korowajczenko, K. T.: Realidad Aumentada Sus Desafíos Y Aplicaciones Para El e- learning. No. June (2012)
9. Cabero, J., Barroso, J, Llorente, C.: La realidad aumentada en la enseñanza universitaria. REDU. Revista de docencia universitaria, **17**(1), 105–118 (2019)
10. Seifert, T., Hervás, C. y Toledo, P.: Diseño y validación del cuestionario sobre percepciones y actitudes hacia el aprendizaje por dispositivos móviles. PixelBit. Revista de Medios y Educación, **54**, 45–64 (2019)
11. Aznar, I., Cáceres, M. P., Trujillo, J. M, Romero, J. M.: Impacto de las apps móviles en la actividad física: Un meta-análisis. Retos: nuevas tendencias en educación física, deporte y recreación, **36**, 52–57 (2019)
12. Cabero, J, Barroso, J.: Los escenarios tecnológicos en Realidad Aumentada (RA): posibilidades educativas. Aula Abierta, **47**(3), 327–336 (2018)
13. Ferguson, R., et al.: Open University Innovation Report 6, p. 2017. The Open University, UK, Milton Keynes (2017)

14. Cabero, J., Llorente, M.C., Marín, V.: Comunidades virtuales de aprendizaje. El Caso del proyecto de realidad aumentada: RAFODIUM. Perspectiva Educacional. Formación de Profesores, **56**(2), 117–138 (2017)
15. Marín, V., Muñoz, V. P.: Trabajar el cuerpo humano con realidad aumentada en educación infantil. Revista Tecnología, Ciencia y Educación, **9**, 148–158 (2018)
16. Akçayır, M., Akçayır, G.: Advantages and challenges associated with augmented reality for education: a systematic review of the literature. Educ. Res. Rev. **20**, 1–11 (2017)
17. Barroso, J., Gallego, O.: Producción de recursos de aprendizaje apoyados en Reali-dad Aumentada por parte de los estudiantes de magisterio. Edmetic. Revista de Edu-cación Mediática y TIC, **6**, 1, 23–38 (2017)
18. Madanipour, P., Cohrssen, C.: Augmented reality as a form of digital technology in early childhood education. Australas. J. Early Childhood **45**(1), 5–13 (2020)
19. Gómez-García, G., Rodríguez-Jiménez, C., Ramos, M.: Virtual reality in physical education area. J. Sport Health Res. **11**, 177–186 (2019)
20. Fabregat-Gesa, R.: Combinando la realidad aumentada con las plataformas de e-elearning adaptativas. Enl@ce: Revista Venezolana de Información, tecnología y conocimiento, **9**(2), 69–78 (2012)
21. Almenara, JC., Osuna, JB.: Posibilidades educativas de la Realidad Aumentada. Revista de nuevos enfoques en investigación educativa (NAER J.), **5**(1), 44–50 (2016)
22. Vidal Ledo, M., Lío Alonso, B., Santiago Garrido, A., Muñoz Hernández, A., Morales Suárez, IDR, Toledo Fernández, AM.: Realidad aumentada. Educación Médica Superior, **31** (2) (2017)
23. Cabero, J., Barroso, J., Gallego, O.: La producción de objetos de aprendizaje en realidad aumentada por los estudiantes. Los estudiantes como prosumidores de información. Tecnología, Ciencia y Educación, **11**, 15–46 (2018)
24. Valderrama-Zapata, C. Garcia-Cavero, R. G., Montehermozo, T. S., Ocejo, P. R., Sánchez, C. G.: La tecnología de la realidad virtual como herramienta educativa en el marco de la pandemia de COVID 19: Una revisión sistemática. Revista Ibérica de Sistemas e Tecnologias de Informação (E53), 233–243 (2022)
25. Calderón, S.J., Tumino, M.C., Bournissen, J.M.: Realidad virtual: impacto en el aprendizaje percibido de estudiantes de Ciencias de la Salud. Revista Tecnología, Ciencia y Educación **16**, 65–82 (2020)
26. Figueras- Ferrer, E.: Reflexiones en torno a la cultura digital contemporánea. Retos futuros en educación superior. Arte, Individuo y Sociedad **33**(2),449–466 (2021)
27. Díaz-López, L., Tarango, J., Romo-González, J.: Realidad Virtual en procesos de aprendizaje en estudiantes universitarios: motivación e interés para despertar vocaciones científicas. Cuadernos de Documentación Multimedia **31**, e68692 (2020)
28. Mariscal, G., Jiménez, E., Vivas-Urias, M.D., Redondo-Duarte, S., Moreno-Pérez, S.: Aprendizaje basado en simulación con realidad virtual. Educ. Knowl. Soc. (EKS) **21**, 15 (2020)
29. Anacona, J., Millán, E., Gómez, C.: Aplicación de los metaversos y la realidad virtual en la enseñanza. Entre ciencia e ingeniería **13**(25), 59–67 (2019)

Analysis of the Development Approach and Implementation of the Service Tokenization Project. Hoteliers Project Ocean Blue Corp - Hotel Best Wester Quito- Ecuador City and Beach

Paúl Alejandro Mena Zapata[1] (✉) and César Augusto Plasencia Robles[2]

[1] Center for Global Governance, Campus Unamuno, University of Salamanca, P° Francisco Tomas y Valiente S/N, 37007 Salamanca, Spain
paul.mena@usal.es

[2] School of Law, Virtual Campus, Universidad Privada del Norte, Lima Central Tower, Av. El Derby 254, 14th Floor, Surco, Lima, Peru
cesar.plasencia@upn.edu.pe

Abstract. This paper explores from the nature of Civil Law and the source theory of legal obligation, towards the evolution of the development of *Smarts Contracts* and *Blockchain* Technology in the implementation of a system of RC20 and RC21 token issuing machine, with fiduciary support as a mechanism of legal security, in order to propose a model of potentialization of destinations with high tourism demand in developing countries, within the framework of design, implementation and commissioning of the destination Hotel Aiden by Best Western Quito DM, Republic of Ecuador City and Beach.

Keywords: Smart Constracts · Smart Tourism · Tokenization · Blockchain · contractware

1 Introduction

By way of introduction, we start by determining the legal nature of subjective rights as a source of contractual obligation, on the basis that they are faculties or powers that people have to demand something from someone or from the State. They can be defined as legal powers, granted by the legal system. The doctrine set forth by Llambías (1997), points out as a prerogative recognized to the individual by the legal system, whose purpose is to generate a relationship of enforceability towards the other persons or subjects of the legal relationship.

The legal nature of obligations is understood as a bond that requires the debtor as a subject obliged to give, do, or not do something in favor of another person called creditor.

S. Ossowski et al. (Eds.): DCAI 2023, LNNS 740, pp. 336–346, 2023.
https://doi.org/10.1007/978-3-031-38333-5_34

Giving rise to an acceptance, which is called obligation susceptible of pecuniary estimation. Thus, and in accordance with Betti's statement[1], the obligation is a patrimonial commitment to another.

Louis Josserand (2014), states that an obligation or personal right is a legal relationship that assigns to one or several persons the position of debtor before others who play the role of creditors and with respect to whom they would be obliged to a positive performance, determined by an obligation to give or to do, and in a negative form of not doing.

2 On the Nature of the Contractual Legal Obligation to *Contractware,* in Contracts for the Provision of Hotel Services

Anguiano (2018) points out that according to Nick Szabo (1996), in *"Extropy", Building Blocks for digItal free Markets*, proposes the modulated functionality of a vending machine, and manages to project the contractual nature towards a model based on the mechanical and digital advances of his time, proposing in conceptual terms the automation of increasingly dynamic contracts, his modulation was based on *hardware and software* systems, which would come to emulate the cycle of beginning and end of the pre- and post-contractual stages, proposing the replacement of human intervention in the contractual act.

Under the modulation proposed by Szabo (1996), the vending machine in its functional cycle of constant operation fulfills two natural and contractual requirements of a service provision process, it delivers the good or service, for a fair price, so that on the one hand it executes the payment and on the other it dispenses or delivers the product, with criteria of security and autonomy in this contractual, constant, dynamic and automatic process.

This simple but at the same time complex legal and IT situation, inspires the development of the present project, because if the dynamics emulated by Szabo focused on *Blockchain* technology, is transferred to the contractual terms of service provision interconnected to the computer code, which through the interface of a token express the will of the parties, the substantial elements of both the legal obligation and the nature of the contract are being integrated.

In this order of ideas, starting from the dynamics of smart contracts with two phenomena of great furor in the dynamic boom of commerce, with different concepts in form, but interrelated in practice, *Blockchain, Smart properties* and *smart tourism*, with a proposal for the conformation of an ecosystem of service provision and legal security.

The implementation of technological mechanisms in the provision of services has become an indispensable tool that not only responds to the social reality of the moment, but also presents itself as a prerogative that can also be used by developing destinations, but with high tourism potential, seen as a whole.

[1] "The obligation is the patrimonial content of the performance owed by the debtor to the creditor. The obligation is not primarily a legal relationship, as are the contract or the legal bond that binds the parties; it is a relationship between two things, between the fact of enrichment and the fact of impoverishment that the law places face to face" BETTI, E. (1955). *Teoría General del Negocio Jurídico.* Buenos Aires: Editorial Universitaria de Buenos Aires.

Blockchain provides us with a relevant technological security, and this is one of the most relevant parameters when using the technology for example for financial markets. This new technological era offers us an internet of value, now the web is not only informative, it will not only circulate with data, but we will be able to create value within the network and within the *Blockchain.*

The way of validating transactions and circulating information allows us to create assets that are unique, referenced within the Blockchain itself and when we transmit an asset we lose it. For example, when an information file is sent to any recipient it can be PDF, it can be cloned and sent to many recipients, the sender has the control and power to do with that file whatever he wants. When using Blockchain technology by using Bitcoin transfer for example what happens is that the issuer loses control and power over that asset that is the Bitcoin, therefore, you can create value, and you can store, or transfer because they become codes that are unique and of which you lose control when you decide to transmit them.

3 Blockchain and Hotel Services

Blockchain is conceived as an IT tool that enables the implementation of "distributed registries", for the purpose of this article, a distributed registry should be understood as a registry that is constructed in such a way that the annotations that integrate it are agreed among the anonymous participants in its creation, with the peculiarity that each of the participants has an updated copy of this registry, which is universal and unique. This distribution pattern can be used for the creation of other registers, which can and must be mathematically and legally represented.

For the purposes of this research, a "distributed registry" is understood as a registry that is constructed in such a way that the annotations or value that comprise it are agreed upon by the anonymous participants in its creation. In addition, each of them has an updated copy of it. This "proof matrix" has been used to optimize costs in cryptocurrency transfer. It can also be used for the creation of other "crypto-registers" as long as the elements that make them up can be mathematically represented, establishing a contractual, dynamic and evidential procedure for the generation of obligations between the parties, both in the use and in the provision of the service.

It is necessary to expose one of the basic and most important components of Blockchain technology, and that is decentralization. Decentralization is one of the fundamental principles of the *Blockchain* process, it consists in the fact that all transactions and operations are recorded in a distributed network, which means that there is no central authority or entity that controls that network. Instead, the information is stored in multiple nodes or devices that are part of that network, and that work together to verify and update the transaction record. Each node in the network has a copy of the entire *Blockchain database*, ensuring that there is no single point of failure or vulnerability.

Decentralization in the *Blockchain* process has several advantages, including security, transparency and resistance to censorship or manipulation. With no central authority in charge of the *Blockchain* network, transactions are more secure and harder to hack. In addition, the transparency in the network makes it easier to track transactions and detect any suspicious activity. With no central entity that can censor or manipulate information on the network, decentralization offers greater resistance to potential abuses of power or hidden interests.

It now corresponds to describe the conceptualization for the present project of the smart property or crypto property-service in the model implemented, once the distributed registry for the exchange of currency has been dynamized, it corresponds to propose which related rights could be transferred within the framework of this implementation system. In general terms, *tokenization* is the mathematical representation of a right of use or property, in a sequence of characters (letters and digits), which are unequivocally incorporated into a distributed registry.

The represented rights of use or *tokens*, with full capacity to transfer the ownership of a right, represented in digital form, but with full capacity for its use and execution, as a simplification of the transmission of a service, under unique and own criteria of non-fungibility.

In this digital environment, the tokens, and their representation in goods and services that grant legal title to the specific act object of the provision, making it susceptible of transmission to new holders under the contractual terms established in the *smart contract*. The virtue of *Blockchain* technology in this interface of emulation of decentralized transfer of goods and services, guarantees the uniqueness of each transaction, and its integration to the block chain of distributed occupancy.

The last substantial element of this ecosystem is implemented in the *Smart contract*, with the core elements of the contractual obligation, towards the development model approach, the *Smart contract*[2] is a computer program that runs autonomously and automatically on the Blockchain. These contracts are designed to automate the process of verifying and executing transactions and to make agreements transparent, secure and tamper-proof. A *Smart contract* is written in a programming language and embedded in the *Blockchain* so that it can be executed anytime and anywhere on the network.

2 "Smart contracts are mainly based on Blockchain technology. Proof of this is that they share many characteristics with it, such as unchangeability, transparency, security or pseudonymity. For this reason only the specific characteristics of smart contracts will be analyzed now, in order to avoid superfluous reiterations. We can mainly point out that smart contracts are autonomous, which means that they do not require the intervention of the parties for the services that comprise them to be executed. They use a computer language that must be translated into the common language, so that any average citizen can understand its content. Otherwise, the contract could be annulled, as there is a vitiated consent" SÁNCHEZ ÁLVAREZ, E. and GARCÍA PACIOS, A. (2021). *Blockchain technology and electronic contracting: critical points of integration of the so-called smart contracts in our contract law system.* In Revista CEFLegal, 246, 71–98.

At this stage of tokenization, the aim is to apply *contractware*[3], which involves the transfer of the contractual dynamics to a computer code, migrating to the nature of the substantial elements of the contract, the obligations of the contracting parties, and the consequences of the observance or non-observance of the contract.

Contractware is a term that refers to the codification of the terms and conditions of a contract in a computer program. It consists of the transformation of a conventional contract into a computer code that can be automatically executed to implement and guarantee the agreed conditions. The term contractware is commonly used in the context of Smart contracts, which are computer programs designed to automatically execute the terms of a contract on a Blockchain. Smart contracts use contractware to incorporate the terms of the contract into computer code that can be executed if previously defined conditions are met (Fig. 1).

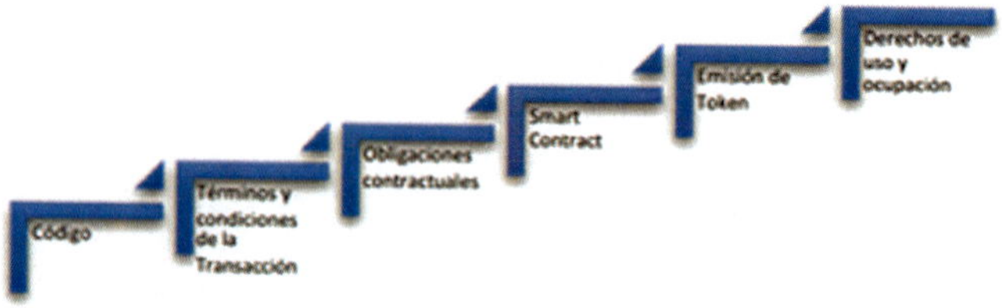

Fig. 1. Modulation of the proposed ecosystem: Source: Own elaboration.

In the graph, the form of transfer of a token is determined, since the smart contract is not limited to stating the transfer agreements, it also verifies the data and executes the legal consequences agreed upon by the contracting parties, thus if the code receives the parties' services (payment in crypto *exchance*) and the token transfer mandate, it refers to each of the parties for the perfection of the contractual model, delivery of the good or service, and payment of the fair price.

4 Activation Cycle of Smart Contracts for the Provision of Hotel Services

In the nature of the contractual act, provision of services, its cycle of conditionality and validity contains three phases: the first is the generation of the contractual will, the second is the perfection of the legal act, and the third is the consummation of the act

[3] "Contractware is the translation of legal (contractual) prose into computer code. The coding incorporates not only the agreements reached but also the consequences that may arise from their fulfillment or non-fulfillment. If, for example, the transfer of a token is formalized, the smart contract is not limited to stating the transfer agreements. It also verifies the data and executes the consequences agreed upon by the signatories. Thus, for example, the code receives the parties' performances (the payment in cryptocurrencies and the token transfer mandate) and forwards to each party the other party's performance. Many authors reject that these sequences of code can be considered authentic contracts, since while simple contracts merely incorporate promises of future performance, intelligent ones execute them".

of the will of the parties. Within the implementation of the decentralized system under *contractware* criteria.

The first phase of the nature of the contractual act, the generation of the contractual will, refers to the process in which the parties involved manifest their intention to establish a contract and express their consent to do so. This phase involves the following aspects:

1. Offer: One party makes a specific and clear proposal to the other party, indicating the terms and conditions under which it is willing to enter into the contract. The offer must be sufficiently precise and complete for the other party to accept or reject it.
2. Acceptance: The other party involved responds to the offer in an affirmative and unreserved manner, accepting the terms and conditions set forth. Acceptance must be a clear and express manifestation of agreement with all the essential elements of the offer.
3. Consent: Both parties must have the legal capacity to enter into the contract and must give their consent freely, without being subjected to coercion or fraud. Mutual consent is essential for the contractual will to be valid and binding.

During this phase, clear and effective communication between the parties is crucial to ensure that both parties understand the terms and conditions of the proposed contract. In addition, any negotiations, modifications or counter-offers also take place at this stage before a final agreement is reached.

In the contractual context, the second perfection of the contract refers to the stage at which the final and binding agreement between the parties is reached. It consists of the following key elements pertaining to civil law.

1. Offer and acceptance: After an offer is made by one party and accepted by the other party, a preliminary agreement between the parties is generated. However, this preliminary agreement is not binding until certain additional requirements are met.
2. Formal requirements: Depending on applicable laws and regulations, it may be necessary for the contract to comply with certain formal requirements in order to be valid and binding. These requirements may include the written form, the signature of the parties or the presence of witnesses.
3. Full consent: Both parties must give full consent free of defects, such as error, fraud or violence. The consent must be valid and free of vices for the contract to be considered valid

   The third phase refers to the consummation of the act of will of the parties. This phase implies that all the conditions and requirements established in the contract are fulfilled, and that the parties perform the agreed obligations.

The consummation of the act of the will of the parties implies the following:

1. Fulfillment of obligations: Each party involved in the contract must carry out and fulfill the agreed obligations. This involves performing the actions, providing the goods or services, or performing any other activity stipulated in the contract.
2. Compliance with deadlines and conditions: The parties must comply with the bonds established in the contract and respect the specific conditions agreed upon. This includes meeting delivery deadlines, scheduled payments, confidentiality clauses, among other agreed terms and conditions.

3. Record of performance: It is common for the parties to record the performance of the contract through documents, reports, invoices, receipts or other records evidencing compliance with contractual obligations.

The consummation of the parties' act of will marks the point at which the performance of the contract is complete and what has been agreed by both parties is satisfied. At this stage, the parties can review whether all conditions have been satisfactorily fulfilled and, if necessary, take appropriate action to resolve any problems or non-compliance.

It is necessary to pre-establish for this development model the conceptualization of "oracle", understanding this as a computer program that in its structure intends to conceive a series of variables conditioned to specific and contemplated cases, unknown at the moment and conceived as future data, which condition the result of the code execution, in the implementation model proposed when the *contracware* is created.

In other words, an "oracle" in the context of computer science and technology refers to a computer program or software component that provides information or answers based on predefined data and rules. Its purpose is to provide a specific output or response based on input variables or set conditions.

The term "oracle" is used metaphorically to refer to the idea of a system or entity that has specialized knowledge or wisdom and can provide reliable and accurate answers to questions or situations posed.

For the present development model, we take as a starting point the rules on electronic commerce referred to by the United Nations Commission on International Trade Law[4], this commission assumes the challenge of the exponential growth of electronic relations in the international traffic of goods and services, through the creation of various instruments based on the principle of functional equivalence between the electronic document and the physical document, and the principle of technological neutrality.

This frame of reference will be in the immediate future the basis for the generation of legislative initiatives in the field of smart contracts, taking as a starting point, electronic

[4] "The United Nations Commission on International Trade Law (UNCITRAL) has developed a set of legislative texts to enable and facilitate the use of electronic means in commercial activities, which have been adopted by more than 100 States. The most widely adopted text is the UNCITRAL Model Law on Electronic Commerce (1996), which sets out the rules for equal treatment of electronic and paper-based information and legal recognition of electronic transactions and processes, based on the fundamental principles of non-discrimination in the use of electronic means, functional equivalence and technological neutrality. The UNCITRAL Model Law on Electronic Signatures (2001) provides further rules on the use of electronic signatures.The United Nations Convention on the Use of Electronic Communications in International Contracts (New York, 2005) takes as its point of departure the earlier texts of UNCITRAL to become the first treaty to give legal certainty to electronic contracting in international trade.More recently, the UNCITRAL Model Law on Electronic Transferable Documents (2017) applies the same principles to enable and facilitate the use of transferable documents and securities in electronic form, such as bills of lading, bills of exchange, checks, promissory notes and warehouse receipts.In 2019, UNCITRAL approved the publication of the Notes on key issues related to cloud computing contracts, while continuing to work on the development of a new instrument on the use and cross-border recognition of electronic identity management services and authentication services (trust services)".

transferable documents, in emulation of securities. In this way, the representation of goods and services associated with international trade will become a reality.

5 Collaborative Economy and *Blockchain* in the Provision of Hotel Services

According to Alfonso Sánchez (2016) the collaborative economy, as a reference of the new systems of production and consumption of services and goods derived from the advances of information technologies, to exchange and share through technological platforms, based on exponential economic development and welfare in a non-linear way and under the applicability of the type of performance of *Product Service System*, which provides the consumer with access to the services they require in a precise way, under the modality of service provision, or otherwise the approach of the right of use of memberships, under trusteeship administration support, as a security guarantee mechanism.

In this context, the implementation project intends to start from the concept of collaborative accommodation service provision and assignment of use of accommodation services autonomously, through *peer-to-peer* platforms[5], singularized to precepts of booking, accommodation, exchange of the service, through the use interface of *blockchain* technology and *tokenization* of hotel service provision in the framework of the exponential development of *smart tourism.*

Peer-to-peer (P2P) platforms are computer systems that allow users to share resources, files or services directly with each other, without the need for a central authority or intermediaries. These platforms are based on direct connection and collaboration between users. Instead of using a centralized server to store and distribute information, P2P platforms allow resources to be distributed across multiple nodes or individual computers. Each node acts as both a resource provider and a consumer, allowing the sharing of files, data, bandwidth, storage capacity or other resources.

P2P platforms have been widely used in various contexts, especially in file sharing. A popular example of P2P platforms can be seen in Skype: The communication application Skype uses a P2P architecture to facilitate calls and message exchange between users. Instead of going through a central server, calls and messages are routed directly between users' devices, enabling faster and more decentralized communication.

Based on the concept of collaborative accommodation, conceived as an alternative to conventional tourism and from the vision of *Smart Tourism and Smart Travel*, this can be enhanced with the application in the implementation of *Blockchain* technology

[5] "P2P technology can be defined as a network in the form of a backbone, composed of nodes that act as clients and servers of other nodes. When a client enters this system", it makes a direct connection to one of the latter, where it collects and stores all the information and content available for sharing. It is then a program whose function is to connect users through a serverless network that facilitates the download53 of music, movies, books, photos and software among all other users, free of charge. These files are shared "from computer to computer" by the mere fact of having access to the system". VARELA PEZZANO, Eduardo Secondo.*Peer-to-peer technologies, author's rights and copyright*. Bogotá: Editorial Universidad del Rosario, 2009.

and *Tokenization* of services under the support of the traceability of trustee assets as a mechanism to guarantee legal security, thus implementing the advantages towards large-scale marketing services and with criteria of supranationality to global markets, as an instrument to generate collaborative economy.

Specific Case

In this conceptual framework, the implementation project of *Tokenization* of hotel services operations called "Blue Ocean Project - Best Wetern Quito city and beach", allows to put in context the first project of marketing of non-fungible *tokens* with trustee support, and with *Blockchain* technology in the Republic of Ecuador. As a model of concept development, implementation, and implementation of this collaborative economy process in developing countries with high tourism potential.

From the functional architecture scheme for the implementation of the project based on *Blockchain* technology whose main objective is to register rooms, control the use of hotel occupancy in the form of Smart Contract, under implementation of ERC-721 non-fungible *token* standard model, which will allow these rooms can be booked through a NFT system. Where it seeks to deploy different *Smart Contracts* based on *tokens* in both ERC20 and ERC721 standard models, whose functionality allows buying and selling NFT both with fiat money and cryptocurrencies.

An NFT (Non-Fungible Token) system refers to a type of digital token that is used to represent the ownership or authenticity of a unique object or asset in a digital environment. Unlike cryptocurrencies such as Bitcoin, which is fungible and can be exchanged for equivalent units, NFTs are unique and indivisible. The key feature of NFTs is that they use *Blockchain* technology, as a decentralized and immutable registry that guarantees the security and traceability of digital assets. Each NFT has a unique identity and is recorded on the blockchain, which allows verifying its authenticity and ownership.

In the first phase of development to determine the technical feasibility of the project, two functional and harmonic alternatives are proposed, the first is to create the NFT in a *Blockchain* compatible with the Ethereum Virtual Machine (EVM), and the second is to create this NFT in several *Blockchain* networks and centralize the management in another *Blockchain*, under criteria of high commercial standardization, for the implementation of the proposed model.

It is recommended to create a central *Smart contract* that registers all operations in a single blockchain, and that added to its integration in several *Blockchain* networks (Ethereum, Tron and others compatible with EVM), centralizes the management in another *Blockchain*. Under a central *Smart contract*, backed in trustee administration as a figure of legal legal security, in other words, that the operations of all blockchains are recorded in a single blockchain.

To implement this solution, a smart contract is proposed in the central *Blockchain*, which interacts with the smart contracts in each of the secondary *Blockchains* where the NFTs will be issued. This centralized smart contract is in charge of keeping a record of all transactions related to the NFTs, it also updates the information of the rooms available for booking, as well as the availability of memberships under the *time sharing* model management modality.

As part of this integration and implementation, when a user makes a reservation through one of the EVM-compatible *blockchains*, an NFT will be created on that specific blockchain. The centralized smart contract will take care of recording this transaction in its registry and update the availability of hotel rooms on all supported and adhering *blockchains*.

Thus, the proposed model, so that users can buy and sell these NFTs, has to aim at the creation of a centralized platform that interacts with the smart contracts of each compatible blockchain. On this platform, users can buy and sell the NFTs using both fiat money and cryptocurrencies, and transactions will be recorded in the centralized smart contract to keep a total record of all transactions.

As it is, the TRC-721 *Smart contract* in TRON will have a similar structure to Ethereum's ERC-721 *Smart contract* that will include a list of the current owners of each NFT, as well as information about the creation date, the value of the NFT and any other relevant information. In addition, this *Smart contract* will have methods that allow for the creation of new NFTs, the transfer of NFTs between accounts, and verification of the authenticity of an NFT. It will also include a method that allows communication with the centralized *Smart contract* on Ethereum to record transactions on both networks.

Finally, to enable *inter-blockchain token* transfer, it is necessary to use a *chain bridge* that acts as a communication channel between two different blockchains. This chain bridge will be responsible for validating and recording the *token* transfer transaction on both blockchains. The interface solution intended by this technical development to implement *inter-Blockchain* token transfer is based on:

a. Creation of the token on the chain of origin: the ERC-721 token must be created on the chain of origin and the corresponding property rights must be assigned, under the establishment of the right of use and its interrelation with the fiduciary guarantee.
b. Chain bridge initialization: a chain bridge must be initialized on both blockchains to enable *inter-Blockchain* transfer of tokens.

6 Perspective of the Right of Use of Hotel Services and Its Legal Relationship to the *Smart Contract*

From the legal perspective, integrating the computerized nature of the technology of implementation of *Smart contracts* into the blockchain, it advocates the existence of a given right that is born with the purpose of producing legal effects. Thus, the *Smart contract* integrated to the blockchain, would execute an act of contractual will drafted in whole or in part in form (or computer language), whose clauses determine the obligations to which the parties submit themselves in whole or in part to an existing legal relationship, the same system being responsible for its automatic execution when the fulfillment of the agreement is verified.

From this perspective, the *Smart contract* proposed for the present project poses a conceptual challenge for the civil legal doctrine. First, the integration of the object of the contract and its regulation in the normative framework, always maintaining the substantial elements of the contract, in form and substance in the terms of the Civil Code, regarding the nature of the obligation and the consideration for the service.

Under this order of ideas, we conceive collaborative hosting for this development as a right of use, which arises from an act of contractual willingness to provide a service, in an activity issued and transferred based on algorithms and computer protocols, based on the right of use represented by the token. For this it is necessary to distinguish between the *right to a token*, or the right of ownership of the token, as in the case of cryptography, to the rights it represents, in this particular case the real right of use in the contractual terms of the *smart contract*, with respect to third parties, and to the right of ownership of the *token* as an asset, as well as to the issuer of the *token*.

Consequently, the present implementation project intends to expose the natural concepts of pre-contractual and contractual acts of a legal relationship of commerce, integrated to the right of use, from a cryptographic and interactive perspective in a system that carries a process of conceptual and commercial technological integration towards destinations in developing countries.

Which is finally, what encourages the exploration and implementation of existing conceptualizations and regulated in the general civil norm in matters of real rights, credit rights, according to its material scope of application, with criteria of cybersecurity in the protection of data of the chain, as of the relations of the *Exchange*, both in primary and secondary markets. Always guaranteeing the traceability of use, and the fiduciary administration as a substantial element of legal security, towards the promoter, the developer and user of the project, in favor of the dynamization and automation of legal traffic in the consumption of services at international level as a destination for promotion in developing countries, all within the framework of protection of consumer rights.

References

Alfonso Sánchez, R.: Collaborative economy: a new market for the social economy. CIRIEC-Spain J. Public Soc. Coop. Econ. (88), 230–258 (2016). https://www.redalyc.org/articulo.oa?id=17449696008. Accessed 16 Mar 2023. ISSN 0213-8093

Anguiano, J.M.: 'Smart Contracts'. Introduction to 'contractware'. Garrigues opine (2018). https://www.garrigues.com/es_ES/noticia/smart-contracts-introduccion-al-contractware

Betti, E.:Teoría General del Negocio Jurídico. Buenos Aires: Editorial Universitaria de Buenos Aires (1955)

Uncitral Guide: Basic Facts and Functions of the United Nations Commission on International Trade Law. United Nations (2013). https://uncitral.un.org/sites/uncitral.un.org/files/media-documents/uncitral/es/12-57494-guide-to-uncitral-s.pdf

Josserand, L.: Teoría General de las Obligaciones, 2nd edn. Ediciones Jurídica de Santiago (2014)

Llambías, J.J.: Tratado de Derecho Civil - Parte General (Volume II). Editorial Perrot (1997)

Sánchez Álvarez, E., García Pacios, A.: Blockchain technology and electronic contracting: critical points of integration of the so-called smart contracts in our contract law system. Rev. CEFLegal **246**, 71–98 (2021)

Szabo, N.: Smart Contracts: Building Blocks for Digital Markets [PDF File] (1996). https://www.fon.hum.uva.nl/rob/Courses/InformationInSpeech/CDROM/Literature/LOTwinterschool2006/szabo.best.vwh.net/smart_contracts_2.html

Varela Pezzano, E.S.: Peer-to-peer technologies, author's rights and copyright. Editorial Universidad del Rosario, Bogotá (2009)

Regulations

UNCITRAL Model Law on Electronic Commerce Guide to Enactment 1996 with new article 5 adopted in 1998

UNCITRAL Model Law on Electronic Signatures with Guide to Enactment 2001

UNCITRAL Model Law on Electronic Transmittable Documents 2017

Implementing a Software-as-a-Service Strategy in Healthcare Workflows

Regina Sousa, Hugo Peixoto, António Abelha, and José Machado(✉)

ALGORITMI/LASI, University of Minho, Braga, Portugal
regina.sousa@algoritmi.uminho.pt, {hpeitoxo,abelha,jmac}@di.uminho.pt

Abstract. The spread of healthcare technology has resulted in a massive amount of data, particularly in the form of laboratory test results, which play an important role in medical diagnosis and treatment. However, managing and interpreting such large amounts of data has proven increasingly difficult, particularly for resource-constrained healthcare facilities. To address this issue, we present a multi-agent system for effective laboratory test result management based on Software-as-a-Service (SaaS) technology. This paper contains a case study that evaluates the system's performance and efficacy. The study's goal is to examine the viability of using a multi-agent system and SaaS technology to manage laboratory test data, highlighting the system's advantages over conventional alternatives. In the age of big data, the deployment of this system could dramatically improve healthcare service efficiency, quality, and cost-effectiveness.

Keywords: Multi-Agent System · Big Data · Real-Time Information Systems · Cloud Paradigm · Software as a Service

1 Introduction and Contextualization

Advancements in healthcare technology have led to an explosion of healthcare data in recent years. Laboratory test results, in particular, have become a critical component in medical diagnosis and treatment. As the world's population ages, the need for medical care increases. According to the World Health Organization (WHO), the number of people over 60 is expected to double by 2050 [14]. To address this demand, healthcare systems have begun to focus more on preventive measures and lifestyle changes that can help improve overall health and well-being rather than just reacting to specific diseases or conditions as they arise. This approach, known as integrative medicine, recognizes the importance of considering the whole person rather than just focusing on specific health issues [3,12]. In addition to preventive measures, there has also been a trend towards personalized medicine, which involves tailoring treatment and prevention strategies to each patient's individual needs and characteristics [18]. However, managing and analyzing these vast amounts of data is becoming increasingly challenging, especially for healthcare providers with limited resources.

S. Ossowski et al. (Eds.): DCAI 2023, LNNS 740, pp. 347–356, 2023.
https://doi.org/10.1007/978-3-031-38333-5_35

Interoperability is crucial for successfully implementing personalized medicine, especially in the context of laboratory test results. The need for interoperability arises from the fact that healthcare institutions often use different Electronic Health Record (EHR) systems that most of the time are not fully compatible [17]. As a result, difficulties arise to ensure seamless patient information sharing, including laboratory test results, across institutions. This issues will most likely have a negative impact in patient care. Without interoperability, healthcare providers may have to rely on outdated or incomplete information, which can lead to misdiagnosis or inappropriate treatment [13]. In addition, patients may have to undergo unnecessary tests or procedures if healthcare providers cannot access their previous results. To address the need for interoperability in personalized medicine, healthcare institutions must adopt common data standards and protocols for exchanging patient data. These standards can help to ensure that patient data is accurate, complete, and easily accessible across different systems and institutions [9].

Multi-Agent Systems (MAS) are essential for interoperability implementation because they provide a distributed computing environment that enables data exchange and communication among various agents. MAS provide an efficient and flexible way to coordinate heterogeneous systems, including different software platforms and programming languages, making interoperability feasible [17]. In contrast, interoperability is essential for MAS because it enables the agents in a system to communicate and collaborate seamlessly. Interoperability helps agents overcome the challenges of heterogeneity, enabling them to communicate and exchange information with each other effectively. Without interoperability, MAS may suffer from communication breakdowns, incompatible data formats, and other integration challenges that can affect the system's overall performance [10]. Overall, MAS and interoperability are complementary concepts essential for the success of complex distributed systems. MAS enable the implementation of interoperability, while interoperability enables the seamless coordination of agents in a MAS. Combining these concepts can help achieve complex goals that cannot be achieved with a single system or agent. By enabling healthcare providers to access and share patient data across different institutions, interoperability can improve the quality of care and help ensure that patients receive the most appropriate treatments based on their needs [6,7].

Unfortunately, developing and implementing traditional Health Information Systems (HIS) is often costly and requires significant software and hardware investments [16]. This can burden small/medium institutions that may rely on more precarious tools instead. Additionally, some of the shortcuts taken can be vulnerable to several security flaws, including patient data breaches, loss of data, incompatibility with other systems, and high maintenance costs. This is where the cloud computing paradigm and the concept of Software as a Service (SaaS) come into play [5]. SaaS offers a cost-effective and secure solution for healthcare organizations, allowing them to access the necessary technology without the upfront investment in hardware and software. It also allows for greater interoperability, as SaaS systems can easily communicate with one another and share data. SaaS presents an excellent opportunity for the healthcare industry to improve patient care and streamline operations. By leveraging the power of the cloud, healthcare organizations can access the latest technology and services without breaking the bank, all while keeping patient data secure [1].

This manuscript presents a case study of the proposed system, focusing on the evaluation of its performance and effectiveness. This study aims to evaluate the feasibility of using a multi-agent system powered by SaaS technology and to demonstrate the advantages of the proposed system over other similar approaches.

The document is structured into five primary segments, commencing with an introduction that identifies the principal issue and outlines the main inspirations and principles that drove this study. Subsequently, there is an analysis of the current state of knowledge, followed by an exposition of the primary outcomes attained. In addition, the results section presents the primary architecture of the multi-agent system. Ultimately, the final two sections are the discussion, where the primary contributions are highlighted, and the conclusion and future work, which concludes the manuscript.

2 State of the Art

In the healthcare sector, Laboratory Information Systems (LIS) are used to manage laboratory operations, including test orders, sample tracking, and result reporting. With the increasing demand for more efficient and cost-effective healthcare services, a paradigm shift is occurring, leading to greater interest in developing SaaS-based LIS that can leverage the benefits of cloud computing and multi-agent systems [19].

SaaS-based LIS are cloud-based applications that can be accessed from anywhere and at any time through the internet. These systems provide a flexible and scalable platform for managing laboratory operations, without the need for expensive hardware and software investments. They also offer enhanced data security and reliability, as the data is stored and backed up in the cloud [20].

Multi-agent systems have been proposed as a promising approach for building intelligent LIS that can facilitate collaboration and coordination among laboratory staff, equipment, and other stakeholders. These systems can use autonomous agents to perform various tasks, such as sample tracking, test result interpretation, and quality control. The agents can communicate with each other and with external systems to exchange information, coordinate workflows, and resolve conflicts [4].

Several SaaS-based LIS that use multi-agent systems have been developed in recent years. For example, LabVantage Solutions offers a cloud-based LIS that uses multi-agent systems to automate laboratory operations and improve quality control. The system uses autonomous agents to monitor and manage laboratory workflows, alert users of any deviations from standard procedures, and optimize testing processes. It also provides real-time analytics and reporting capabilities, allowing laboratory managers to track performance metrics and make data-driven decisions [8].

Clinidata is a cloud-based LIS developed by Maxdata. This system uses multi-agent systems to automate laboratory operations and improve quality control. The system uses autonomous agents to monitor and manage laboratory workflows, alert users of any deviations from standard procedures, and optimize testing processes. It also provides real-time analytics and reporting capabilities, allowing laboratory managers to track performance metrics and make data-driven decisions [11].

Modulab is another example of a SaaS-based LIS that uses multi-agent systems. This system is developed by Werfen, a leading provider of in vitro diagnostic solutions. Modulab uses autonomous agents to automate laboratory workflows, including sample processing, quality control, and result reporting. The agents can communicate with each other and with external systems to coordinate workflows and resolve conflicts. The system also provides real-time analytics and reporting capabilities, allowing laboratory managers to track performance metrics and optimize laboratory operations [21].

Other Software like Epic Systems, Cerner, Meditech, and NextGen Healthcare can be considered a good example. All have the ability to integrate and exchange data with other systems in real-time using standardized communication protocols and data formats.

In summary, SaaS-based LIS that use multi-agent systems offer a promising approach for managing laboratory operations in a cost-effective and efficient manner. These systems leverage the benefits of cloud computing and autonomous agents to automate workflows, improve quality control, and enhance collaboration and communication among laboratory staff and stakeholders [15]. The examples provided above demonstrate the potential benefits of this approach and highlight the need for further research and development in this field.

In addition, to leveraging cloud computing and multi-agent systems, SaaS-based LIS that use data standards have been shown to offer significant advantages over traditional LIS. Data standards, such as Health Level Seven International (HL7), facilitate the exchange of laboratory data between different systems and organizations, enabling interoperability and data sharing. By using data standards, SaaS-based LIS can facilitate the integration of laboratory data with other healthcare systems, such as electronic health records (EHRs), and support clinical decision-making [2].

By using data standards, SaaS-based LIS that use multi-agent systems can facilitate the exchange of laboratory data between different systems and organizations, support clinical decision-making, and improve patient care. However, the implementation of data standards in SaaS-based LIS can also pose some challenges, such as the need for data mapping and the potential for data loss or corruption. Therefore, further research and development are needed to address these challenges and ensure the effective use of data standards in SaaS-based LIS.

Indeed, as shown above several implementations and approaches are already in place. Nevertheless, the system proposed hereby wants to go a step forward, and integrate multiple of this LIS as well as different other healthcare information systems using a MAS approach based in healthcare data exchange standards, making professional's life easier and but, more important, combining such data in an transparent way can lead to an optimal patient care.

3 Results

Before developing a multi-system response to the problem in hands, the solution's global architecture must be designed. In fact, given its complexity, it is essential to understand the global orchestration of the information flow and each component in order to preserve the integrity of data and information along the way. The proposed architecture reflects a multi-agent approach in which the components collaborate to implement the system. Figure 1 depicts a system that mimics a monolithic architecture. However, each component may be replicated and redundant to ensure uninterrupted service. Connecting and integrating diverse data sources is the first step in the process. In this instance, we will concentrate on the LIS case study, although the architecture is flexible enough to accommodate any other data type.

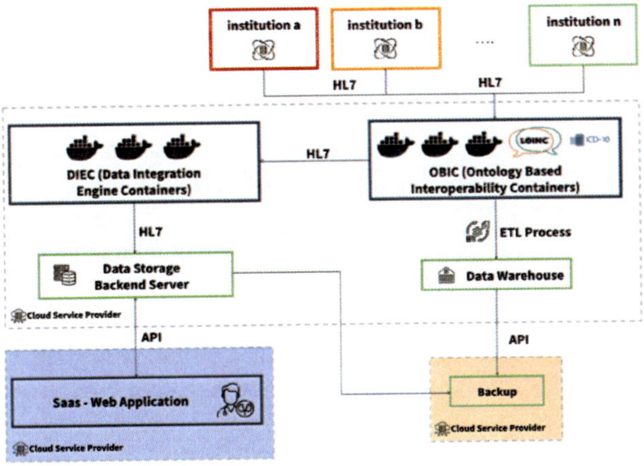

Fig. 1. Overall Architecture.

Analysing the problem, four distinct components can be identified. Specifically, it was considered:

- **Incoming Data Sources:** First, there are the institutions that will supply the infrastructure. This means that not only may each institution furnish this workflow with the data it deems pertinent to this project, but more importantly, the system is prepared to handle a wide range of data types.
- **Data Normalization and Standardization:** Regarding each income observed by the institutions, the MAS can take data inside the Ontology-Based Interoperability Container (OBIC) and output data in compliance with

healthcare ontology, such as Logical Observation Identifiers Names and Codes (LOINC) or ICD-10. Moreover, this data is split according to the primary needs: the Backend Database creation or Data Warehouse exploration for making some of the relevant data accessible for Machine and Data Mining purposes. As one may see in more detail later in this manuscript, several steps to ensure data anonymisation and quality are taken into the Extract Transform Load (ETL) process before storing it in the data warehouse.
- **Saas Web App:** This component is best characterized as the user interface for data consulting and information retrieval. In this phase, users can interact with the data and start decision-making based on the most accurate data available.
- **Backup:** The final component ensures service continuity by storing all data from the Data Storage Backend Server and data warehouse in a secure way. This backup is entirely customizable to ensure best practices or specific requirements.

Its architecture seeks not only to meet the day-to-day demands of physicians and other healthcare professionals with the incorporation of the User Interface, but also to ensure that the data can be made openly accessible and anonymous to promote Knowledge Extraction for other studies and trials.

3.1 Technical Description

HL7 Integration Agent. In the realm of intern system communication, HL7 serves as the go-to standard for enabling seamless and streamlined data exchange between disparate suppliers. The popularity of HL7 can be attributed to its ability to standardize communication in a clear and effective manner, as well as the widespread adoption of this standard among the majority of suppliers. As the healthcare industry continues to evolve, there is an increasing emphasis on developing specialized technicians with a deep understanding of HL7, thereby contributing to the growing consensus on its use. HL7 messages are typically transmitted over a network using a protocol such as Transmission Control Protocol(TCP)/Internet Protocol(IP). The messages themselves are structured using an ASCII-based format, which is designed to be both human-readable and machine-readable. Each HL7 message is composed of segments, which are groups of related data elements. Each segment starts with a three-letter code that identifies the type of data contained within that segment. For example, the PID segment contains patient identification information, while the OBX segment contains observation results. Both of these segments are present in the messages exchanged in this MAS. Within each segment, data elements are identified by a two or three-letter code, followed by the actual data value. For example, the code "PID-3" refers to the patient's medical record number, while the code "OBX-5" refers to the result of a laboratory test. HL7 messages can be sent in a variety of ways, including point-to-point connections, message queues, and web services. In our scenario we propose a queue system based in multi-agent implementation. OBIC opens several TCP/IP ports and waits for the incoming of messages. This ports are distributed by several Mirth Connect containers, in order for the traffic to be fully balanced and ensuring that if one of the containers goes down, the all system can continue to work as expected. The container that receives the message sent, proceeds to return an answer to the sender, called acknowledgment (ACK). The ACK in HL7 message exchange is a response message that the receiving system sends to signify that it has received and processed the original message. The original message's message control ID (a distinctive number issued by the sender) and message type are both included in the ACK message. The segments and data elements used in the original message's structure are also used in the ACK message. But, rather than conveying genuine medical information, the data pieces in the ACK message are utilized to offer details about the status of the original message. In order to assist the sender in identifying and resolving the problem, the receiver may add error data in the ACK message if it experiences a problem while processing the original message. Ultimately, the ACK message is a crucial component of HL7 message exchange since it informs the sender of the message's status and enables effective and efficient communication across healthcare systems.

Standardization Agent. Regarding the system's key components, OBIC and the Data Integration Engine Containers (DIEC) are also able to communicate with one another via HL7, thus ensuring uninterrupted syntactic interoperability. To achieve semantic interoperability - the pinnacle of interoperability - a sophisticated ontology engine is employed, with proper application of this engine serving as a guarantee. In the case study presented, analytical results are converted to the LOINC standard, a coding system that provides substantial benefits for the creation of a free, accessible, and real-time updated data repository.

Data Transformation Agent. The data generated by the system is bifurcated into two separate streams to fulfill different objectives. Firstly, a real-time updated repository is created, which provides an open and accessible platform for the storage of data. This repository is designed to facilitate data sharing and enable data-driven insights to support clinical decision-making trough needed studies. Secondly, the data is consolidated and stored in a database that serves as a LIS provided as a service. This LIS database is optimized for efficient data retrieval, management, and analysis, with a focus on streamlining laboratory workflows and improving the quality of patient care.

In the first step and so that the data warehouse can be built there is a complex ETL process that is applied, throughout a python agent. Among all the data processing process the following methods stand out: Anonymization, Uppercase, Date Format, Missing values, Noisy Data, Outliers.

The second step feeds a database, in this case a relational model database, that holds the multi-institutional LIS. This database can be supported by different database management systems and its choice can and should be made according to the different existing constraints, namely licensing costs, processing and storage capacities.

Backup Agent. The last component of this MAS is the backup agent. All organization's data management plan must include a data backup procedure since it offers a way to restore data in the case of loss, corruption, or disaster, and the system proposed hereby should be no exception. A typical data backup procedure should have these steps:

- Identify what data needs to be backed up. In this case all the data from the data warehouse as well as the relational database is saved;
- Select a backup method and frequency: in this approach one considered weekly full backups and incremental backups, twice a day;
- Determine the backup location: for this step a different cloud location was chosen; - Test the backup process: every week a Q&A is performed to evaluate the backups quality.

4 Discussion

The MAS proposed hereby is a system, as seen in the previous section, composed of multiple autonomous agents that interact with each other and with their environment to achieve specific goals. When compared to the approaches researched and found in the State-of-the-art, the proposed solution can be better for several reasons:

- **Flexibility:** It can adapt to changes in the environment and adjust behavior accordingly, making it more flexible than any other system.
- **Robustness**: If one agent in this approach fails, the other agents can still continue to function, making the system more resilient and robust.
- **Scalability:** As the number of institutions in place increases, the system can scale up without requiring significant changes to the overall architecture, making it easier to deploy and manage.
- **Standardization:** The use of international data standards enables all entities involved to understand the information that is being exchanged.
- **Cooperation:** Agents can collaborate and work together to achieve a common goal, allowing for more complex tasks to be completed that would be difficult or impossible for a single agent.
- **Specialization:** Different agents can be specialized to perform specific tasks or functions, which can result in more efficient and effective overall system performance.

Overall, the proposed architecture, with its approach can outperform other systems in situations where there are multiple agents that need to interact with each other and their environment to achieve a common goal. This approach is innovative and has the potential to significantly improve the efficiency and quality of healthcare by enabling easier and more effective communication between different healthcare systems and professionals.

In summary, the system presented has its own characteristics and advantages in terms of robust flexibility, scalability, cooperation, and specialization. The choice of system to be used depends on the specific needs of each healthcare setting and the priorities of the users, and the most generic appears to be the most advantageous.

5 Conclusions and Future Work

The study presented here demonstrates the effectiveness of a multi-agent system (MAS) approach for addressing common healthcare stakeholder issues. The researchers designed a flexible, robust, and scalable architecture based on healthcare standards, which was then applied in a case study focused on laboratory results. This study's results showed promising outcomes compared to other solutions, highlighting the potential for achieving interoperability within and outside healthcare institutions.

The present study explores the potential benefits of implementing a Software-as-a-Service (SaaS) strategy in healthcare workflows. SaaS not only reduces the initial investment for all stakeholders involved but also provides an opportunity for smaller stakeholders to participate in the overall workflow. The approach enables all stakeholders to contribute to data collection, paving the way for future knowledge extraction. More significantly, the SaaS approach enables data sharing among healthcare professionals, ultimately leading to better care for patients.

In the future the main objectives should focus on using the same approach to different scenarios and case studies, for example using data from Intensive Care Unities, or even a more wider approach using data from Electronic Health Records.

Acknowledgements. This work has been supported by FCT (Fundação para a Ciência e Tecnologia) within the R&D Units Project Scope: UIDB/00319/2020.

References

1. Asija, R., Nallusamy, R.: Healthcare SaaS based on a data model with built-in security and privacy. Int. J. Cloud Appl. Comput. (IJCAC) **6**(3), 1–14 (2016)
2. Chao, P.C., Sun, H.M.: Multi-agent-based cloud utilization for the it office-aid asset distribution chain: an empirical case study. Inf. Sci. **245**, 255–275 (2013)
3. Coalition, PM: Personalized medicine: the future of healthcare (2017). https://www.personalizedmedicinecoalition.org/policy/pmc-resources/personalized-medicine-the-future-of-healthcare
4. Dolcini, G., Sernani, P.: A multi-agent architecture for health information systems. In: Advanced Methods and Technologies for Agent and Multi-Agent Systems, vol. 252, p. 375 (2013)
5. Guha, S., Kumar, S.: Emergence of big data research in operations management, information systems, and healthcare: past contributions and future roadmap. Prod. Oper. Manag. **27**(9), 1724–1735 (2018)
6. Iqbal, S., Altaf, W., Aslam, M., Mahmood, W., Khan, M.U.G.: Application of intelligent agents in health-care. Artif. Intell. Rev. **46**, 83–112 (2016)

7. Jemal, H., Kechaou, Z., Ayed, M.B., Alimi, A.M.: A multi agent system for hospital organization. Int. J. Mach. Learn. Comput. **5**(1), 51 (2015)
8. Labware: An online comprehensive guide to understanding LIMS (2023). https://www.labware.com/lims-guide
9. Loreto, P., Fonseca, F., Morais, A., Peixoto, H., Abelha, A., Machado, J.: Improving maternity care with business intelligence. In: 2017 5th International Conference on Future Internet of Things and Cloud Workshops (FiCloudW), pp. 170–177. IEEE (2017)
10. Machado, J., Abelha, A., Novais, P., Neves, J., Neves, J.: Quality of service in healthcare units. Int. J. Comput. Aided Eng. Technol. **2**(4), 436–449 (2010)
11. Maxdata: Laboratórios (2023). https://www.maxdata.pt/pt/laboratorios
12. AoM National: Personalized medicine and health information technology (2014). https://www.nap.edu/catalog/18996/personalized-medicine-and-health-information-technology
13. Oliveira, D., et al.: OpenEHR modelling applied to complementary diagnostics requests. Procedia Comput. Sci. **210**, 265–270 (2022)
14. World Health Organization: Ageing and health (2022). https://www.who.int/news-room/fact-sheets/detail/ageing-and-health
15. Othmane, B., Hebri, R.S.A.: Cloud computing & multi-agent systems: a new promising approach for distributed data mining. In: Proceedings of the ITI 2012 34th International Conference on Information Technology Interfaces, pp. 111–116. IEEE (2012)
16. Reis, Z.S.N., Maia, T.A., Marcolino, M.S., Becerra-Posada, F., Novillo-Ortiz, D., Ribeiro, A.L.P.: Is there evidence of cost benefits of electronic medical records, standards, or interoperability in hospital information systems? Overview of systematic reviews. JMIR Med. Inform. **5**(3), e7400 (2017)
17. Sousa, R., Ferreira, D., Abelha, A., Machado, J.: Step towards monitoring intelligent agents in healthcare information systems. In: Rocha, Á., Adeli, H., Reis, L.P., Costanzo, S., Orovic, I., Moreira, F. (eds.) WorldCIST 2020. AISC, vol. 1161, pp. 510–519. Springer, Cham (2020). https://doi.org/10.1007/978-3-030-45697-9_50
18. Twilt, M.: Precision medicine: the new era in medicine. EBioMedicine **4**, 24–25 (2016)
19. Wager, K.A., Lee, F.W., Glaser, J.P.: Health Care Information Systems: A Practical Approach for Health Care Management. Wiley, Hoboken (2021)
20. Wang, X., Tan, Y.: Application of cloud computing in the health information system. In: 2010 International Conference on Computer Application and System Modeling (ICCASM 2010), vol. 1, pp. V1-179–V1-182 (2010). https://doi.org/10.1109/ICCASM.2010.5619051
21. Werfen (2023). https://www.werfen.com/pt/pt-pt/hemostase/modulab

Recognizing Emotions from Voice: A Prototype

Manuel Rodrigues(✉) and Guilherme Andrade

LASI/ALGORITMI Centre, University of Minho, Guimarães, Portugal
{mfsr,d2263}@di.uminho.pt

Abstract. Over time, the interaction between humans and machines has become increasingly important for both personal and commercial use. As technology continues to permeate various aspects of our lives, it is essential to seek healthy progress and not only improve, but also maintain the benefits that technology brings. While this relationship can be approached from many angles, this discussion focuses on emotions. Emotions remain a complex and enigmatic concept that remains a mystery to scientists. As such, it is crucial to pave the right way for the development of technology that can help understand emotions. Some indicators, such as word use, facial expressions and voice, provide important information about mental states. This work focuses on voice and proposes a comprehensive process for automatic emotion recognition by speech. The pipeline includes sound capture and signal processing software, algorithms for learning and classification within the Semi-Supervised Learning paradigm and visualisation techniques to interpret the results. For classifying the samples, a semi-supervised approach using Neural Networks is adopted to reduce the reliance on human emotion labelling, which is often subjective, difficult and costly. Empirical results carry more weight than theoretical concepts, given the complexity and uncertainty inherent in human emotions, but prior knowledge in this domain is not ignored.

Keywords: Automatic Speech Emotion Recognition · Semi Supervised learning · Human emotion · Unlabelled dataset

1 Introduction

Emotions are a fundamental aspect of human nature and have the power to influence our actions and shape our behaviours in society. However, their complex nature is not fully understood, as different mental states can overlap and result in different perspectives for each individual. Over time, researchers have attempted to categorise these mental states, with Paul Ekman initially focusing on six universally recognised emotions and Robert Plutchik proposing a psycho-evolutionary classification approach based on a few primary emotions giving rise to more complex sets. However, even today's experts may face challenges in accurately identifying emotions.

Currently, we are witnessing a revolution in deep learning, which has demonstrated remarkable success in solving a wide range of problems when carefully planned and supported by advanced hardware. The latest advances in hardware have made it possible to

S. Ossowski et al. (Eds.): DCAI 2023, LNNS 740, pp. 357–367, 2023.
https://doi.org/10.1007/978-3-031-38333-5_36

train increasingly complex deep learning models and improve the training speed of simpler models, allowing extended experimentation with different variants of algorithms and potentially obtaining more efficient and accurate results. In addition, deep learning has been used with various architectures for various applications, such as finance, computer vision, medicine and stocks, among others.

Describing emotions has been a complex over time. There are many subtle clues that expose our mental state, even when we are not aware of it. Using the abstraction capacity of machine learning algorithms, with minimal human intervention, these signs can be detected. This research has the potential to lead to many important applications that are currently costly or impractical, such as medical diagnosis, recommendation systems, social anomaly and fraud detection, and human resource management in emotionally intense situations. By taking advantage of the capabilities of artificial intelligence, these tasks could become cheaper and even affordable for everyday use. Previous studies by Rodrigues et al. (2012, 2020) and Gonçalves et al. (2015) have demonstrated the importance and potential of these applications.

Various solutions can be implemented using a generic or custom approach without it being considered intrusive. The primary objective of this work is to establish a robust methodology for any of these solutions and subsequently pave the way for multiple paths for specific tasks that may require minor adjustments. The primary objectives can be briefly outlined as follows:

- Use of a standard audio speech dataset (for this purpose IEMOCAP will be used) (IEMOCAP);
- Creation of an adaptable methodology, which is not specific to any particular domain;
- Creation of abstract representations of multiple data instances;
- Classification based on pre-defined mental states.

2 State of the Art and Literature Review

Automatic emotion recognition (ASER) involves many different approaches of distinct nature. From a general perspective, it is relevant to distinguish between multimodal and monomodal emotion recognition: the former is usually based on more than one source of information: voice, facial expression, body language, among other types of data; the latter uses only one type of data. The latter will be the one used in this work. As only voice data is used, a comparison is made with other works also focusing on a monomodal approach having voice as data source.

Fully supervised approaches remain the most relevant ones, compared to SSL approaches. A simplified overview of supervised approaches is therefore necessary to understand classification patterns, safeguarding that, the comparison of real results can be complicated, since different ways of sampling the dataset, different types of classification and different evaluation metrics originate from a given goal set by each research work, which leads to a variety of parameters, making a direct comparison not as reliable as it normally could be.

In Xu, Zhang, Cui and Zhang (2021), a relatively small convolutional neural network in terms of parameters is used with feature fusion and multiscale area attention, extending in Li et al. (2020), using Log Mel Spectrograms, which can achieve 79.34% Weighed

Accuracy (WA) and 77.54% Unweighted Accuracy (UA) in IEMOCAP. Only a certain part of IEMOCAP is used, namely four classes: sad, neutral, angry and happy + excited. Subsequently, a filtering of the data is performed: the dataset is divided into two parts, a represented part and an improvised part; only the improvised part is used. An ablation test is performed to gauge the impact of each refinement technique applied to the model, such as attention variation or Vocal Tract Length Perturbation (VLTP) for data augmentation, allowing to verify the improvement made by the implementation choices.

In Jalal et al. (2020), two groups of experiments are performed, which can be distinguished mainly by the type of neural network, Bi Long Short Term Memory with attention (BLSTMATT), described by Milner et al. (2019), and Convolutional Self-Attention (CSA), described by Jalal et al. (2019), both with higher complexity than the previous work mentioned. In the standard IEMOCAP benchmark, with the BLSTMATT approach, a value of 80.1% UA and 73.5% WA is obtained, while with the CSA approach, the best values go up to 76.3% UA and 69.4% WA. Finally, combining the two approaches into one, with a manipulation being performed on the outputs of each network, a peak value of 80.5% UA combined with 74% WA is obtained. As mentioned earlier in other works, four classes are used, sad, neutral, angry and happiness + excitement, but in this setup improvised and represented speech samples are used. The dataset is separated into training/test in a speaker-independent way, where the actors in one set are different from the actors in the other.

Fully supervised approaches are reaching increasingly high levels of performance and are slowly evolving into highly reliable systems. Of course, the reliability of such a system always depends on the end goal of the task.

In specific tasks, the gap between a fully supervised and a semi-supervised paradigm is no longer black and white (Andrade et al., 2022). Thus, SSL presents itself as a more viable solution for the future, since the amount of collected data is increasing despite the fact that its processing is increasingly expensive. To take advantage of the abundance of unlabeled audio data, many researchers have defined original approaches in the SSL paradigm.

In Zhao et al. (2020), a GAN system is used in conjunction with a classifier to add some significant information from unlabeled and generated data to the labelled set. To complement the system, in many works it is chosen to implement some techniques originated in Adversarial Training (Goodfellow et al., 2015) and Virtual Adversarial Training (Miyato et al., 2018), implementing a Smooth Semi Supervised Generative Adversarial Network (SSSGAN) and a Virtual Smooth Semi Supervised Generative Adversarial Network (VSSSGAN), respectively.

As the features used to represent speech are not similar to those of images, this may move the algorithm implementation away from stable architectures based on the Deep Convolutional Generative Adversarial Network (DCGAN) (Radford et al., 2016).

Considerably good results are however achieved in IEMOCAP (also with four classes, including improvised and represented speech), with 59.3% and 58.7% UAR, on 2400 labelled data, in the proposed SSSGAN and VSSSGAN methods, respectively, claiming to surpass the state of the art in this context. This is most likely one of the few works performed on SGANs in a monomodal baseline with speech-only data, which translates the lack of real practical resources for the task proposed in this paper.

3 Datasets

Although there is a huge amount of audio data currently available, in order for it to be correctly used in machine learning models there is the need for processing and labelling of it. Regarding ASER (Automatic Speech Emotion Recognition), several datasets are available, including spontaneous speech - characterized by naturalness, being essential for real-world scenarios, allowing neural networks to better approximate the emotional representation; actuated speech, whose main advantage is that there is total control over the experimental factors, for example the emotion to be expressed and the text, but may interfere with the emotions themselves; finally induced speech, which seeks to combine the advantages of the previous two, inducing an emotional response through fabricated situations. It may however be subject to problems of authenticity. Despite these challenges, efforts to provide audio speech data have allowed to improve machine learning frameworks. In this work, the IEMOCAP dataset is the main data source due to its high quality and quantity of actuated and induced speech.

While during the training process, each chunk represents an isolated instance, at test time, an aggregation of all chunks per utterance is performed, averaging the prediction produced by the network. A higher level of overlap between chunks was shown to lead to more robust classification and better overall performance (Xu, Zhang and Zhang, 2021). Lin and Busso (2021) took a very interesting chunking approach in which a more dynamic solution is presented, with chunks of varying length from instance to instance, which leads to different levels of overlap in each utterance and creates adaptive behavior for speech processing.

Each of these chunks is then converted into a Log Mel Spectrogram, where 40 bands are used, a common parameter used in this type of transformation for ML purposes (Parthasarathy and Busso, 2019), with a STFT window of 2048 and a hop length of 512, giving rise to a representation of 1 channel of dimensions 40×59. Note that their chunk-like approach extends the number of training samples. Depending on the (randomly) selected statements, the data set, from the Neural Network point of view, becomes larger, usually approaching the value of 13000 training samples. The same does not happen with the test set, since the clustering step in classification fixes its length.

4 Proposed Model

The use of a GAN system introduces a significant dependency on the chosen architecture and hyperparameters, as these models are sensitive to structural variations such as network depth, learning rate, and batch size (Lucic et al., 2018). The SGAN algorithm is no different in this regard. Optimizing hyperparameters for a task can be a resource-intensive and potentially unfeasible process, depending on factors such as available data, computational resources, time constraints, and personnel limitations. Therefore, it is often advantageous to incorporate conditions and baselines established by other researchers to ensure a logical starting point and streamline the experimentation process.

The model begins by receiving a Log Mel Spectrogram input with dimensions $40 \times 59 \times 1$. Subsequently, three 2D convolutional layers are applied to the input, each consisting of 32 filters. After each convolutional layer, a LeakyReLU activation function, as proposed by Xu et al. (2015), is applied. Following this, a Spatial Dropout layer with a ratio of 0.3, based on the work of Tompson et al. (2015), is introduced.

The process is then repeated with another set of identical convolutional layers, but this time with 64 filters each. Again, a Spatial Dropout layer with the same ratio is incorporated. The final convolutional block consists of a 3 × 3 layer, followed by two 1 × 1 layers, both with 64 filters and LeakyReLU activation. A Global Average Pooling layer is then applied, leading to Dense units that produce the logits for both the classifier and discriminator.

For the classifier, a standard approach is employed, where a softmax activation is applied to obtain the class probabilities. In the case of the discriminator, the kernel trick mentioned in Salimans et al. (2016) is employed, enabling the sharing of layers and units between the classifier and discriminator.

The unsupervised model utilizes the logits prior to the application of the softmax function and calculates a normalized sum of the exponential outputs. In practice, an aggregation function Lambda layer is used on the logits output by the Dense units, as defined by:

$$D(x) = \frac{Z(x)}{Z(x) + 1}, \text{ where } Z(x) = \sum^{K} \exp[l_k(x)]$$

The model has a small number of parameters due to the size of the labelled dataset as in (Alom et al., 2019).

To address the potential issue of overfitting to a limited number of labelled instances, the model incorporates Spatial Dropout and a final block of Convolutional layers with 1 × 1 kernel sizes, followed by Global Average Pooling (GAP) as proposed by Lin et al. (2014). These techniques are employed to enhance the model's generalization ability and reduce its susceptibility to overfitting on a small labelled dataset.

It is important to mention that Weight Normalization (Salimans and Kingma, 2016) is applied to all the layers in the model. Weight Normalization is a technique that reparametrizes the weights to enhance the optimization problem's conditioning. It has demonstrated promising outcomes in generative models such as GANs. The schematic representation of the shared architecture for the Discriminator and Classifier, including the layers up to the final logits Dense layer, can be found in Table 1.

To complete the GAN system, the Generator component needs to be defined. As per the standard procedure, a latent dimension of 100 is sampled from a Gaussian distribution. The initial processing of the noise input involves 3 × 4 × 256 Dense units to shape it into an image. Subsequently, four similar blocks are employed, each consisting of a Strided Convolution layer with 128 filters, Batch Normalization (Ioffe and Szegedy, 2015) with a momentum of 0.8, and LeakyReLU activation function. This is followed by two blocks, each comprising a Convolutional layer with 128 filters, Batch Normalization, and LeakyReLU activation, for a more refined processing of the latent features. Finally, a Convolutional layer with 1 filter is applied, with Weight Normalization also applied. The kernel shapes are not specified as they are adapted to achieve the final shape of 40 × 59 × 1. Please refer to Table 2 for a visual representation of the model.

To train each of the models, two different optimizer objects are utilized. Both optimizers are Adam optimizers with a learning rate of 0.0001 and a beta value of 0.5.

Additionally, stochastic weight averaging (Izmailov et al., 2019) is incorporated with an average decay parameter set to 0.999. The batch size for the discriminator is set to 128. Within each batch, samples of the same nature (supervised samples, unsupervised real samples, and unsupervised fake samples) are grouped together and fed into the model accordingly. For the Generator, the batch size is increased to 256. The number of training epochs is set to 300, where each epoch covers the entire dataset (both labelled and unlabeled) divided by the batch size. All modules are initialized with random values drawn from a normal distribution with a mean of 0 and a standard deviation of 0.05.

4.1 Evaluation Metrics

In order to evaluate and compare the obtained results, three metrics were recorded: cross entropy loss, UAR (also known as WA - Weighed Accuracy) and UA (Unweighted Accuracy). Among these metrics, the WA is commonly used because it provides a better understanding of model performance in unbalanced datasets. It is important to use consistent metrics when comparing results, as they may vary between different ASER research papers due to variations in dataset conditions and experimental objectives. In some cases, researchers choose to use continuous measures of valence, arousal and dominance instead of categorical emotional labels, as training models with continuous measures have shown advantages over categorical labels (Khorrami et al., 2016). However, this shift to continuous measures may affect human readability, as interpreting continuous values in relation to associated emotions may be more difficult.

Table 1. Discriminator/Classifier architecture for the GAN system

Model Architecture
40 × 59 × 1 image (Spectrogram) 32 3 × 3 conv2D Padding Stride = (1, 1) weightnorm lReLU 32 3 × 3 conv2D Padding Stride = (1, 1) weightnorm lReLU 32 3 × 3 conv2D Padding Stride = (2, 2) weightnorm lReLU
SpatialDropout(0.3)
64 3 × 3 conv2D Padding Stride = (1, 1) weightnorm lReLU 64 3 × 3 conv2D Padding Stride = (1, 1) weightnorm lReLU 64 3 × 3 conv2D Padding Stride = (2, 2) weightnorm lReLU
SpatialDropout(0.3)
64 3 × 3 conv2D Stride = (2, 2) weightnorm lReLU 64 1 × 1 conv2D Stride = (1, 1) Padding weightnorm lReLU 64 1 × 1 conv2D Stride = (1, 1) Padding weightnorm lReLU
Global Average Pooling Dense weightnorm

Table 2. Generator architecture for the GAN system.

Model Architecture
100 × 1 image (Noise)
3 × 4 × 256 Dense batchnorm lReLU
128 4 × 4 conv2DTranspose Stride = (2, 2) Padding batchtnorm lReLU
128 4 × 4 conv2DTranspose Stride = (2, 2) Padding batchtnorm lReLU
128 4 × 4 conv2DTranspose Stride = (2, 2) Padding batchtnorm lReLU
128 4 × 4 conv2DTranspose Stride = (2, 2) Padding batchtnorm lReLU
128 4 × 4 conv2D Stride = (1, 1) batchtnorm lReLU
128 4 × 4 conv2D Stride = (1, 1) batchtnorm lReLU
1 3 × 3 conv2D Stride = (1, 1) weightnorm

5 Tests and Results

Each testing instance is determined using a 5-fold cross-validation approach, along with varying numbers of labelled instances. The available labelled instances can range from 300, 600, 1200, 2400, up to a fully labelled dataset. These partial values represent approximately 2%, 4%, 9%, and 18% of the labelled data within the entire dataset. To ensure a fair assessment of performance, the results are averaged across multiple tests, giving equal weight to each run of the procedure. Additionally, a simple ablation test is conducted where the generator and discriminator components are deactivated, transforming the task into a fully supervised scenario. This allows for an evaluation of the added value that the GAN system brings to the model.

The initial set of experiments commences with the smallest number of labelled samples in order to establish a baseline for further progress. This test is expected to yield the lowest performance among all the semi-supervised learning (SSL) experiments. By utilizing only 2% of the dataset as labelled samples, this particular test is considered crucial and insightful as it simulates scenarios with extremely limited processed data. The first experiment, conducted with 300 supervised samples, achieves a UAR (Unweighted Average Recall) of 53.059%, UA (Unweighted Accuracy) of 49.391%, and a cross entropy loss of 1.5880238. Based on these initial findings, it is anticipated that subsequent tests will demonstrate significantly higher levels of performance.

In the second round of tests, with 600 labelled samples, the model hits 55.597% UAR and 52.967% UA, with a cross entropy loss of 1.3667043. Going from around 2% of labelled data to around 4% has resulted in the improvement of the evaluation metrics, UAR and UA, by at least 2.5%.

With 1200 labels, the model hits 58.329% UAR and 56.067% UA, with a cross entropy loss of 1.1982944. At this point, almost 10% of labelled data is used which sets a mark on SSL paradigm based, as it can bee seen a reference for testing such algorithms. As the number of labelled instances goes higher, it's expected that the results of it converges slower and slower to the results that would've been obtained if all the samples were labelled. So 10% would be a good average indicator for low sample training performance on DL models. As expected, the improvement is still considerably high.

With 2400 labels, the model hits 62.855% UAR with 61.089% UA with a cross entropy loss of 1.0183306. There is a steady improvement on every metric, where doubling the amount of labelled data is resulting in a more or less consistent growth of the model's effectiveness. Naturally, the test instance with higher number of labelled samples achieves the best results. The compilation of all test results described are shown in Figs. 1 and 2.

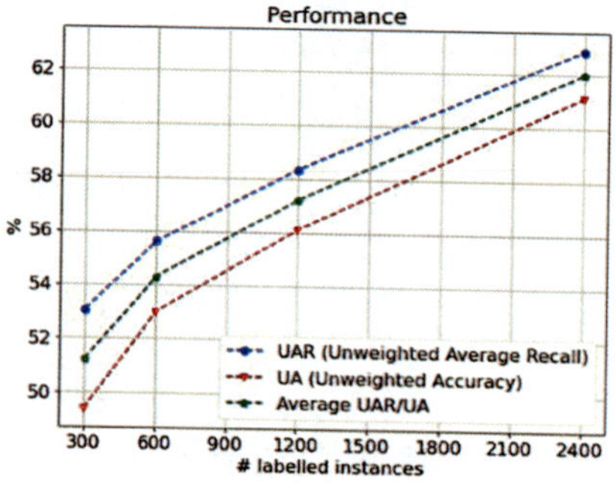

Fig. 1. Performance metrics of UAR and UA along with an average.

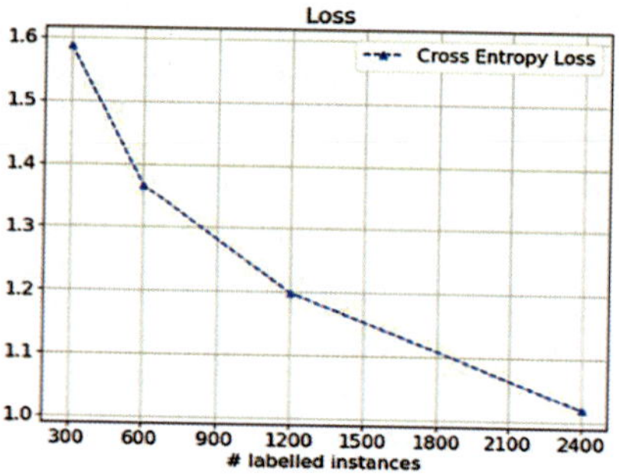

Fig. 2. Cross entropy loss across tests.

By examining these tests alone, there are indications of a performance decrease, suggesting that there may be an optimal number of labelled instances for a given dataset within the context of this procedure, which allows for acceptable performance to be achieved.

The results obtained using this approach are competitive and even slightly surpass the state-of-the-art results presented by Zhao et al. (2020) in the context of an SGAN system on the IEMOCAP dataset. This could be attributed to several factors. Firstly, the utilization of Log Mel Spectrograms combined with Convolutional layers demonstrates enhanced feature extraction capabilities, which, when combined with the efficiency of training CNNs, favours this approach. Secondly, the aggregation function employed, where the average of predictions is taken as the final prediction for a collection of chunks from the same utterance, alters the classification scheme. Therefore, direct comparisons between models under different conditions may not be entirely valid. This can also apply to the training process, where the division of chunks inflates the dataset size by multiple factors. Lastly, the generator's impact may be significant, as the naturalness of a convolutional generator in generating single-channel representations allows for easier feature selection. Additionally, the generator is trained using the Feature Matching loss, a loss function demonstrated by Salimans et al. (2016) to improve the performance of SGAN systems.

Along the experiments, as the number of labelled instances changes, the behavior of the GAN system is more or less consistent, where they tend to converge to the same values of loss. This would be expected since the unsupervised task associated doesn't really change, data wise, as every single sample is utilized. Varying the number of unlabelled samples fed to the GAN system would be an interesting testing point to verify the impact of number of samples, which could potentially translate to changes

Table 3. Comparison of the results obtained on this work, compared to the state of the results for GAN on the IEMOCAP.

UAR (%)				
	300	600	1200	2400
Zhao et al. (2020)	52.3	55.4	57.8	59.3
This work	53.1	55.6	58.3	62.9

in the supervised vertent as well. Table 3 shows a comparison with the values of UAR obtained in Zhao et al. (2020), to understand how algorithm and procedures can change the output performance.

6 Conclusions

The field of SSL is very young. Only in the last few years, with the growth of technology, software and hardware wise, resources were allocated towards SSL experiments as the margin of experimentation grew larger and larger. Competitive results with fully supervised approaches are being obtained on the most diverse tasks, which is motivating young researchers to dig deeper on the use cases of this framework. However, there is a rather complex nature, intrinsic to semi supervised approaches, that makes its introduction harder. A lot of work is needed to refine it to levels of reliability of SL and we are slowly moving towards it.

After this initial stabilizing, the aim is to deploy real world uses for such models. Unlabeled data is abundant in many scenarios, which could turn SSL into a very attractive approach in areas like health, education.

As an improving point, the priority would come from the theoretical and practical understanding of GANs on SSL. The expression "leveraging unlabeled data" needs to be translated into something more concrete, so the development towards solid systems on this paradigm, on the context, can start in a more systematic way. With more experimentation, more empirical cues will be found, making the task of developing such a model more consistent.

The multiple different phases on the development of ML systems need to be more evident, which by itself allows its maintenance in a more simplified, efficient and effective way. The proposed framework showed that an interesting level of accuracy can be obtained, regarding emotion detection, leading the way for newer and even more accurate systems in emotion detection using voice.

Acknowledgement. This work has been supported by FCT – Fundação para a Ciência e Tecnologia within the R\&D Units Project Scope: UIDB/00319/2020;

References

Andrade, G., Rodrigues, M., Novais, P.: A survey on the semi supervised learning paradigm in the context of speech emotion recognition. In: Arai, K. (ed.) IntelliSys 2021. Lecture Notes in Networks and Systems, vol. 295, pp. 771–792. Springer, Cham (2022). https://doi.org/10.1007/978-3-030-82196-8_57

Alom, M.Z., et al.: A state-of-the-art survey on deep learning theory and architectures. Electronics **8**, 292 (2019)

Gonçalves, S., Rodrigues, M., Carneiro, D., Fdez-Riverola, F., Novais, P.: Boosting learning: non-intrusive monitoring of student's efficiency. In: Mascio, T.D., Gennari, R., Vittorini, P., De la Prieta, F. (eds.) Methodologies and Intelligent Systems for Technology Enhanced Learning. AISC, vol. 374, pp. 73–80. Springer, Cham (2015). https://doi.org/10.1007/978-3-319-19632-9_10

Goodfellow, I.J., Shlens, J., Szegedy, C.: Explaining and harnessing adversarial examples (2015)

IEMOCAP. https://sail.usc.edu/iemocap/

Ioffe, S., Szegedy, C.: Batch normalization: accelerating deep network training by reducing internal covariate shift (2015)

Izmailov, P., Podoprikhin, D., Garipov, T., Vetrov, D., Wilson, A.G.: Averaging weights leads to wider optima and better generalization (2019)

Jalal, A., Milner, R., Hain, T.: Empirical interpretation of speech emotion perception with attention based model for speech emotion recognition (2020)

Khorrami, P., Le Paine, T., Brady, K., Dagli, C., Huang, T.S.: How deep neural networks can improve emotion recognition on video data. In: 2016 IEEE International Conference on Image Processing (ICIP), pp. 619–623 (2016)

Li, Y., Kaiser, L., Bengio, S., Si, S.: Area attention (2020)

Lin, M., Chen, Q., Yan, S.: Network in network (2014)

Lin, W.-C., Busso, C.: Chunk-level speech emotion recognition: A general framework of sequence-to-one dynamic temporal modelling. IEEE Trans. Affect. Comput. 1 (2021)

Lucic, M., Kurach, K., Michalski, M., Gelly, S., Bousquet, O.: Are gans created equal? a large-scale study (2018)

Miyato, T., Maeda, S., Koyama, M., Ishii, S.: Virtual adversarial training: a regularization method for supervised and semi-supervised learning (2018)

Parthasarathy, S., Busso, C.: Semi-supervised speech emotion recognition with ladder networks (2019)

Radford, A., Metz, L., Chintala, S.: Unsupervised representation learning with deep convolutional generative adversarial networks (2016)

Rodrigues, M., Fdez-Riverola, F., Novais, P.: An approach to assessing stress in e-learning students (2012)

Rodrigues, M., Monteiro, V., Fernandes, B., Silva, F., Analide, C., Santos, R.: A gamification framework for getting residents closer to public institutions. J. Ambient Intell. Human. Comput. **11** (2020)

Salimans, T., Goodfellow, I., Zaremba, W., Cheung, V., Radford, A., Chen, X.: Improved techniques for training GANs (2016)

Salimans, T., Kingma, D. P.: Weight normalization: a simple reparameterization to accelerate training of deep neural networks (2016)

Tompson, J., Goroshin, R., Jain, A., LeCun, Y., Bregler, C.: Efficient object localization using convolutional networks (2015)

Xu, M., Zhang, F., Zhang, W.: Head fusion: Improving the accuracy and robustness of speech emotion recognition on the IEMOCAP and RAVDESS dataset. IEEE Access **9**, 74539–74549 (2021)

Xu, B., Wang, N., Chen, T., Li, M.: Empirical evaluation of rectified activations in convolutional network (2015)

Zhao, H., Yufeng, X., Zhang, Z.: Robust semisupervised generative adversarial networks for speech emotion recognition via distribution smoothness. IEEE Access **8**, 106889–106900 (2020)

Author Index

S. Ossowski et al. (Eds.): DCAI 2023, LNNS 740, pp. 369–370, 2023.
https://doi.org/10.1007/978-3-031-38333-5